ORDINARY DIFFERENTIAL EQUATIONS

By

L. S. PONTRYAGIN

Translated from the Russian

by

LEONAS KACINSKAS and WALTER B. COUNTS

Library of Congress

ADDISON-WESLEY PUBLISHING COMPANY, INC.

READING, MASSACHUSETTS · PALO ALTO · LONDON

ADIWES INTERNATIONAL SERIES
IN MATHEMATICS

A. J. LOHWATER
Consulting Editor

Printed in the United States of America

Library of Congress Catalog Card No. 62-17075

PREFACE

This book has been written on the basis of lectures which I delivered at the department of mathematics and mechanics of Moscow State University. In drawing up the program for my lectures, I proceeded on the belief that the selection of material must not be random nor must it rest exclusively on established tradition. The most important and interesting applications of ordinary differential equations to engineering are found in the theory of oscillations and in the theory of automatic control. These applications were chosen to serve as guides in the selection of material. Since oscillation theory and automatic control theory without doubt also play a very important role in the development of our contemporary technical culture, my approach to the selection of material for the lecture course is, if not the only possible one, in any case a reasonable one. In attempting to give the students not only a purely mathematical tool suitable for engineering applications, but also to demonstrate the applications themselves, I included certain engineering problems in the lectures. In the book they are presented in §13, 27, and 29. I consider that these problems constitute an integral organic part of the lecture course and, accordingly, of this book.

In addition to the material presented in the lectures, I have included in the book more difficult problems which were investigated in student seminars. They are contained in §19 and 31. The material contained in §24, 25, and 30 was only partially presented in the lectures. For the convenience of the reader, in the last chapter, the sixth, are presented certain facts from linear algebra in the form in which they are used in this book.

In closing, I wish to express my gratitude to my students and to my closest co-workers V. G. Boltyanskiy, R. V. Gamkrelidze, and E. F. Mishchenko, who helped me in the preparation and delivery of the lectures and in writing and editing this book. I want also to note the decisive influence upon my scientific interests exerted by the outstanding Soviet specialist in the field of oscillation theory and automatic control theory, Aleksandr Aleksandrovich Andronov, with whom for many years I have had a friendly relationship. His influence has substantially affected the character and direction of this book.

Moscow
16 July 1960

L. S. Pontryagin

FOREWORD

This book constitutes a mildly radical departure from the usual one-semester first course in differential equations. There was a time when almost all first courses in differential equations were devoted to an exhaustive treatment of the methods and artifices by which certain elementary equations can be solved explicitly. Fortunately, much of the material of the "classical" first course in differential equations has been moved back into the elementary calculus, where it finds its proper place among the so-called techniques of integration.

The disadvantages of offering a methods course in differential equations immediately following a second course in the calculus have been recognized for a long time: Students, whose interest and tastes in mathematics had begun to take shape, are confronted with the least challenging course in the curriculum, and even the engineering students, to whose interests the methods course is supposedly dedicated, are given little idea of the theory of differential equations. In recent years, a number of books have done much to improve the level of the first course in differential equations, though many of these books are in the nature of treatises, more suitable for the two-semester course at the graduate level.

The present volume is designed for a one-semester course in the junior or senior year, preferably after the course in calculus. It makes no pretense of being a treatise; the methods and artifices which belong to the calculus are omitted, though a new importance is given the linear equation with constant coefficients. An important feature of this book is the chapter on stability theory (Chapter 5) and the introduction to the Lyapunov theory. The engineering student will be challenged by the nontrivial treatment of such topics as the Watt regulator for a steam engine and the vacuum-tube circuit. The use of matrices and linear algebra will complement the one-semester course in linear algebra which is appearing more frequently at the junior level.

The Publisher

CONTENTS

CHAPTER 1

INTRODUCTION

This chapter is devoted primarily to the definition of those concepts which will be studied subsequently. What is a system of ordinary differential equations, what do we mean by a solution of it, and how many of these solutions exist? These are the basic questions which we shall attempt to answer in this chapter. The number of solutions is determined by theorems of existence and uniqueness, which will not be proved here, but only formulated. The proofs of these and of a number of other theorems of the same type are given in the fourth chapter, but theorems previously formulated in the first chapter are repeatedly used, so that their meaning is thus clarified. In addition to such basic information, solutions of differential equations of several of the simplest types are given in the first chapter. At the end of the chapter complex differential equations and their complex solutions are studied, and elementary facts concerning systems of linear differential equations are given.

1. First-order differential equations. Equations in which the unknowns are functions of one or several variables and which contain not only the functions themselves, but also their derivatives, are called *differential equations.* If the unknown functions are functions of several variables, then the equations are called *partial* differential equations; in the opposite case, i.e., for the case of functions of only one independent variable, the equations are called *ordinary* differential equations. In this book we shall deal only with the latter.

In applications to physics the time is taken as the independent variable, which is conventionally designated by the letter t; throughout this book the independent variable will be designated by t. Unknown functions will be designated by x, y, z, and so on. Derivatives of functions with respect to t will as a rule be designated by dots: $\dot{x} = dx/dt$, $\ddot{x} = d^2x/dt^2$, and so on. When this is inconvenient or impossible, we shall denote the order of a derivative by an upper index in parentheses; for example, $x^{(n)} = d^nx/dt^n$.

First we shall study the *first-order differential equation.* This equation may be written in the form

$$F(t, x, \dot{x}) = 0. \tag{1}$$

Here t is the independent variable, x the unknown function, $\dot{x} = dx/dt$ the derivative, and F a given function of three variables. The function F need not be defined for all values of its arguments; therefore we speak of the *domain of definition* B of the function F or simply, the *domain* B of F; here we have in mind a domain in the space of the three variables x, y, z. Equation (1) is called a *first-order* equation because it contains only the first derivative $\dot{x}$ of the unknown function x. A function $x = \varphi(t)$ of the independent variable t, defined on a certain interval $r_1 < t < r_2$ (the cases $r_1 = -\infty, r_2 = +\infty$ are not excluded), which, when substituted for x in equation (1), reduces (1) to an identity on the entire interval $r_1 < t < r_2$, is called a *solution* of equation (1). The interval $r_1 < t < r_2$ is called the *interval of definition* of the solution $\varphi(t)$. It is evident that substitution of $x = \varphi(t)$ in (1) is possible only when the function $\varphi(t)$ has a first derivative (and, in particular, is continuous) on the entire interval $r_1 < t < r_2$. For the substitution of $x = \varphi(t)$ into equation (1) to be possible, it is also necessary that the point with coordinates $(t, \varphi(t), \dot{\varphi}(t))$ belong to the domain B of the function F for any value of t in the interval $r_1 < t < r_2$.

Relation (1) connects the three variables t, x, $\dot{x}$. In certain cases it determines $\dot{x}$ as a single-valued, implicit function of the independent variables t, x. In this case (1) is equivalent to a differential equation of the form

$$\dot{x} = f(t, x). \tag{2}$$

Equation (2) is said to be *solved explicitly for the derivative;* in certain respects it is more amenable to study than the general differential equation (1). It is such explicit equations which we shall now study. We shall not assume that (2) has been obtained as a result of solving (1) for $\dot{x}$, but shall proceed from the function $f(t, x)$ as a given function of the two independent variables t, x.

In order to visualize the situation geometrically, we introduce for study the tx-plane P. We shall plot t, as an independent variable, along the axis of abscissas, and x, as a dependent variable, along the axis of ordinates. The function f appearing in (2) need not be defined for all values of t and x, or, in geometric language, need not be defined at all points of the plane P, but only at points of a certain set Γ of P (Fig. 1). We shall assume that the set Γ is a domain. This means that for every point p in Γ there is some circle of positive radius with center at p also contained in Γ. Concerning the function f, it will be assumed that both the function itself and its partial derivative, $\partial f/\partial x$, are continuous functions of t and x in Γ. A solution $x = \varphi(t)$ of equation (2) may be thought of geometrically in P as a curve with the equation $x = \varphi(t)$. This curve has a tangent at every point and lies entirely in the domain Γ; it is called an *integral curve* of the differential equation (2).

Existence and uniqueness theorem. In algebra it is known that a large role is played by theorems which give the number of solutions to a given equation or system of equations. One such example is the *fundamental*

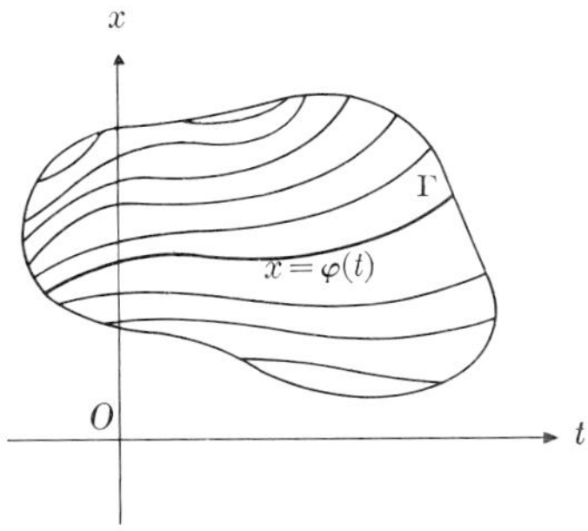

FIGURE 1

theorem of algebra, which asserts that a polynomial of the nth degree always has exactly n roots (counted according to multiplicity). In exactly the same way, in the theory of differential equations the important theoretical problem is how many solutions the differential equation has. It turns out that every differential equation has a *continuum* of solutions and this is why the question to be posed does not concern the *number* of solutions, but rather *how* the set of all solutions of a given differential equation can be described. The answer to this question is given by the *existence and uniqueness theorem* (Theorem 1), which is presented without proof in this section. The proof will be given considerably later (see §20).

THEOREM 1. Let

$$\dot{x} = f(t, x) \tag{3}$$

be a differential equation. We shall assume that the function $f(t, x)$ is defined in a certain domain Γ of the plane P of the variables t, x. We shall assume that the function f and its partial derivative $\partial f/\partial x$ are continuous in the entire domain Γ. The theorem asserts that

(1) For every point (t_0, x_0) of the domain Γ there exists a solution $x = \varphi(t)$ of equation (3) which satisfies the condition

$$\varphi(t_0) = x_0; \tag{4}$$

(2) If two solutions $x = \varphi(t)$ and $x = \chi(t)$ of equation (3) coincide for one value $t = t_0$, that is, if

$$\varphi(t_0) = \chi(t_0),$$

then these solutions are identically equal for all values of t for which they are defined.

The numbers t_0, x_0 are called the *initial values* for the solution $x = \varphi(t)$, the relation (4) represents the *initial conditions* for this solution, and we

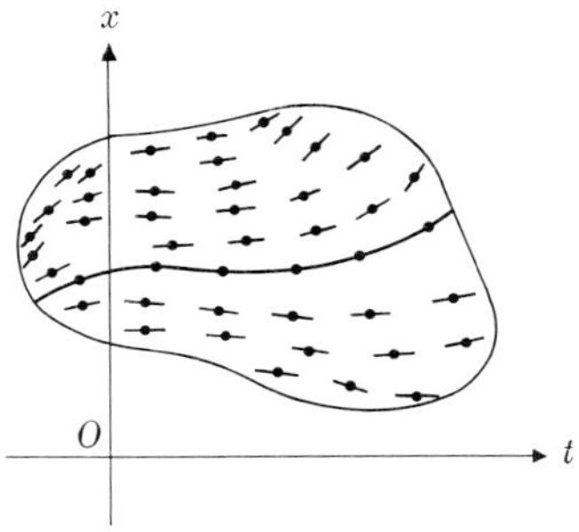

FIGURE 2

shall also say that the solution $x = \varphi(t)$ *satisfies the initial conditions* (4) or that it has *initial values* t_0, x_0. The assertion that the solution $x = \varphi(t)$ satisfies the initial conditions (4) (or has initial values t_0, x_0) assumes that the interval $r_1 < t < r_2$, where the solution $x = \varphi(t)$ is defined, contains the point t_0.

Thus Theorem 1 asserts that the coordinates of any point (t_0, x_0) of the domain Γ are initial values for some solution of equation (3) and that two solutions with common initial values coincide.

The geometrical meaning of Theorem 1 consists in the fact that *through every point* (t_0, x_0) *of* Γ *passes one and only one integral curve of equation* (3) (see Fig. 1).

We have interpreted geometrically every solution $x = \varphi(t)$ of equation (3) in the form of the graph of the function $\varphi(t)$. We now give a geometric interpretation of equation (3) itself. Through every point (t, x) of Γ we shall draw a straight line $l_{t,x}$ with slope $f(t, x)$. We obtain the *direction field* (or tangent field) corresponding to equation (3) and thus the geometric interpretation of this equation.

The connection between the geometrical interpretation of the equation and the geometrical interpretation of its solutions consists in the fact (Fig. 2) that *any integral curve* $x = \varphi(t)$ *is tangent to the straight line* $l_{t,\varphi(t)}$ at each of its points $(t, \varphi(t))$.

EXAMPLES

1. To illustrate the significance of Theorem 1 (in this case, of its second part), we shall solve the differential equation

$$\dot{x} = \alpha x, \tag{5}$$

where α is a real number. Here

$$f(t, x) = \alpha x,$$

so that the function f in fact depends only on the variable x. The domain of f coincides with the entire plane P. Both the function $f(t, x) = \alpha x$ and its derivative $\partial f(t, x)/\partial x = \alpha$ are continuous functions of t and x in the entire plane P. Thus Theorem 1 is applicable to equation (5). By direct substitution into equation (5) it is verified that each of the functions

$$x = ce^{\alpha t}, \tag{6}$$

where c is an arbitrary real number, is a solution of equation (5). We shall show that by assigning all possible values for c, we shall obtain *all* solutions of equation (5). Let $x = \varphi(t)$ be an arbitrary solution of this equation. We shall show that by proper choice of the number c we have $\varphi(t) = ce^{\alpha t}$. Let t_0 be a certain point of the interval of existence of the solution $x = \varphi(t)$, and let $x_0 = \varphi(t_0)$. Let us assume that $c = x_0 e^{-\alpha t_0}$. Then the solutions $x = \varphi(t)$ and $x = ce^{\alpha t} = x_0 e^{\alpha(t-t_0)}$ of equation (5) have the same initial values (t_0, x_0), and therefore coincide by virtue of the second part of Theorem 1. Thus, formula (6) exhausts the set of all solutions of differential equation (5).

2. We shall give a mathematical description of the process of decay of radioactive matter. The quantity of matter not yet decayed at the instant t we shall denote by $x(t)$. Then the quantity of matter which has decayed over the small interval of time t to $t + h$ is determined by the formula $\alpha h x(t)$, where α is a coefficient which depends on the properties of the radioactive matter and is slightly dependent on h; more accurately, it tends to a definite limit β as $h \to 0$. Thus we have

$$x(t) - x(t + h) = \alpha h x(t).$$

Dividing this relation by h and passing to the limit as $h \to 0$, we obtain

$$\dot{x}(t) = -\beta x(t).$$

We see that the function $x(t)$ satisfies the very simple differential equation examined in Example 1, so that

$$x(t) = ce^{-\beta t}.$$

To determine the constant c it is sufficient to specify any initial values. If, for example, it is known that at the instant $t = 0$ there was a quantity of matter x_0, then $c = x_0$, and we have

$$x(t) = x_0 e^{-\beta t}.$$

The rate of decay is expressed here by the value β having the dimension 1/sec or $(\text{sec})^{-1}$. Instead of the value β, the rate of decay is often characterized by the so-called *half-life*, i.e., the time required for half of the existing matter to decay. We shall designate the half-life by T and establish the connection between the values β and T. We have

$$\frac{x_0}{2} = x_0 e^{-\beta T},$$

whence

$$T = \frac{1}{\beta} \ln 2.$$

2. Some elementary integration methods. The main problem facing us when we deal with a differential equation is the problem of finding its solutions. In the theory of differential equations, just as in algebra, the question of what it means to find the solution of an equation may be understood in various ways. In algebra the original aim was to find a general formula involving radicals for the solution of equations of any degree. Such were the formulae for the solution of a quadratic equation, Cardan's formula for the solution of a cubic equation, and Ferrari's formula for the solution of an equation of the fourth degree. Later, it was established that for equations of degree higher than the fourth, a general formula for solution in radicals does not exist. The possibility remained of an approximate solution of equations with numerical coefficients and also the possibility of relating the dependence of the roots of an equation on its coefficients. The evolution of the concept of solution in the theory of differential equations was approximately the same. The original aim was to solve, or, as it was said, "to integrate" differential equations by means of "quadratures," i.e., the attempt was to write the solution in terms of the elementary functions and their integrals. Later, when it became clear that a solution in this sense exists only for very few types of equations, main emphasis of the theory was transferred to the study of general laws of the behavior of solutions. In this section we shall develop integration methods by quadratures for certain first-order differential equations.

(A) We shall solve the equation

$$\dot{x} = f(t), \tag{1}$$

the right-hand side of which depends only on the independent variable t. We shall assume that the function $f(t)$ is defined and continuous on the interval $r_1 < t < r_2$. Under this assumption, equation (1) satisfies the conditions of Theorem 1, and the domain Γ for this equation is a strip in the tx-plane P which is determined by the inequalities $r_1 < t < r_2$. Let t_0 be an arbitrary point of the interval $r_1 < t < r_2$; we assume

$$\varphi_0(t) = \int_{t_0}^{t} f(\tau)\, d\tau.$$

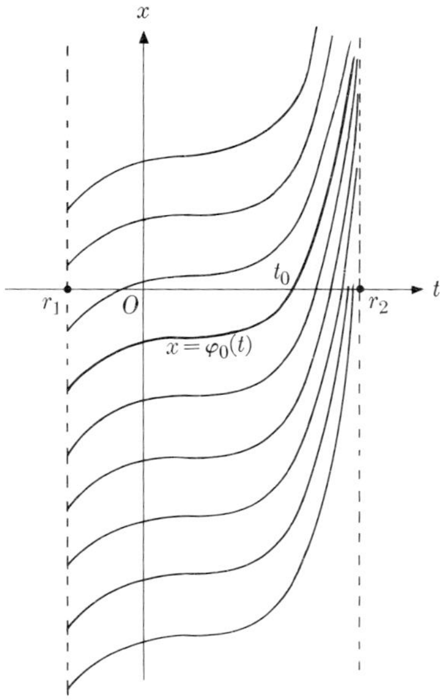

FIGURE 3

The function $\varphi_0(t)$ is defined on the interval $r_1 < t < r_2$. By inspection, an arbitrary solution of the equation (1) is given by the formula

$$x = \varphi(t) = \varphi_0(t) + c, \tag{2}$$

where c is an arbitrary constant. The right-hand side of (2) is, as is known, the *indefinite integral* of the function $f(t)$, so that (2) may be written in the form

$$x = \int f(t)\, dt.$$

It is seen by direct inspection that the function (2) satisfies equation (1). Further, the graph of every solution (2) for an arbitrary c is obtained from the graph of the solution $x = \varphi_0(t)$ by using a vertical-parallel translation by the quantity c (Fig. 3). From this it is evident that through every point of Γ passes a curve defined by formula (2). Hence, by Theorem 1 it follows that (2) actually encompasses the set of *all* solutions of (1).

(B) We shall solve the equation

$$\dot{x} = g(x), \tag{3}$$

the right-hand side of which depends only on the unknown function x. We shall assume that the function $g(x)$ is defined and has a continuous derivative on the interval $a_1 < x < a_2$. Then Theorem 1 is applicable to equation (1), and a strip in the tx-plane P which is determined by the inequalities $a_1 < x < a_2$ serves as the domain Γ. For the sake of simplicity, we assume in addition that on the interval $a_1 < x < a_2$ the function $g(x)$ does not vanish and consequently does not change sign. Let x_0 be an arbitrary point of the interval $a_1 < x < a_2$; we assume

$$G_0(x) = \int_{x_0}^{x} \frac{d\xi}{g(\xi)}. \tag{4}$$

The function $G_0(x)$ is defined on the interval $a_1 < x < a_2$, and its derivative on this interval is never zero; therefore the function $G_0(x)$ has an inverse, i.e., there exists a function $\psi_0(t)$ such that

$$G_0(\psi_0(t)) = t. \tag{5}$$

Consequently, an arbitrary solution of equation (3) is given by the formula

$$x = \psi(t) = \psi_0(t - c), \tag{6}$$

where c is an arbitrary constant. The function $\psi(t)$ is monotonic and assumes all values belonging to the interval $a_1 < x < a_2$.

We shall first prove that the function (6) is a solution of equation (3). From (5) it follows that

$$G_0(\psi(t)) = G_0(\psi_0(t - c)) = t - c. \tag{7}$$

Differentiating this relation with respect to t, we obtain

$$G_0'(\psi(t))\dot{\psi}(t) = 1,$$

hence [see (4)]

$$\dot{\psi}(t) = g(\psi(t)).$$

Since the function $\psi_0(t)$ is obtained as the inverse of the monotonic function $G_0(x)$, which is defined on the *entire* interval $a_1 < x < a_2$, the function $\psi_0(t)$ [and consequently $\psi(t)$] is monotonic and assumes all values on the interval $a_1 < x < a_2$. Since, further, the integral curve (6) is obtained from the curve $x_0 = \psi_0(t)$ by a horizontal-parallel translation (Fig. 4), a curve of the form (6) passes through every point of the strip Γ. Thus by Theorem 1, (6) contains the set of *all* solutions of equation (3).

Note. The relation (7) shows that the function $\psi(t)$ is the inverse of the function $G_0(x) + c$, which is the indefinite integral of the function $1/g(x)$. Thus all solutions $x = \psi(t)$ of equation (3) are described by the formula

$$\int \frac{dx}{g(x)} = t. \tag{8}$$

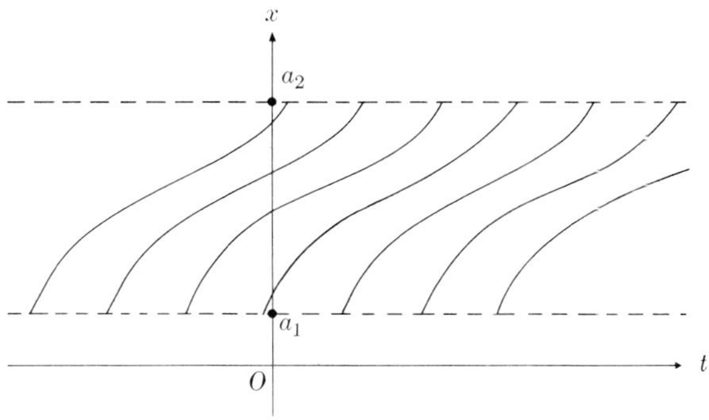

FIGURE 4

If the function $t = G_0(x) + c$ is taken as the unknown function, then we obtain for it the differential equation

$$\frac{dt}{dx} = \frac{1}{g(x)},$$

which is equivalent to equation (3). It is solved by the method presented in (A), which gives (8).

(C) We shall solve the equation

$$\dot{x} = f(t)g(x), \tag{9}$$

which is called an *equation with separable variables*. We shall assume that the function $f(t)$ is defined and continuous on the interval $r_1 < t < r_2$ and that the function $g(x)$ is defined and has a continuous derivative on the interval $a_1 < x < a_2$. Then Theorem 1 is applicable to equation (9), and the rectangle determined by the inequalities

$$r_1 < t < r_2, \qquad a_1 < x < a_2$$

serves as its domain Γ. For the sake of simplicity, we shall assume that $g(x)$ does not vanish on the interval $a_1 < x < a_2$. For the solution of (9) we form two auxiliary equations:

$$\frac{du}{dt} = f(t), \tag{10}$$

$$\frac{dx}{du} = g(x). \tag{11}$$

Equations (10) and (11) are solved by the rules given in (A) and (B). Let $u = \varphi_0(t)$ be some solution of equation (10) and $x = \psi_0(u)$ some solution of equation (11) defined on $b_1 < u < b_2$ and assuming all values belonging to the interval $a_1 < x < a_2$ [see (B)]. Hence an arbitrary solution of equation (9) can be written in the form

$$x = \chi(t) = \psi_0(\varphi_0(t) - c), \tag{12}$$

where c is an arbitrary constant. The interval of variation of t for every solution (12) must be such that values of $u = \varphi_0(t) - c$ belong to the interval $b_1 < u < b_2$.

We shall prove first that the function (12) is a solution of (9). Since the function $\psi_0(u)$ satisfies (11), $\psi_0'(\varphi_0(t) - c) = g[\psi_0(\varphi_0(t) - c)] = g(\chi(t))$. Further, $\dot{\varphi}(t) = f(t)$ [see (10)]. Differentiating equation (12), we obtain

$$\dot{\chi}(t) = \psi_0'(\varphi_0(t) - c)\dot{\varphi}_0(t) = g(\chi(t))f(t).$$

Thus the function $x = \chi(t)$ satisfies (9).

We shall now show that formula (12) gives *all* solutions of equation (9). Let (t_0, x_0) be an arbitrary point of Γ. Since the function $x = \psi_0(u)$ assumes all values on the interval $a_1 < x < a_2$, there exists a value u_0 such that $\psi_0(u_0) = x_0$. Let us set $c = \varphi_0(t_0) - u_0$; then by (12) we have $\chi(t_0) = \psi_0(u_0) = x_0$. Thus through every point (t_0, x_0) of Γ passes a curve of the form (12) and, by Theorem 1, formula (12) contains all solutions of equation (9).

Note. Formula (12) may be written in the form

$$\int \frac{dx}{g(x)} = \int f(t)\, dt$$

[compare (A) and (B)].

(D) We shall solve the equation

$$\dot{y} = h\left(\frac{y}{t}\right), \tag{13}$$

in which the right-hand side depends only on the ratio of the variables y and t. Such an equation is called *homogeneous*. We shall assume that $h(x)$ is defined and has a continuous derivative on the interval $a_1 < x < a_2$. Then Theorem 1 is applicable to equation (13), and Γ consists of all points of the ty-plane P satisfying the inequality

$$a_1 < \frac{y}{t} < a_2 \qquad (t \neq 0).$$

For the sake of simplicity, we also assume that on the interval $a_1 < x < a_2$ the function $h(x) - x$ does not vanish. We shall solve (13) by making a *change of variables,* that is, instead of the unknown function y we shall introduce an unknown function x by setting

$$y = xt. \tag{14}$$

By replacing the function y in equation (13) by its expression (14) we obtain for the new unknown x the equation

$$\dot{x}t + x = h(x),$$

or, what is the same thing,

$$\dot{x} = \frac{h(x) - x}{t}.$$

This equation is an equation with separable variables and may be solved by the method shown in (C).

(E) We shall solve the equation

$$\dot{y} = a(t)y + b(t). \tag{15}$$

This equation is called *linear*, because the unknown function and its derivative appear linearly in it. We shall assume that the functions $a(t)$ and $b(t)$ are defined and continuous on the interval $r_1 < t < r_2$. Then for equation (15) Theorem 1 is applicable, and the domain Γ is defined by the inequalities

$$r_1 < t < r_2.$$

If $b(t) \equiv 0$, then equation (15) is called *homogeneous*.

To the *nonhomogeneous* equation (15) corresponds the homogeneous equation

$$\dot{x} = a(t)x, \tag{16}$$

which we shall study first. Equation (16) is an equation with separable variables and it can be solved by the method shown in (C). However, the function $g(x)$ in the given equation is equal to x and can vanish. Therefore, for the solution of equation (16) it is necessary to investigate separately the domains $x > 0$ and $x < 0$ and also the solution $x = 0$. Instead, we shall write the solution directly by means of the formula

$$x = c \exp\left[\int_{t_0}^{t} a(\tau)\, d\tau\right], \tag{17}$$

where $r_1 < t_0 < r_2$ and c is an arbitrary constant. Substitution of (17) into (16) shows that (17) gives a solution; we shall show that it contains the set of *all* solutions. Let (ϑ_0, x_0) be an arbitrary point of the strip $r_1 < t < r_2$. In order that the solution (17) have the initial values (ϑ_0, x_0)

it is sufficient that the constant c satisfy the condition

$$x_0 = c \exp\left[\int_{t_0}^{\vartheta_0} a(\tau)\, d\tau\right],$$

from which c is uniquely determined.

The nonhomogeneous equation (15) is solved by the method of *variation of parameters*. We set

$$y = c \exp\left[\int_{t_0}^{t} a(\tau)\, d\tau\right] \tag{18}$$

and assume that c is not a constant quantity but depends on t. Substitution of (18) into (15) gives

$$\dot{c} \exp\left[\int_{t_0}^{t} a(\tau)\, d\tau\right] = b(t). \tag{19}$$

Thus

$$\dot{c} = b(t) \exp\left[-\int_{t_0}^{t} a(\tau)\, d\tau\right].$$

This differential equation for c is solved by the method presented in (A).

The method of variation of parameters is in this case a method of *introduction of a new unknown*. That is, instead of the unknown function y, we introduce, by (18), a new unknown function c. By finding an arbitrary solution of (19), we also find an arbitrary solution of (15) by formula (18).

Examples

1. We shall solve the equation

$$\dot{x} = \frac{2}{t^2 - 1}. \tag{20}$$

Here, as in (A), the right-hand side,

$$f(t) = \frac{2}{t^2 - 1},$$

depends only on the independent variable t, but it has discontinuities at the points $t = 1$ and $t = -1$.

Thus the domain Γ in the tx-plane P consists of a strip, $-1 < t < 1$, and two half-planes, $t < -1$ and $t > 1$. In order to solve (20) by the rules indicated in (A), it is necessary to divide the interval $-\infty < t < +\infty$ into three intervals, $-\infty < t < -1$, $-1 < t < 1$, $1 < t < +\infty$, and in each of them to take the indefinite integral of $f(t)$. To carry out the

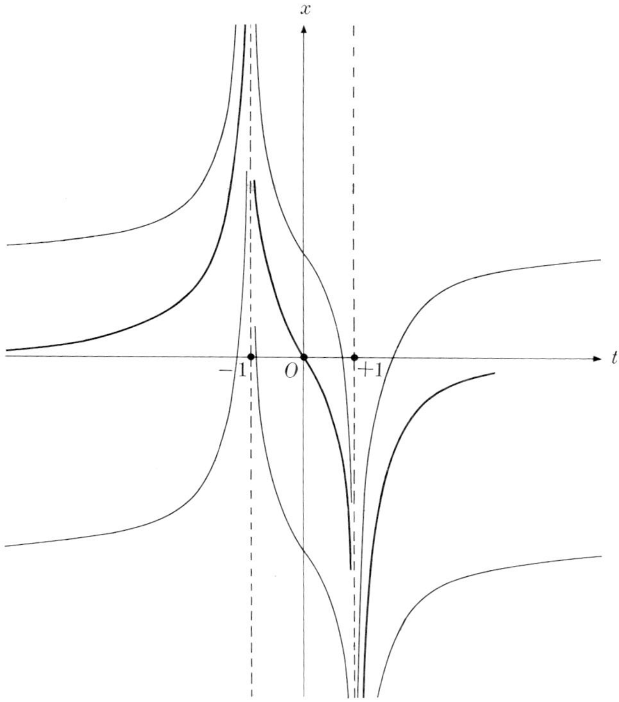

FIGURE 5

integration, we expand the function $f(t)$ into partial fractions:

$$\dot{x} = \frac{1}{t-1} - \frac{1}{t+1} \cdot$$

For each of the three intervals, the solution of equation (20) is written in the form

$$x = \int \left(\frac{1}{t-1} - \frac{1}{t+1}\right) dt = \ln |t-1| - \ln |t+1| + c$$
$$= \ln \left|\frac{t-1}{t+1}\right| + c. \qquad (21)$$

As was indicated in §1, a continuous function which satisfies the equation and is defined *on the interval* is called a solution of the differential equation; thus formula (21) for a fixed c defines not one but *three* solutions of equation (20), the first of which is defined on the interval $-\infty < t < -1$, the second on the interval $-1 < t < 1$, and the third on the interval $1 < t < \infty$ (Fig. 5).

2. We shall solve the equation

$$\dot{x} = \frac{x^2 - 1}{2}. \tag{22}$$

The right-hand side here depends only on x; it is defined, continuous, and differentiable for all values of x, and for this reason the domain Γ, for equation (22), coincides with the entire plane P. The function

$$g(x) = \frac{x^2 - 1}{2},$$

on the right-hand side of equation (22), vanishes at the points $x = -1$ and $x = 1$; for the solution of equation (22) according to the rules indicated in (B), therefore, it is necessary to split Γ into the three domains determined by the inequalities

$$-\infty < x < -1, \qquad -1 < x < +1, \qquad 1 < x < \infty, \tag{23}$$

and in addition to investigate the obvious solutions $x \equiv 1$, $x \equiv +1$. In each of the domains (23), the solution $x = \psi(t)$ is determined implicitly from the equation

$$\int \frac{2\,dx}{x^2 - 1} = t,$$

or else from the equation

$$\ln \left| \frac{x - 1}{x + 1} \right| = t - c_1.$$

This equation is equivalent to the equation

$$\left| \frac{x - 1}{x + 1} \right| = e^{t - c_1} = c_2 e^t, \tag{24}$$

where c_2 is an arbitrary positive constant. On the intervals $-\infty < x < -1$ and $1 < x < \infty$ equation (24) may be written in the form

$$\frac{x - 1}{x + 1} = c_2 e^t,$$

and on the interval $-1 < x < 1$ in the form

$$\frac{x - 1}{x + 1} = -c_2 e^t;$$

both cases can be embraced by a single formula

$$\frac{x - 1}{x + 1} = c e^t, \tag{25}$$

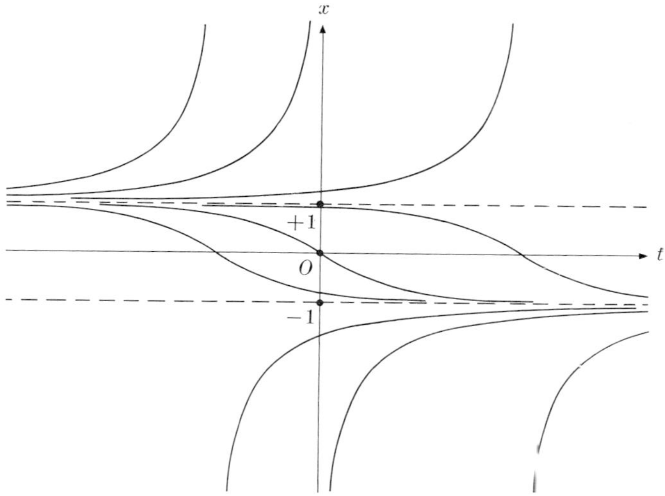

FIGURE 6

where c is an arbitrary constant other than zero. Solving equation (25) for x, we have

$$x = \frac{1 + ce^t}{1 - ce^t}, \tag{26}$$

where $c \neq 0$.

For a fixed negative c formula (26) gives one solution, located in the strip $-1 < x < 1$. For a fixed positive c formula (26) defines two solutions, of which one, defined on the interval $-\infty < t < -\ln c$, is located in the half-plane $x > 1$, and the other, defined on the interval $-\ln c < t < \infty$, is located in the half-plane $x < -1$. In addition to the solutions located in the domains (23), equation (22) has two more solutions, $x \equiv 1$ and $x \equiv -1$, which are obtained formally from (26) for $c = 0$ and $c = \infty$. Since one of the solutions found (Fig. 6) passes through every point of P, it follows from Theorem 1 that other solutions do not exist. Thus formula (26) includes the set of all solutions of equation (22) if c is allowed to take all real values, including ∞.

3. Example 2 is a typical one for an equation of the form

$$\dot{x} = g(x), \tag{27}$$

where $g(x)$ is a continuously differentiable function defined for all values x, $-\infty < x < \infty$. Theorem 1 is applicable to equation (27), Γ being the entire plane P. We note first that, if a is a zero of the function $g(x)$, that is, if $g(a) = 0$, then the function $x \equiv a$ is a solution of equation (27). Thus to every zero of $g(x)$ corresponds a solution whose graph is a horizontal

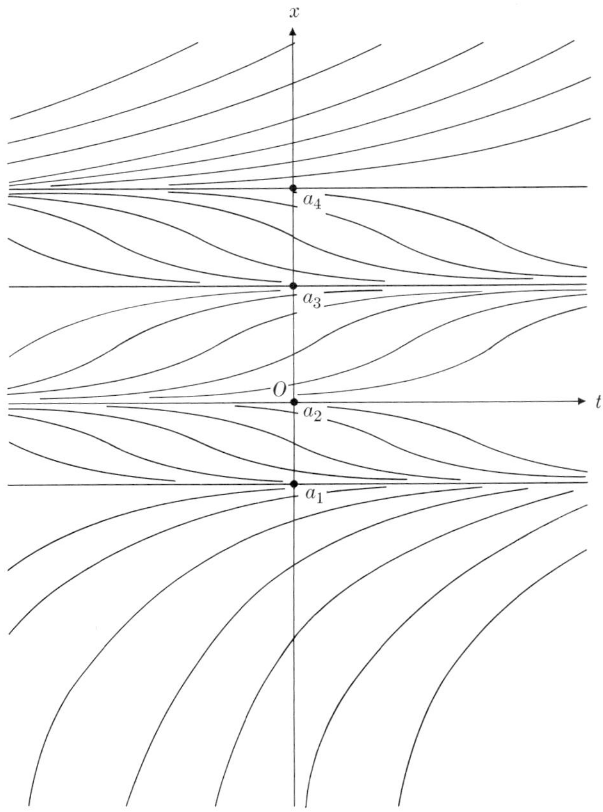

FIGURE 7

straight line $x = a$ in the plane P. For the sake of simplicity, we shall assume that the zeros of $g(x)$ do not have accumulation points. Then P will be decomposed by solutions of the form $x = a$ into a series of strips, including possibly one or two half-planes of the form $x < a'$ and $x > a''$. In each of these strips equation (27) may be solved by the method indicated in (B).

We shall note certain essential properties of these solutions. Let a_1 and a_2, $a_1 < a_2$, be two successive zeros of $g(x)$, so that on the interval $a_1 < x < a_2$ the function $g(x)$ does not vanish. Let $x = \psi_0(t)$ be any solution of (27) in the domain $a_1 < x < a_2$. As was noted in (B), all other solutions in this domain are obtained by the formula $x = \psi_0(t - c)$, i.e., by a horizontal-parallel translation of the solution $\psi_0(t)$. Since on the interval $a_1 < x < a_2$ the function $g(x)$ does not vanish, the solution $\psi_0(t)$ is monotonic. In order to be definite, we shall assume that $g(x) > 0$

for $a_1 < x < a_2$. Then the solution $\psi_0(t)$ is a monotonically increasing function. From Theorem 1 it is comparatively easy to derive the fact that the solution $\psi_0(t)$ can be continued to the interval $-\infty < t < \infty$, whence

$$\lim_{t\to-\infty} \psi_0(t) = a_1, \quad \lim_{t\to+\infty} \psi_0(t) = a_2.$$

We shall now describe the behavior of the solutions in the domain $x > a$, where a is the maximum zero of the function $g(x)$ (if such a value exists). It is obvious that $g(x) \neq 0$ for $x > a$; to be precise, we shall assume that $g(x) > 0$ for $x > a$. Let $\psi_0(t)$ be some solution of (27) passing into the domain $x > a$. By virtue of (B) all solutions in this domain are given by the formula $x = \psi_0(t - c)$. From Theorem 1 it is comparatively easy to derive the fact that the solution $\psi_0(t)$ can be continued over the interval $-\infty < t < r$, whence

$$\lim_{t\to-\infty} \psi_0(t) = a, \qquad \lim_{t\to r} \psi_0(t) = +\infty;$$

here r may assume finite values or the value $+\infty$. Thus in the half-plane $x > a$ two distinct cases are possible: every solution is defined on the interval $-\infty < t < +\infty$ or every solution is defined on the interval $-\infty < t < r$, where the number r is finite and depends on the solution selected. Since a solution of the form described above passes through every point of the plane P (Fig. 7), the solutions described above constitute, by Theorem 1, the set of all solutions of equation (27).

4. We shall show that, if the right-hand side of the equation does not have a continuous derivative, then the second part of Theorem 1 (uniqueness) need not apply. Let us consider the equation

$$\dot{x} = 3x^{2/3}. \tag{28}$$

The right-hand side of equation (28) is defined and continuous for all values of x, but its derivative $2x^{-1/3}$ has a discontinuity at the point $x = 0$. If the set of all points for which $x \neq 0$ is taken as the domain Γ, then Theorem 1 is applicable to equation (28) in this domain, and in each of the half-planes $x > 0$, $x < 0$, equation (28) can be solved by the method indicated in (B). Solving equation (28) by this method, we obtain

$$x^{1/3} = t - c. \tag{29}$$

Part of the graph of the function (29) (for $t < c$) is in the half-plane $x < 0$ and part (for $t > c$) is in the half-plane $x > 0$. It is immediately verified, however, that the function (29) [that is, $x = (t - c)^3$] is a solution of (28) for *all* values of t in the interval $-\infty < t < +\infty$. At the same time,

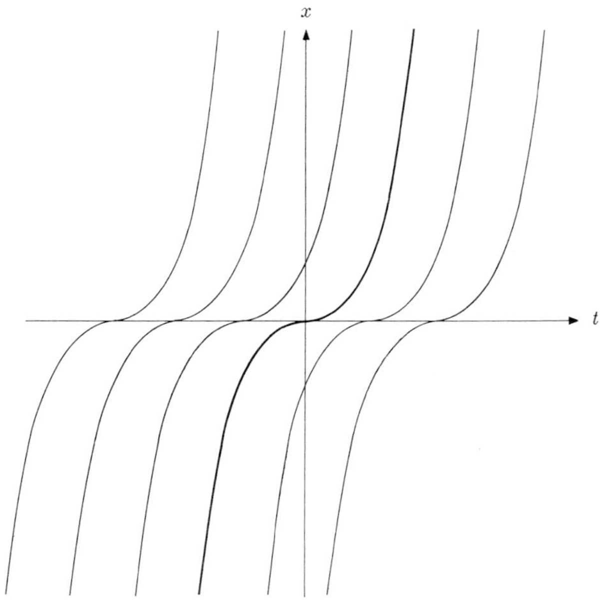

FIGURE 8

$x \equiv 0$ is also a solution of equation (28). Thus through every point $x = 0$, $t = c$ of the straight line $x = 0$ pass *two* solutions (Fig. 8): the solution (29) and the solution $x = 0$. We see that the second part of Theorem 1 (uniqueness) does not hold for equation (28).

3. Formulation of the existence and uniqueness theorem. In §1 we studied a first-order differential equation and formulated an existence and uniqueness theorem for this equation. The theory of ordinary differential equations deals also with more general systems of equations. Usually a system of ordinary differential equations consists of as many equations as there are unknown functions in the system; in this connection, all the unknown functions are functions of one and the same independent variable. In all cases the existence and uniqueness theorem is the fundamental theoretical proposition which makes it possible to approach the study of a given system of differential equations.

The existence and uniqueness theorem is formulated and proved for a system of equations which superficially appears to be of a rather particular type. But in fact, systems of comparatively general type reduce to this system of equations. Systems of differential equations of the particular type discussed here we shall in the future call *normal,* although this term is by no means generally accepted.

The system

$$\dot{x}^i = f^i(t, x^1, x^2, \ldots, x^n), \qquad i = 1, \ldots, n, \tag{1}$$

of ordinary differential equations is called a *normal system*. In this system t is the independent variable, $x^1, \ldots, x^n$ are the unknown functions, and $f^1, \ldots, f^n$ are functions of $n + 1$ variables defined in a certain domain Γ of the $(n + 1)$-dimensional space of the variables $t, x^1, \ldots, x^n$. Here and in the sequel, a "domain" of euclidean space is understood to be an *open set*, i.e., a set which together with every point contains a certain sphere of positive radius with its center at this point. It will always be assumed that the functions

$$f^i(t, x^1, x^2, \ldots, x^n), \qquad i = 1, \ldots, n, \tag{2}$$

are continuous in the domain Γ; it will also be assumed that their partial derivatives

$$\frac{\partial f^i(t, x^1, x^2, \ldots, x^n)}{\partial x^j}, \qquad i, j = 1, \ldots, n, \tag{3}$$

are continuous in Γ. It should be noted that the partial derivatives (3), the continuity of which is assumed, are taken only with respect to $x^1, \ldots, x^n$, but not with respect to the independent variable t.

In the presently fashionable theory of ordinary differential equations, the continuity of the derivatives (3) is not usually required, and this requirement is replaced by a weaker one, namely a Lipschitz condition for the functions (2). I consider that this generalization is nonessential and there is no need to dwell on it (this condition will be formulated below).

By a *solution* of system (1) we shall mean a system of continuous functions

$$x^i = \varphi^i(t), \qquad i = 1, \ldots, n, \tag{4}$$

which are defined on some interval $r_1 < t < r_2$ and which satisfy (1). The interval $r_1 < t < r_2$ is called the *interval of definition* of the solution (4) (the cases $r_1 = -\infty$, $r_2 = +\infty$ are not excluded). It is assumed that the system of functions (4) satisfies the system of equations (1) if after substitution of the functions (4) into (1) in place of $x^1, \ldots, x^n$, the relations (1) reduce to identities in t on the entire interval $r_1 < t < r_2$. For this substitution to be possible, it is necessary that the functions (4) have derivatives at every point of the interval $r_1 < t < r_2$ and that the right-hand sides of (1) be defined for all values of their arguments. Thus the point with coordinates

$$t, \varphi^1(t), \ldots, \varphi^n(t)$$

must belong to Γ for all values of t in the interval $r_1 < t < r_2$. We shall now give a formulation of the existence and uniqueness theorem for the normal system (1).

THEOREM 2. Let (1) be a normal system of ordinary differential equations. Here the right-hand sides of equations (1) are defined in a certain domain Γ, and the functions (2) and (3) are continuous in this domain. Then, for every point

$$t_0, x_0^1, x_0^2, \ldots, x_0^n \tag{5}$$

of Γ, there exists a solution

$$x^i = \varphi^i(t), \qquad i = 1, \ldots, n, \tag{6}$$

of the system (1), which is defined on some interval containing the point t_0 and which satisfies the conditions

$$\varphi^i(t_0) = x_0^i, \qquad i = 1, \ldots, n. \tag{7}$$

Further, if there exist two solutions

$$\begin{aligned} x^i &= \psi^i(t), \qquad i = 1, \ldots, n, \\ x^i &= \chi^i(t), \qquad i = 1, \ldots, n, \end{aligned} \tag{8}$$

of (1) which satisfy the conditions

$$\psi^i(t_0) = \chi^i(t_0) = x_0^i, \qquad i = 1, \ldots, n, \tag{9}$$

each defined on an interval of values of t containing t_0, then these solutions coincide wherever both are defined.

The values (5) are called *initial values* for the solutions (6) and (8), and (7) and (9) are called *initial conditions* for these solutions. We shall say hereafter that our solutions *have the initial values* (5) or *satisfy the initial conditions* (7) and (9). Thus the existence and uniqueness theorem for a normal system can be formulated briefly as follows:

For any initial values (5) there always exists a solution of system (1) with these initial values, which is defined on a certain interval containing the point t_0. Further, if there are two solutions with identical initial values (5), each of which is defined on a proper interval containing t_0, then these solutions coincide on the common part of these intervals.

The existence and uniqueness theorem permits us to formulate and solve the problem of the *maximum interval of existence of a solution* with given initial values. In order to attack this problem we shall assume that we have two solutions with the same initial values (5), the first solution being defined on the interval $r_1 < t < r_2$ and the second solution on the interval

$s_1 < t < s_2$. Both these intervals contain the point t_0, that is, they intersect. If one of the intervals were contained in the other, for example, the first one in the second, then consideration of the first would be superfluous, since, by virtue of uniqueness, the first solution coincides in this case with the second wherever it is defined. It can happen, however, that neither of the two intervals is contained in the other; for example, it may be that $r_1 < s_1 < r_2 < s_2$. In this case, the uniqueness implies that the two solutions coincide on the interval $s_1 < t < r_2$; the first solution is defined only on the half-interval $r_1 < t \leq s_1$ and the second only on the half-interval $r_2 \leq t < s_2$. On the interval $r_1 < t < s_2$, which contains both original intervals, no solution has as yet been formally defined. This solution, however, is very easy to define. It is only necessary to consider that on the interval $r_1 < t < r_2$ this solution coincides with the first one, and on the interval $s_1 < t < s_2$ it coincides with the second. Thus we arrive at a solution defined on the interval $r_1 < t < s_2$ which contains both of the original intervals, and there is no need to investigate the original solutions which are defined on the subintervals of the interval $r_1 < t < s_2$. It is evident that if there is a finite number of solutions with common initial values, then it is possible to construct the solution on the interval which contains all the intervals of existence of the original solutions. Also, *all* solutions with common initial values can be easily consolidated into one, and a concept of the maximum interval of existence of a solution with given initial values can be attained. We shall formulate and prove the corresponding proposition:

(A) For any initial values (5), there exists a solution

$$x^i = \hat{\varphi}^i(t), \qquad i = 1, \ldots, n,$$

of the system (1), defined on a certain interval $m_1 < t < m_2$, such that for any solution $x^i = \varphi^i(t)$, $i = 1, \ldots, n$, of system (1) with initial values (5), its interval of definition is contained in the interval $m_1 < t < m_2$. By virtue of the uniqueness theorem, the solution $x^i = \varphi^i(t)$ coincides with the solution $x^i = \hat{\varphi}^i(t)$ along its entire interval of definition, and for this reason there is no need to consider it. The interval $m_1 < t < m_2$ will be called the *maximum interval of existence* for the initial values (5).

We shall prove the existence of a maximum interval for given initial values (5). To each solution of system (1) with initial values (5) corresponds its own interval of definition. The set of all right-hand endpoints of these intervals we shall denote by R_2 and the set of all left-hand endpoints by R_1. The upper bound of the set R_2 we shall denote by m_2 (in particular, m_2 may be $+\infty$), and the lower bound of the set R_1 we shall denote by m_1 (in particular, m_1 may be $-\infty$). We shall now show that $m_1 < t < m_2$ is a maximum interval for the initial values (5). We shall first construct the solution $x^i = \hat{\varphi}^i(t)$, $i = 1, \ldots, n$, with initial values

(5) which is defined on the interval $m_1 < t < m_2$. Let t^* be an arbitrary point of this interval, that is, $m_1 < t^* < m_2$. To be precise we assume that $t^* \geq t_0$. Since m_2 is an upper bound of the set R_2, there exists a solution $x^i = \psi^i(t)$, $i = 1, \ldots, n$, of (1) with initial values (5), whose interval of definition $r_1 < t < r_2$ contains the point t^*, and we set $\hat{\varphi}^i(t^*) = \psi^i(t^*)$, $i = 1, \ldots, n$. The value obtained for the solution $\hat{\varphi}^i(t)$ at the point $t = t^*$ does not depend on the choice of the solution $x^i = \psi^i(t)$, $i = 1, \ldots, n$. Indeed, if instead of the solution $x^i = \psi^i(t)$, $i = 1, \ldots, n$, we took $x^i = \chi^i(t)$, $i = 1, \ldots, n$, with initial values (5) and interval of definition $s_1 < t < s_2$ which also contains the point t^*, then by virtue of the uniqueness (see Theorem 2) we would have $\psi^i(t^*) = \chi^i(t^*)$, $i = 1, \ldots, n$. Thus the functions $\hat{\varphi}^i(t)$, $i = 1, \ldots, n$, are uniquely defined on the entire interval $m_1 < t < m_2$. At the same time they constitute a *solution* of (1) with initial values (5). Indeed, near each point t^* of the interval $m_1 < t < m_2$ the system of functions $\hat{\varphi}^i(t)$, $i = 1, \ldots, n$, coincides by construction with some solution of (1), and for this reason the functions $\hat{\varphi}^i(t)$, $i = 1, \ldots, n$, are continuous and constitute a solution of (1). It remains to be shown that the interval $m_1 < t < m_2$ is maximal. Let $x^i = \varphi^i(t)$, $i = 1, \ldots, n$, be a certain solution of system (1) with initial values (5) which is defined on the interval $r_1 < t < r_2$. Then r_1 is an element of the set R_1 and r_2 is an element of the set R_2, and therefore $m_1 \leq r_1$, $r_2 \leq m_2$, that is, the interval $r_1 < t < r_2$ is contained in the interval $m_1 < t < m_2$.

We shall formulate here without proof one more existence theorem, which we shall later obtain as a simple corollary of Theorem 2.

THEOREM 3. Let

$$\dot{x}^i = \sum_{j=1}^{n} a_j^i(t)x^j + b^i(t), \qquad i = 1, \ldots, n \tag{10}$$

be a normal linear system of equations. Here the coefficients $a_j^i(t)$ and the free terms $b^i(t)$ are continuous functions of t defined on a certain interval $q_1 < t < q_2$. Then for arbitrary initial values

$$t_0, \overset{1}{x_0}, \overset{2}{x_0}, \ldots, \overset{n}{x_0}, \qquad q_1 < t_0 < q_1, \tag{11}$$

there exists a solution of (10) with these initial values which is defined on the *entire* interval $q_1 < t < q_2$.

In other words, the maximum interval of existence of the solution of the linear system (10) is the *entire* interval $q_1 < t < q_2$ [for the arbitrary initial conditions (11)].

For the case when the coefficients and free terms of system (10) are defined on the entire straight line, i.e., when $q_1 = -\infty$, $q_2 = +\infty$ for any initial values, there exists a solution of (10) defined on the entire infinite interval $-\infty < t < +\infty$.

The solutions of a normal system (1) can be interpreted geometrically as *integral curves* in the $(n+1)$-dimensional space with coordinates $t, x^1, \ldots, x^n$ (compare §1). The equations of an integral curve have the form

$$x^i = \varphi^i(t), \qquad i = 1, \ldots, n, \tag{12}$$

where (12) is a solution of the system.

The system (1) itself may be interpreted with the aid of the direction field in an $(n+1)$-dimensional space (compare §1).

Examples

1. We shall solve the normal linear system of equations

$$\dot{x} = -\omega y, \qquad \dot{y} = \omega x. \tag{13}$$

The domain Γ for this system is the entire space (t, x, y) By direct inspection, we see that the system of functions

$$x = c_1 \cos(\omega t + c_2), \qquad y = c_1 \sin(\omega t + c_2), \tag{14}$$

where c_1 and c_2 are arbitrary constants, is a solution of (13). To show that, by proper selection of the constants c_1 and c_2, an arbitrary solution can be obtained from formula (14), we shall prescribe initial values t_0, x_0, y_0 and show that among the solutions (14) there is a solution with these initial values. We obtain for the constants c_1 and c_2 the conditions

$$c_1 \cos(\omega t_0 + c_2) = x_0, \qquad c_1 \sin(\omega t_0 + c_2) = y_0. \tag{15}$$

Let ρ and φ be polar coordinates of the point (x_0, y_0), so that

$$x_0 = \rho \cos \varphi, \qquad y_0 = \rho \sin \varphi.$$

Then equations (15) may be rewritten in the form

$$c_1 \cos(\omega t_0 + c_2) = \rho \cos \varphi, \qquad c_1 \sin(\omega t_0 + c_2) = \rho \sin \varphi.$$

By setting

$$c_1 = \rho, \qquad c_2 = \varphi - \omega t_0,$$

it is evident that we shall satisfy conditions (15). Thus, through every point (t_0, x_0, y_0) passes a solution given by formula (14). By virtue of Theorem 2 (uniqueness), formula (14) contains the set of *all* solutions.

2. Let us solve the equation

$$\dot{x} = x^2 \cos t \tag{16}$$

with separable variables. The domain Γ for this equation is the entire plane (t, x). For $x > 0$ and for $x < 0$ this equation can be solved by the method presented in (C) of §2. For each of these half-planes we have

$$\int \frac{dx}{x^2} = \int \cos t \, dt,$$

or

$$-\frac{1}{x} = \sin t - c.$$

Thus we obtain

$$x = \frac{1}{c - \sin t}. \tag{17}$$

In addition to the solutions described by this formula, we have the trivial solution

$$x = 0. \tag{18}$$

We shall show that formulas (17) and (18) include the set of all solutions of equation (16). Let (t_0, x_0) be arbitrary initial values. If $x_0 = 0$, then the solution (18) has these initial values. If $x_0 \neq 0$, however, then the constant c has the value

$$c = \sin t_0 + \frac{1}{x_0}.$$

The solution (18) is defined on the interval $(-\infty, +\infty)$, and this interval is the maximum interval of existence of the solution for the corresponding initial values. In exactly the same way, for $|c| > 1$ formula (17) determines one solution defined on the interval $(-\infty, \infty)$, and this interval is maximal. For a fixed constant c which satisfies the inequality $|c| \leq 1$, formula (17) gives not one solution, but an *infinite set of solutions*. Each individual solution in this case is defined on the interval $r_1 < t < r_2$, where r_1 and r_2 are two consecutive zeros of the function $\sin t - c$. This interval is the maximum interval of existence of the solution for the corresponding initial values, for, as t approaches the endpoints of the interval, the function (17) tends to infinity (Fig. 9).

3. We shall show that if the right-hand sides (2) of the system of equations (1) are k times continuously differentiable, i.e., have continuous derivatives of the kth order (including derivatives of mixed type) with respect to all variables $t, x^1, \ldots, x^n$, then the $(k+1)$st derivative of the solution (4) of (1) exists and is continuous.

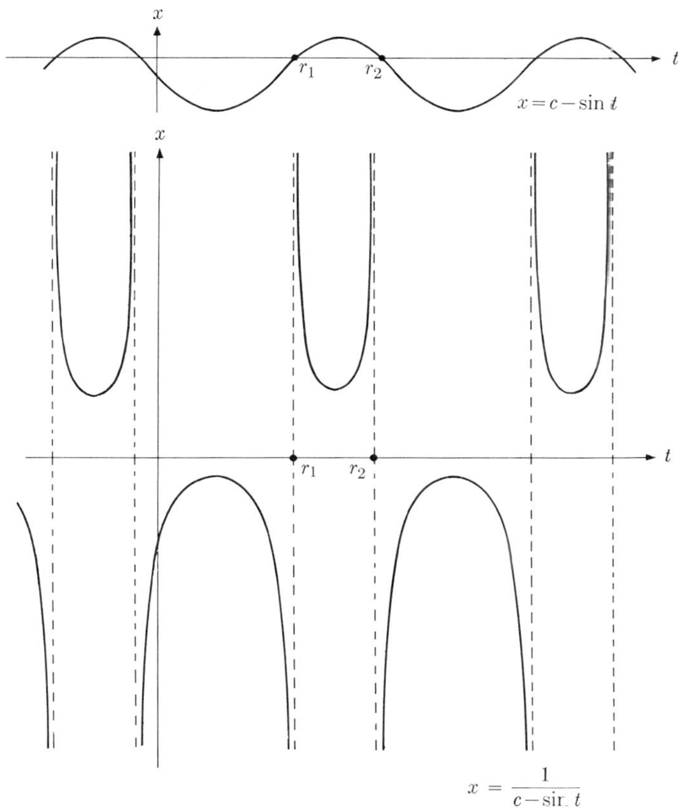

FIGURE 9

Indeed, for (4), we have the identity

$$\dot{\varphi}^i(t) = f^i(t, \varphi^1(t), \ldots, \varphi^n(t)), \qquad i = 1, \ldots, n. \tag{19}$$

If the right-hand sides (2) have continuous first derivatives, then the right-hand side of (19) has a continuous derivative with respect to t, so that $\ddot{\varphi}^i(t)$ exists and is continuous. By differentiating (19) k times, we establish (in order) the existence and continuity of all derivatives of the 2nd, 3rd, . . . , $(k + 1)$st order of the functions $\varphi^i(t)$.

4. Reduction of a general system of differential equations to a normal system. In the preceding section we formulated an existence and uniqueness theorem for a normal system of differential equations. Here it will be shown how quite general systems of differential equations may be reduced to normal systems, and at the same time an existence and uniqueness theorem will be established for these general systems.

We shall first give the concept of a system of differential equations in the general form.

In the case of one unknown function x of an independent variable t it is customary to consider one equation, which can be written in the form

$$F(t, x, \dot{x}, \dots, x^{(n)}) = 0. \tag{1}$$

Here t is an independent variable, x is the unknown function, and F is a given function of $n + 2$ variables. The function F need not be defined for all values of its arguments; for this reason we speak of the domain of definition B of the function F. Here we have in mind the $(n + 2)$-dimensional domain B in which the coordinates of a point are the variables $t, x, \dot{x}, \dots, x^{(n)}$. If the highest-order derivative entering the differential equation is equal to n, then we speak of an nth-order equation. By a *solution* of equation (1) we shall mean a continuous function $x = \varphi(t)$ of the variable t, which is defined on a certain interval $r_1 < t < r_2$, such that substitution of $\varphi(t)$ for x in (1) reduces (1) to an identity in t on the interval $r_1 < t < r_2$. It is clear that substitution of $x = \varphi(t)$ in (1) is possible only when $\varphi(t)$ has derivatives up to the nth order, inclusive, on the entire interval of existence $r_1 < t < r_2$. For the substitution of $x = \varphi(t)$ into relation (1) to be possible, it is also necessary that the point with coordinates $\{t, \varphi(t), \dot{\varphi}(t), \dots, \varphi^{(n)}(t)\}$ belong to the domain B of F for any t in the interval $r_1 < t < r_2$.

If there are two unknown functions of one independent variable, then two differential equations are considered which together form a *system of equations*. This system can be written in the form

$$\begin{aligned} F(t, x, \dot{x}, \dots, x^{(m)}, y, \dot{y}, \dots, x^{(n)}) &= 0, \\ G(t, x, \dot{x}, \dots, x^{(m)}, y, \dot{y}, \dots, y^{(n)}) &= 0. \end{aligned} \tag{2}$$

Here t is an independent variable, x and y are the two unknown functions, and F and G are two functions, each of $m + n + 3$ variables, which are defined in a certain domain B. If the highest-order derivative of x entering into (2) is equal to m, and the highest-order derivative of y entering into (2) is equal to n, then the number m is called the *order* of the system (2) *with respect to* x, the number n is the *order* of the system (2) *with respect to* y, and the number $m + n$ is called the *order* of the system (2). A pair of continuous functions $x = \varphi(t)$ and $y = \psi(t)$ defined on a certain interval $r_1 < t < r_2$ will be called a *solution* of (2) if they have the property that substituting them into (2) reduces (2) to an identity in t on the entire interval $r_1 < t < r_2$. As in the case of one equation, it is assumed that conditions are satisfied which make possible the substitution of $x = \varphi(t)$, $y = \psi(t)$ into (2).

Systems of differential equations with three or more unknown functions of one independent variable are defined similarly. If the unknown functions of a system of differential equations are the functions $x^1, \ldots, x^n$, and the highest-order derivative of x^i entering the system is q_i, $i = 1, \ldots, n$, then the number q_i is called the *order* of the system *with respect to* x^i, and the number $q = q_1 + q_2 + \cdots + q_n$ is called the *order* of the system. Thus, the normal system (1) of §3 is of order n.

If (1) can be solved for $x^{(n)}$, then equation (1) may be rewritten in the form

$$x^{(n)} = f(t, x, \dot{x}, \ldots, x^{(n-1)}). \tag{3}$$

In exactly the same way, if (2) can be solved for $x^{(m)}$ and $y^{(n)}$, then this system can be rewritten in the form

$$\begin{aligned} x^{(m)} &= f(t, x, \dot{x}, \ldots, x^{(m-1)}, y, \dot{y}, \ldots, y^{(n-1)}), \\ y^{(n)} &= g(t, x, \dot{x}, \ldots, x^{(m-1)}, y, \dot{y}, \ldots, y^{(n-1)}). \end{aligned} \tag{4}$$

Equation (3) and system (4) are said to be *solved for the highest derivatives.* Systems with an arbitrary number of functions which can be solved for the highest derivatives are defined analogously. In particular, the normal system (1) of §3 can be solved for the highest derivatives. Later we shall concern ourselves almost exclusively with systems which can be solved for the highest derivatives.

We shall now show that any nth-order system of differential equations which is solved for the highest derivatives may be reduced to a normal nth-order system. To begin, we shall show how one nth-order equation is reduced to an nth-order normal system.

(A) Let

$$y^{(n)} = f(t, y, \dot{y}, \ldots, y^{(n-1)}) \tag{5}$$

be an nth-order differential equation which is solved for the highest derivative. Here t is an independent variable and y is an unknown function of the variable t. Further, $f(t, y, \dot{y}, \ldots, y^{(n-1)})$ is a given function of $n + 1$ variables $t, y, \dot{y}, \ldots, y^{(n-1)}$, which is defined in a certain domain Γ of an $(n + 1)$-dimensional space. With respect to the function

$$f(t, y, \dot{y}, \ldots, y^{(n-1)}),$$

we shall assume that it is continuous in Γ and that its partial derivatives

$$\frac{\partial f(t, y, \dot{y}, \ldots, y^{(n-1)})}{\partial y^{(k)}}, \qquad k = 0, 1, \ldots, n - 1,$$

(where it is assumed that $y^{(0)} = y$) are also continuous in Γ. To change equation (5) into a normal system of equations, new unknown functions

$x^1, x^2, \ldots, x^n$ of the independent variable t are introduced by means of the equalities

$$x^1 = y, \qquad x^2 = \dot{y}, \ldots, \qquad x^n = y^{(n-1)}. \tag{6}$$

Thus equation (5) is equivalent to the system

$$\begin{aligned} \dot{x}^1 &= x^2, \\ \dot{x}^2 &= x^3, \\ &\vdots \\ \dot{x}^{n-1} &= x^n, \\ \dot{x}^n &= f(t, x^1, x^2, \ldots, x^n). \end{aligned} \tag{7}$$

Theorem 2 implies that for every point $t_0, y_0, \dot{y}_0, \ldots, y_0^{(n-1)}$ of Γ there exists a solution $y = \psi(t)$ of equation (5) which satisfies the initial conditions

$$\psi^{(k)}(t_0) = y_0^{(k)}, \qquad k = 0, 1, \ldots, n-1,$$

or, as it is called, a solution with initial conditions

$$t_0, y_0, \dot{y}_0, \ldots, y_0^{(n-1)}. \tag{8}$$

Further, any two solutions with initial values (8) coincide on the common part of their intervals of definition. If equation (5) is linear, i.e., if f is linear with respect to the variables $y, \dot{y}, \ldots, y^{(n-1)}$, and if its coefficients are defined and continuous on the interval $q_1 < t < q_2$, then for any initial values $t_0, y_0, \dot{y}_0, \ldots, y_0^{(n-1)}$, where $q_1 < t_0 < q_2$, a solution $y = \psi(t)$ exists which is defined on the entire interval $q_1 < t < q_2$.

We shall prove that equation (5) is equivalent to the system (7). We shall assume that y satisfies (5) and prove that the functions $x^1, \ldots, x^n$ defined by (6) satisfy (7). By differentiating the relations (6) which introduce the new unknown functions $x^1, \ldots, x^n$, we obtain

$$\dot{x}^k = y^{(k)}, \qquad k = 1, \ldots, n-1, \tag{9}$$

$$\dot{x}^n = y^{(n)}. \tag{10}$$

Replacing the right-hand sides of (9) on the basis of (6) and the right-hand side of (10) on the basis of (5), which is satisfied by y, we obtain (7). Let us assume, conversely, that the functions $x^1, \ldots, x^n$ satisfy (7); we shall then take x^1 for y and show that y satisfies (5). Setting $x^1 = y$ in the first of the equations of (7), we obtain $x^2 = \dot{y}$. Substituting $\dot{y}$ for x^2 in the second equation of (7), we obtain $x^3 = \ddot{y}$. Continuing this construction further, we arrive at the relations (6). Finally, substituting in the last of the equations of system (7) each function $x^1, \ldots, x^n$ by those of (6), we obtain equation (5) for y.

Since the function f is defined in Γ, the right-hand sides of (7) are also defined in Γ under the condition that the change of coordinates obeys (6). For system (7) the conditions of Theorem 2 are satisfied in Γ. Thus it is possible to select arbitrarily the initial values $t_0, x_0^1, x_0^2, \ldots, x_0^n$ in Γ. This set of initial values is transformed by (6) into a set of initial values $t, y_0, \dot{y}_0, \ldots, y_0^{(n-1)}$ for equation (5).

If (5) is linear, then (7) is also linear. From this follows the final part of (A), by Theorem 3. Thus proposition (A) has been proved.

The method described in (A) makes it possible to reduce to a normal system any system of differential equations which is solved for the highest derivatives. In order not to encumber the presentation with formulas, we shall investigate in the following proposition (B) a fourth-order system consisting of two equations.

(B) Let

$$\begin{aligned} \ddot{u} &= f(t, u, \dot{u}, v, \dot{v}), \\ \ddot{v} &= g(t, u, \dot{u}, v, \dot{v}) \end{aligned} \tag{11}$$

be a system of two second-order equations. Here t is an independent variable, and u and v are the unknown functions. We shall reduce (11) to a normal system by introducing new unknowns x^1, x^2, x^3, x^4 according to the formulas

$$x^1 = u, \qquad x^2 = \dot{u}, \qquad x^3 = v, \qquad x^4 = \dot{v}.$$

By this substitution, (11) is transformed into the system

$$\begin{aligned} \dot{x}^1 &= x^2, \\ \dot{x}^2 &= f(t, x^1, x^2, x^3, x^4), \\ \dot{x}^3 &= x^4, \\ \dot{x}^4 &= g(t, x^1, x^2, x^3, x^4). \end{aligned} \tag{12}$$

If it is assumed that the functions f and g, on the right-hand sides of (11), are defined in a domain Γ of a five-dimensional space, where $t, u, \dot{u}, v, \dot{v}$ are the coordinates of a point, these functions being continuous and having continuous first-order partial derivatives with respect to $u, \dot{u}, v, \dot{v}$, then the system (12) is normal and satisfies the conditions of Theorem 2 in Γ. From this it easily follows that for any point $t_0, u_0, \dot{u}_0, v_0, \dot{v}_0$ of Γ there exists a solution $u = \varphi(t)$, $v = \psi(t)$ of (11) which satisfies the initial conditions

$$\begin{aligned} \varphi(t_0) &= u_0, \qquad \dot{\varphi}(t_0) = \dot{u}_0, \\ \psi(t_0) &= v_0, \qquad \dot{\psi}(t_0) = \dot{v}_0. \end{aligned}$$

In addition, all solutions with identical initial conditions coincide on the common part of their intervals of existence.

The proof of (B) is carried out in the same manner as the proof of (A).

If one nth-order equation, given in the form

$$F(t, y, \dot{y}, \ldots, y^{(n)}) = 0, \tag{13}$$

is not solved for the highest-order derivative $y^{(n)}$ of the unknown, then the question immediately arises of its solvability with respect to $y^{(n)}$. We may assume that this question does not relate to the field of differential equations, but rather to the theory of functions. There arise, however, some questions, which are investigated in the theory of differential equations, of the following character. Let us assume that equation (13) is quadratic with respect to the variable $y^{(n)}$. Then it defines a *two-valued* function $y^{(n)}$ of the remaining variables. Where the two values are really distinct, we actually arrive at two different equations of the form (5), but where the two values of variable $y^{(n)}$ defined by equation (13) coincide, splitting them into two equations of form (5) is impossible, and it is necessary to study equation (13). The study of such equations leads to the concept of *singular solutions* of the differential equation and to the study of equations on surfaces. These questions, however, will not be studied in this book.

EXAMPLES

1. We shall solve the equation

$$\ddot{x} + \omega^2 x = 0, \qquad \omega = \text{const.} \tag{14}$$

By direct inspection we see that the function

$$x = r \cos(\omega t + \alpha), \qquad r \geq 0, \tag{15}$$

where r and α are constant, satisfies this equation. We shall show that (15) contains the set of all solutions. Let $x = \varphi(t)$ be any solution of (14). By Theorem 3 [see the end of (A)], it may be assumed that the solution $x = \varphi(t)$ is defined for all values of t. Let us set $\varphi(0) = x_0$, $\dot{\varphi}(0) = \dot{x}_0$. We see easily that r and α can be selected in such a way that $r \cos \alpha = x_0$, $-r\omega \sin \alpha = \dot{x}_0$ hold.

If these equalities are satisfied, then the solutions (15) and $\varphi(t)$ have identical initial values $0, x_0, \dot{x}_0$, and therefore coincide [see (A)].

The function (15) describes the *harmonic oscillatory process* for the *harmonic oscillator*. The positive constant r is called the *amplitude* of the oscillation (15), and α is its *initial phase* or simply its *phase*. Equation (14)

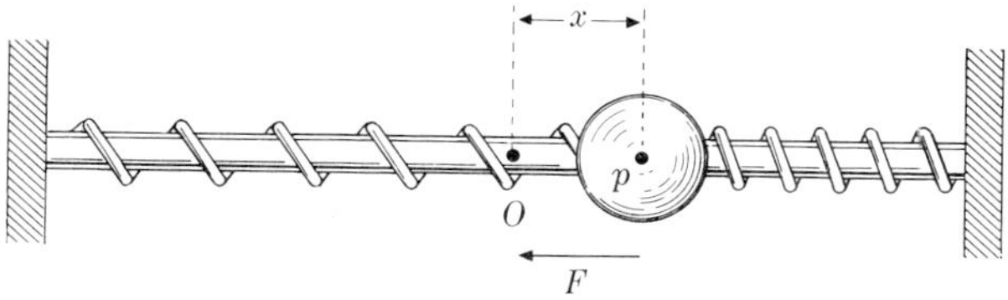

FIGURE 10

is called the equation of the harmonic oscillator. The number ω is called the *frequency* of oscillation, although in reality the number of oscillations per second is determined by the formula

$$\nu = \frac{\omega}{2\pi}\cdot$$

2. We shall investigate the motion of a point p with mass m along a horizontal straight line l under the action of a force F which is attracting it toward the point O on the same straight line and is proportional to the distance between the points p and O. To form the equation of motion of the point p we shall take a coordinate system with the point O as origin on the line l. The variable coordinate of the point p we shall designate by $x = x(t)$. Then, by Newton's second law, the equation of motion of the point p will have the form

$$m\ddot{x} = F = -kx.$$

This equation is usually written in the form

$$m\ddot{x} + kx = 0. \tag{16}$$

Physically, the force F can be realized by a spring of some kind (Fig. 10). The number k is called the *coefficient of elasticity* of the spring. According to formula (15), the solution of (16) has the form

$$x = r\cos\left(\sqrt{\frac{k}{m}}\,t + \alpha\right), \qquad r \geq 0.$$

Thus, the frequency of oscillation $\omega = \sqrt{k/m}$ of the point p is determined by its mass m and by the elasticity k of the spring; it does not depend on the initial conditions. The amplitude of the oscillation r and its initial phase α depend on the initial conditions, i.e., on the location x_0 of the point p and on its velocity $\dot{x}_0$ at the instant $t = 0$.

3. We shall formulate and give an approximate solution of the equation of the *mathematical pendulum*. The mathematical pendulum is represented by a point p of mass m which under the influence of gravity moves on the

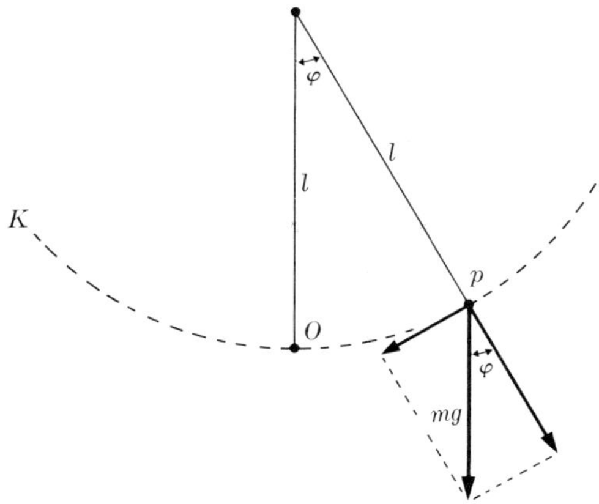

FIGURE 11

circumference K of radius l in the vertical plane. The quantity l is called the *length* of the pendulum. On the circumference K we shall introduce an angular coordinate by taking the lowest point O of the circumference as the origin of coordinates (Fig. 11). The variable coordinate of the point p we shall denote by $\varphi = \varphi(t)$. The point p is subject to the gravitational force $P = mg$, directed vertically downward. The component of this force along the normal to the circumference is balanced by the normal component of the inertial force (i.e., centrifugal force with opposite sign) and by the reaction of the connection (of the circumference or of the thread which forces the point to move along the circumference); the component along the tangent to the circumference at point p in the direction of increase of the angle φ is equal to $-mg \sin \varphi$. Thus the equation of motion of the point p has the form $ml\ddot{\varphi} = -mg \sin \varphi$, or

$$l\ddot{\varphi} + g \sin \varphi = 0. \tag{17}$$

This equation is nonlinear and its solution presents great difficulties.

If it is assumed that the coordinate φ of p is very close to zero during the motion, then in equation (17), φ may be substituted approximately for $\sin \varphi$ and we obtain an "approximate" linear equation of the pendulum:

$$l\ddot{\varphi} + g\varphi = 0.$$

Its solution has the form [see (15)]

$$\varphi = r \cos\left(\sqrt{\frac{g}{l}}\, t + \alpha\right).$$

Thus the frequency of "small oscillations" of the pendulum is determined by the formula $\omega = \sqrt{g/l}$.

The frequency of oscillation of the solutions of equation (17) depends on the amplitude of the oscillations and decreases when the amplitude of the oscillations increases.

The number ν of small oscillations of a pendulum per second is determined by the formula

$$\nu = \frac{\omega}{2\pi} = \frac{1}{2\pi}\sqrt{\frac{g}{l}}\,.$$

For example, the length of a *seconds pendulum*, i.e., of a pendulum which performs one oscillation per second ($\nu = 1$), is determined by the formula

$$l = \frac{g}{4\pi^2} \approx 0.25 \text{ m}.$$

5. Complex differential equations. Up to this point we have considered only real equations and their real solutions. In certain cases, however, as in the case of the solution of linear equations with constant coefficients, it is easier to first find *complex solutions* of the real equation and then to select from them the real solutions. To present this approach we must introduce the concepts of a complex function of a real variable and of a complex system of differential equations.

(A) A *complex function* $\chi(t)$ of a real variable t is said to be defined if, to each t on a certain interval $r_1 < t < r_2$, there corresponds a complex number

$$\chi(t) = \varphi(t) + i\psi(t),$$

where $\varphi(t)$ and $\psi(t)$ are real functions of the real variable t. The function $\varphi(t)$ is called the *real part* of the complex function $\chi(t)$ and function $\psi(t)$ is called its *imaginary part*. A complex function $\chi(t)$ is said to be *continuous* if the functions $\varphi(t)$ and $\psi(t)$ are continuous. In exactly the same way, the complex function $\chi(t)$ is called *differentiable* if the functions $\varphi(t)$ and $\psi(t)$ are differentiable; the derivative $\dot{\chi}(t)$ of a complex function $\chi(t)$ is defined by the formula

$$\dot{\chi}(t) = \dot{\varphi}(t) + i\dot{\psi}(t).$$

By direct inspection we see that the usual rules for differentiation of the sum, product, and quotient of two complex functions of a real variable are valid.

(B) Let

$$\dot{z}^j = h^j(t, z^1, \ldots, z^n), \qquad j = 1, \ldots, n \tag{1}$$

be a normal system of differential equations. We assume that the functions $h^j(t, z^1, \ldots, z^n)$ on the right-hand sides of the equations are defined for complex values of the variables $z^1, \ldots, z^n$. We can confine ourselves, for example, to the case when these functions are polynomials in the variables $z^1, \ldots, z^n$ with coefficients which are real or complex continuous functions of a real variable t defined and continuous on the interval $q_1 < t < q_2$. Under these conditions, it is quite legitimate to pose the problem of finding *complex solutions* of system (1). The system

$$z^j = \chi^j(t), \qquad j = 1, \ldots, n, \tag{2}$$

of complex functions of a real variable t which are defined on some interval $r_1 < t < r_2$ will be called a *solution* of (1) if substitution of the functions (2) for the variables z^j leads to a system of identities in t on this interval. Since we are assuming that the right-hand sides of (1) are polynomials in $z^1, \ldots, z^n$, they are defined for all values of these variables. Hence, the following existence and uniqueness theorem is valid for the system (1).

Let

$$t_0, z_0^1, z_0^2, \ldots, z_0^n$$

be an arbitrary system of initial values. Here $z_0^1, \ldots, z_0^n$ are arbitrary complex numbers, and t_0 is an arbitrary real number satisfying the condition $q_1 < t_0 < q_2$. Then there exists a solution

$$z^j = \chi^j(t), \qquad j = 1, \ldots, n,$$

of (1) which satisfies the initial conditions

$$\chi^j(t_0) = z_0^j, \qquad j = 1, \ldots, n.$$

Any two solutions with identical initial conditions coincide on the common part of their intervals of definition.

If (1) is linear, i.e., if the polynomials h^j are of first degree, then for any initial values there exists a solution of (1) defined on the entire interval $q_1 < t < q_2$.

This existence and uniqueness theorem for a normal system of complex equations follows directly from Theorem 2 after every complex unknown function z^j is split into its real and imaginary parts. Indeed, let us assume

$$z^j = x^j + iy^j, \qquad j = 1, \ldots, n, \tag{3}$$

and make a change of the variables z^j, $j = 1, \ldots, n$, in system (1) according to formulas (3); then we shall have

$$\dot{x}^j + i\dot{y}^j = f^j(t, x^1, \ldots, x^n, y^1, \ldots, y^n) \\ + ig^j(t, x^1, \ldots, x^n, y^1, \ldots, y^n), \tag{4}$$

where f^j and g^j are real functions of real arguments which satisfy the relations

$$f^j(t, x^1, \ldots, x^n, y^1, \ldots, y^n) + ig^j(t, x^1, \ldots, x^n, y^1, \ldots, y^n) = h^j(t, x^1 + iy^1, \ldots, x^n + iy^n).$$

From (4) it follows that

$$\begin{aligned} \dot{x}^j &= f^j(t, x^1, \ldots, x^n, y^1, \ldots, y^n), \qquad j = 1, \ldots, n, \\ \dot{y}^j &= g^j(t, x^1, \ldots, x^n, y^1, \ldots, y^n), \qquad j = 1, \ldots, n. \end{aligned} \tag{5}$$

Thus, in place of the normal system (1) of complex equations, we substitute a normal system (5) of real variables. Since the right-hand sides of equations (1) are polynomials in $z^1, \ldots, z^n$, the right-hand sides of (5) are polynomials in $x^1, \ldots, x^n, y^1, \ldots, y^n$. Since the coefficients of the polynomials h^j are continuous functions of t on the interval $q_1 < t < q_2$, the coefficients of the polynomials f^j and g^j are also continuous functions on the same interval. Thus, the right-hand sides of (5) are defined and satisfy the hypotheses of Theorem 2 in the domain Γ, which is defined by the single condition $q_1 < t < q_2$ imposed on t, while the remaining variables $x^1, \ldots, x^n$ and $y^1, \ldots, y^n$ remain arbitrary. Setting

$$\begin{aligned} z_0^j &= x_0^j + iy_0^j, \qquad & j = 1, \ldots, n, \\ \chi^j(t) &= \varphi^j(t) + i\psi^j(t), \qquad & j = 1, \ldots, n, \end{aligned}$$

we arrive at the problem of determining the solution of (5) under the initial conditions

$$\begin{aligned} \varphi^j(t_0) &= x_0^j, \qquad & j = 1, \ldots, n, \\ \psi^j(t_0) &= y_0^j, \qquad & j = 1, \ldots, n. \end{aligned}$$

By Theorem 2, this solution exists, and any two solutions with identical initial conditions coincide on the common part of their intervals of definition.

If (1) is linear, then (5) is also linear, and for this reason the final part of proposition (B) follows from Theorem 3.

It should be noted that system (1), whose right-hand sides consist of polynomials in $z^1, \ldots, z^n$, can be *real*, i.e., the coefficients of these polynomials can be real functions of t. Nevertheless, we can also treat (1) as a complex system; namely, we can find its complex solutions, assuming that the functions $z^1, \ldots, z^n$ are complex. This approach to real equations is used because in certain cases it is easier to find the complex solutions of real equations than the real solutions. In this case the complex solutions of a real system of equations are found first, and then the real solutions are selected from the complex solutions; i.e., only those solutions are considered whose imaginary part vanishes. Linear equations with constant coefficients will be solved later in exactly this way.

Just as in the real case, we can reduce quite general systems of differential equations to normal systems in the complex case. Thus in the complex case we have propositions analogous to propositions (A) and (B) of §4. Here we shall formulate only the existence theorems for an nth-order differential equation.

(C) Let

$$z^{(n)} = f(t, z, \dot{z}, \ldots, z^{(n-1)}) \tag{6}$$

be an nth-order equation whose right-hand side is a polynomial in the variables $z, \dot{z}, \ldots, z^{(n-1)}$ and whose coefficients are continuous real or complex functions of t defined on the interval $q_1 < t < q_2$. If t_0, z_0, $\dot{z}_0, \ldots, z_0^{(n-1)}$ are now arbitrary initial values, where $z_0, \dot{z}_0, \ldots, z_0^{(n-1)}$ are arbitrary complex numbers, and if t_0 is a real number satisfying $q_1 < t_0 < q_2$, then there exists a solution $z = \varphi(t)$ of equation (6) which satisfies the initial conditions

$$\varphi(t_0) = z_0, \qquad \dot{\varphi}(t_0) = \dot{z}_0, \ldots, \varphi^{(n-1)}(t_0) = z_0^{(n-1)}.$$

Any two solutions with identical initial conditions coincide on the common part of their intervals of definition. If (6) is linear, i.e., if the polynomial f is of the first degree, then for arbitrary initial values there exists a solution defined on the entire interval $q_1 < t < q_2$.

In §7 and below, the complex function $e^{\lambda t}$ of the real variable t, where λ is a complex number, will play an important role. We shall define this function here and prove some of its properties.

(D) Let $w = u + iv$ be an arbitrary complex number; let us assume that

$$e^{w} = e^{u} (\cos v + i \sin v). \tag{7}$$

It is easy to see that the relation $e^{\overline{w}} = \overline{e^{w}}$ holds. Below we shall prove the formula

$$e^{w_1} e^{w_2} = e^{w_1 + w_2}. \tag{8}$$

The well-known *Euler formulas* follow directly from (7):

$$\cos v = \frac{e^{iv} + e^{-iv}}{2}, \qquad \sin v = \frac{e^{iv} - e^{-iv}}{2i}.$$

Let $\lambda = \mu + i\nu$ be a complex number. By (7) we have

$$e^{\lambda t} = e^{\mu t} (\cos \nu t + i \sin \nu t).$$

We shall show that for complex values of λ the following differentiation formula is valid

$$\frac{d}{dt}e^{\lambda t} = \lambda e^{\lambda t}, \tag{9}$$

which is well known for real values of the parameter λ.

Formula (7), taken here as the *definition* of the function e^w of the complex variable w, can be proved if the function e^w is defined by the series

$$e^w = 1 + w + \frac{w^2}{2!} + \frac{w^3}{3!} + \cdots + \frac{w^n}{n!} + \cdots.$$

We shall, however, assume that the function e^w is defined by (7).

We shall prove formula (8). Setting

$$w_1 = u_1 + iv_1, \qquad w_2 = u_2 + iv_2$$

yields

$$\begin{aligned} e^{w_1}e^{w_2} &= e^{u_1}(\cos v_1 + i \sin v_1)e^{u_2}(\cos v_2 + i \sin v_2) \\ &= e^{u_1+u_2}(\cos(v_1 + v_2) + i \sin(v_1 + v_2)) = e^{w_1+w_2}. \end{aligned}$$

We shall now prove formula (9). We shall consider first the case of a pure imaginary number $\lambda = i\nu$. We have

$$\begin{aligned} \frac{d}{dt}e^{i\nu t} = \frac{d}{dt}(\cos \nu t + i \sin \nu t) &= -\nu \sin \nu t + i\nu \cos \nu t \\ &= i\nu(\cos \nu t + i \sin \nu t) = i\nu e^{i\nu t}. \end{aligned}$$

Further, for an arbitrary $\lambda = \mu + i\nu$, by the formula for the differentiation of a product, we have

$$\begin{aligned} \frac{d}{dt}e^{\lambda t} &= \frac{d}{dt}(e^{\mu t} \cdot e^{i\nu t}) = \frac{d}{dt}(e^{\mu t})e^{i\nu t} + e^{\mu t}\frac{d}{dt}(e^{i\nu t}) \\ &= \mu e^{\mu t}e^{i\nu t} + i\nu e^{\mu t}e^{i\nu t} = (\mu + i\nu)e^{\mu t + i\nu t} = \lambda e^{\lambda t}. \end{aligned}$$

Examples

1. We shall investigate the complex equation

$$\dot{z} = \lambda z, \tag{10}$$

where $z = x + iy$ is a complex unknown function of a real variable t, and $\lambda = \mu + i\nu$ is a complex number. From (9) it follows that

$$z = ce^{\lambda t} \tag{11}$$

is a solution of equation (10) for an arbitrary complex constant c. We shall show that formula (11) contains the set of all solutions. For this, as in Example 1 of §1, it would be possible to use the uniqueness theorem, but here we shall use Theorem 3 to show how it can be used to simplify somewhat the calculations. In this case these simplifications are not particularly significant, but later an analogous method can give more substantial results. Thus, let $z = \chi(t)$ be an arbitrary solution of (10). By virtue of Theorem 3 [see the last part of proposition (C)], we may assume that this solution is defined for all values of t. Setting $\chi(0) = z_0$, we see that the solution $z = \chi(t)$ has as its initial values the numbers 0 and z_0. It is clear that the solution

$$z = z_0 e^{\lambda t},$$

which is obtained from (11) by setting $c = z_0$, has the same initial values. If it is assumed that $c = re^{i\alpha}$, where $r > 0$ and α are real numbers, then (11) may be written in the form

$$z = re^{\lambda t + i\alpha}. \tag{12}$$

We now split equation (10) into its real and imaginary parts. We have

$$\dot{x} + i\dot{y} = (\mu + i\nu)(x + iy) = (\mu x - \nu y) + i(\nu x + \mu y)$$

or

$$\dot{x} = \mu x - \nu y, \qquad \dot{y} = \nu x + \mu y. \tag{13}$$

Thus the system (13) of two real equations is equivalent to one complex equation (10), and therefore an arbitrary solution $x = \varphi(t)$, $y = \psi(t)$ of (13) is related to the arbitrary solution (12) of (10) by the relation

$$\varphi(t) + i\psi(t) = re^{\lambda t + i\alpha} = r\big(e^{\mu t}\cos(\nu t + \alpha) + i\sin(\nu t + \alpha)\big),$$

whence we obtain

$$x = \varphi(t) = re^{\mu t}\cos(\nu t + \alpha), \qquad y = \psi(t) = re^{\mu t}\sin(\nu t + \alpha). \tag{14}$$

Thus, by using complex functions and equations, we have found the solution (14) of the system (13) of real equations.

2. We shall present one more example of splitting a complex equation into two real ones. Let

$$\dot{z} = z^2 + iz$$

be a complex equation, where $z = x + iy$ is a complex unknown function of the real variable t. We have

$$\dot{x} + i\dot{y} = (x + iy)^2 + i(x + iy) = (x^2 - y^2 - y) + i(2xy + x)$$

and therefore

$$\dot{x} = x^2 - y^2 - y, \qquad \dot{y} = 2xy + x.$$

6. Some properties of linear differential equations. A system of differential equations is called *linear* if all unknown functions and their derivatives, taken together, enter the equations of the system linearly. Thus the system of linear equations of the most general type can be written in the form

$$\sum_{j,k} a_{ijk}(t)(x^j)^{(k)} + b_i(t) = 0, \qquad i = 1, \ldots, n. \tag{1}$$

Here $x^1, \ldots, x^n$ are unknown functions of the independent variable t, and the coefficients $a_{ijk}(t)$ and the free terms $b_i(t)$ of the equations are functions of t. If all free terms of (1) are identically zero, then the system is called *homogeneous*. To each linear system corresponds a homogeneous linear system obtained from it by discarding the free terms. Thus to the linear system (1) corresponds the homogeneous linear system

$$\sum_{j,k} a_{ijk}(t)(y^j)^{(k)} = 0, \qquad i = 1, \ldots, n. \tag{2}$$

We shall note several immediate properties of linear systems. In formulating these, it will be assumed that all coefficients and free terms of the linear system are defined and continuous on the interval $q_1 < t < q_2$; all solutions under consideration will be assumed to be defined on the entire interval $q_1 < t < q_2$.

(A) If $y^i = \varphi^i(t)$ and $y^i = \psi^i(t)$, $i = 1, \ldots, n$ are two solutions of the linear homogeneous system (2), and if c_1 and c_2 are two arbitrary numbers, then the system of functions

$$y^i = c_1\varphi^i(t) + c_2\psi^i(t), \qquad i = 1, \ldots, n,$$

is also a solution of the homogeneous system (2). An analogous assertion is also valid for three or more solutions of (2).

(B) If $x^i = \psi^i(t)$ and $x^i = \chi^i(t)$, $i = 1, \ldots, n$, are two solutions of (1), then the system of functions

$$y^i = \chi^i(t) - \psi^i(t), \qquad i = 1, \ldots, n,$$

is a solution of the system of homogeneous equations (2). Further, if $y^i = \varphi^i(t)$, $i = 1, \ldots, n$, is a solution of the homogeneous system (2), and $x^i = \psi^i(t)$, $i = 1, \ldots, n$, is a solution of (1), then the system of functions

$$x^i = \varphi^i(t) + \psi^i(t), \qquad i = 1, \ldots, n,$$

is also a solution of (1).

(C) We shall assume that the free terms of (1) may be written in the form of sums

$$b^i(t) = \alpha c_i(t) + \beta d_i(t), \qquad i = 1, \ldots, n.$$

We shall consider along with system (1) two systems of equations:

$$\sum_{j,k} a_{ijk}(t)(x^j)^{(k)} + c_i(t) = 0, \qquad i = 1, \ldots, n, \tag{3}$$

$$\sum_{j,k} a_{ijk}(t)(x^j)^{(k)} + d_i(t) = 0, \qquad i = 1, \ldots, n. \tag{4}$$

If $x^i = \psi^i(t)$, $i = 1, \ldots, n$, is a solution of (3) and $x^i = \chi^i(t)$, $i = 1, \ldots, n$, is a solution of (4), then the system of functions

$$x^i = \alpha\psi^i(t) + \beta\chi^i(t), \qquad i = 1, \ldots, n,$$

is a solution of (1).

CHAPTER 2

LINEAR EQUATIONS WITH CONSTANT COEFFICIENTS

Systems of ordinary differential equations with constant coefficients constitute a large and important class of ordinary differential equations which may be solved completely with the aid of elementary functions. In view of the fact that the solution of these equations does not, in principle, present any great difficulties, they are often considered to be of no great interest for theory, and in textbooks they are usually relegated to the position of simple exercises appended to the general theory of linear equations. Linear equations with constant coefficients, nevertheless, have numerous engineering applications, since the performance of many technical devices is described in an adequate manner by these equations. It is precisely the engineering applications which bring forward a series of new problems of a theoretical nature in the theory of linear equations with constant coefficients. To the solution of these theoretical problems many studies of an applied nature have been devoted, and some of these are reflected in the present chapter. Thus in this chapter we employ the *operational notation* which is customary in engineering practice and which is very convenient for the solution of systems of equations by the method of elimination. The problem of the *stability of solutions* of systems of linear equations, which is very important in the theory of automatic control, is studied. Further, we shall develop the so-called *method of complex amplitude*, which is a convenient means for determining particular steady-state solutions and is widely applied in electrical engineering.

Rather than confining ourselves to the solution of the purely mathematical problems arising from applications, we shall present here in very short dogmatic form an exposition of the theory of electrical circuits. The design of electrical circuits gives a good and important, from the engineering point of view, illustration of the mathematical methods developed in this chapter.

In addition, the present chapter includes a study of the phase plane of second-order linear systems, which is preceded by a very elementary study of phase spaces of (generally speaking, nonlinear) autonomous systems. Phase spaces of autonomous systems also find important applications in engineering.

Because of what has been said above, the chapter on linear equations with constant coefficients occupies a considerably larger place in this book than is customary in textbooks on the theory of ordinary differential equations. The presentation of all the material of the present chapter is very elementary, with the sole exception of §14, where matrices in Jordan form are used. Everything done with the aid of matrices in Jordan form may be skipped, as has been indicated in detail in §14, since it is not used later in the book.

7. The linear homogeneous equation with constant coefficients. Case of simple roots. In this section and in the following we shall solve the linear homogeneous nth-order equation with constant coefficients, i.e., the equation

$$z^{(n)} + a_1 z^{(n-1)} + \cdots + a_{n-1}\dot{z} + a_n z = 0, \tag{1}$$

where z is an unknown function of the independent variable t, and the coefficients $a_1, \ldots, a_n$ are constants (real or complex). First, we shall find all complex solutions of this equation, and then (in the case where coefficients $a_1, \ldots, a_n$ are real) separate from them the real solutions. Equation (1) can be written in the form

$$z^{(n)} = -a_1 z^{(n-1)} - \cdots - a_{n-1}\dot{z} - a_n z, \tag{2}$$

so that the existence and uniqueness theorem can be applied to it [see proposition (C), §5]. Later on, we shall use only the uniqueness, since solutions of equation (2) will be found explicitly, and their existence will be established by this fact; uniqueness itself will be used to prove that *all* the solutions have been found.

In engineering applications of ordinary differential equations with constant coefficients, an important role is played by the *operational calculus.* We shall use here the *symbolic* (or *operational*) *notation* which is the basis of the operational calculus. The essence of this notation consists of the fact that the derivative of an arbitrary function $z = z(t)$ with respect to the time t is designated not by $(d/dt)z$, but by pz, so that the letter p placed to the left of the function is a *differentiation symbol* with respect to t. If we apply to the differentiation symbol p certain algebraic operations, we are led to the notation

$$\frac{d^k}{dt^k} z = p^k z.$$

By using this notation, we can write

$$a_0 z^{(n)} + a_1 z^{(n-1)} + \cdots + a_{n-1}\dot{z} + a_n z$$
$$= a_0 p^n z + a_1 p^{n-1} z + \cdots + a_{n-1} p z + a_n z.$$

If in the right-hand side of the last equation we take the function z out of the parentheses, we obtain the equality

$$a_0 z^n + a_1 z^{(n-1)} + \cdots + a_{n-1}\dot{z} + a_n z = (a_0 p^n + a_1 p^{n-1} + \cdots + a_{n-1} p + a_n) z.$$

Thus we come to a formal definition.

(A) Let

$$L(p) = a_0 p^n + a_1 p^{n-1} + \cdots + a_{n-1} p - a_n$$

be an arbitrary polynomial with constant coefficients (real or complex) with respect to the symbol p, and let z be a certain real or complex function of a real variable t. We set

$$L(p)z = a_0 z^{(n)} + a_1 z^{(n-1)} + \cdots + a_{n-1}\dot{z} + a_n z. \tag{3}$$

If $L(p)$ and $M(p)$ are two arbitrary polynomials in the symbol p (or, as we say, in the *differentiation operator* p) and z, z_1, z_2 are functions of t, then we have the identities

$$L(p)(z_1 + z_2) = L(p)z_1 + L(p)z_2,$$
$$(L(p) + M(p))z = L(p)z + M(p)z,$$
$$L(p)(M(p)z) = (L(p)M(p))z.$$

By the notation introduced, equation (1) can be written in the form

$$L(p)z = 0, \tag{4}$$

where

$$L(p) = p^n + a_1 p^{n-1} + \cdots + a_{n-1} p + a_n.$$

(B) Let $L(p)$ be an arbitrary polynomial with respect to the symbol p. Then

$$L(p)e^{\lambda t} = L(\lambda)e^{\lambda t}. \tag{5}$$

We shall prove formula (5). We have

$$pe^{\lambda t} = \lambda e^{\lambda t}$$

[see formula (9), §5]. From this it follows that $p^k e^{\lambda t} = \lambda^k e^{\lambda t}$. Hence formula (5) follows immediately [see (3)].

It follows directly from formula (5) that the function $e^{\lambda t}$ is a solution of equation (4) if and only if the number λ is a root of the polynomial $L(p)$. The polynomial $L(p)$ is called the *characteristic polynomial* of equation (4). In the case when it has no multiple roots, the set of all solutions of (4) is described by the following theorem.

THEOREM 4. If the characteristic polynomial $L(p)$ of the equation

$$L(p)z = 0 \tag{6}$$

[see (1) and (4)] has no multiple roots, if its roots are

$$\lambda_1, \lambda_2, \ldots, \lambda_n,$$

and if we set

$$z_1 = e^{\lambda_1 t}, \qquad z_2 = e^{\lambda_2 t}, \quad \ldots, \quad z_n = e^{\lambda_n t}, \tag{7}$$

then for any complex constants $c^1, c^2, \ldots, c^n$, the function

$$z = c^1 z_1 + c^2 z_2 + \cdots + c^n z_n \tag{8}$$

is the solution of equation (6). This solution is the general solution in the sense that every solution of equation (6) can be obtained from (8) by proper choice of the constants $c^1, c^2, \ldots, c^n$. Here the constants $c^1, c^2, \ldots, c^n$ (which are called *integration constants*) are defined uniquely for every given solution z.

Thus the functions (7) constitute the so-called *fundamental system of solutions* of (6) [see §18, (C)].

We note that the functions (7) are defined over the entire axis $-\infty < t < +\infty$.

Proof. It follows from (5) that every function of the system (7) is a solution of (6), and, because (6) is linear and homogeneous, it follows [see §6, (A)] that, for any complex constants $c^1, c^2, \ldots, c^n$, (8) gives the solution of (6). We shall show that if $z_* = z_*(t)$ is an arbitrary solution of (6), then it can be written in the form (8). By proposition (C), §5, we can consider the solution z_* to be defined on the entire axis $-\infty < t < \infty$. Let us set

$$z_*(0) = z_0, \qquad \dot{z}_*(0) = \dot{z}_0, \quad \ldots, \quad z_*^{(n-1)}(0) = z_0^{(n-1)}.$$

We shall now show that constants $c^1, c^2, \ldots, c^n$ can be chosen in such a way that the solution $z(t)$, defined by (8), satisfies the same initial conditions

$$z(0) = z_0, \qquad \dot{z}(0) = \dot{z}_0, \quad \ldots, \quad z^{(n-1)}(0) = z_0^{n-1}. \tag{9}$$

Substituting the function z from (8) into the equations (9), we obtain

$$c^1 z_1^{(s)}(0) + \cdots + c^n z_n^{(s)}(0) = z_0^{(s)}; \qquad s = 0, 1, \ldots, n-1. \tag{10}$$

The relations (10) represent a system of equations in the unknowns $c^1, c^2, \ldots, c^n$. In order that (10) may be solved, it is sufficient that the determinant of the matrix

$$\begin{pmatrix} z_1(0) & z_2(0) & \cdots & z_n(0) \\ \dot{z}_1(0) & \dot{z}_2(0) & \cdots & \dot{z}_n(0) \\ \ddot{z}_1(0) & \ddot{z}_2(0) & \cdots & \ddot{z}_n(0) \\ \vdots & & & \\ z_1^{(n-2)}(0) & z_2^{(n-2)}(0) & \cdots & z_n^{(n-2)}(0) \\ z_1^{(n-1)}(0) & z_2^{(n-1)}(0) & \cdots & z_n^{(n-1)}(0) \end{pmatrix} \tag{11}$$

does not vanish.

It is immediately evident that the matrix (11) has the form

$$\begin{pmatrix} 1 & 1 & \cdots & 1 \\ \lambda_1 & \lambda_2 & \cdots & \lambda_n \\ \lambda_1^2 & \lambda_2^2 & \cdots & \lambda_n^2 \\ \vdots & & & \\ \lambda_1^{n-1} & \lambda_2^{n-1} & \cdots & \lambda_n^{n-1} \end{pmatrix}$$

and therefore its determinant (Vandermonde's determinant) is different from zero because all the numbers $\lambda_1, \lambda_2, \ldots, \lambda_n$ are mutually distinct. However, we shall give another (direct) proof that the determinant of (11) is different from zero. Later this proof will also be extended to the case of multiple roots.

If the determinant of matrix (11) were zero, then we could find a linear dependence between its rows. Let us assume that this linear dependence does exist. This means that there exist numbers $b_{n-1}, b_{n-2}, \ldots, b_0$, not all zero, such that, if we multiply the rows of the matrix (11) by these numbers and add, we obtain a row of zeros. If we calculate the nth term of this zero row, we obtain

$$b_{n-1}z_k(0) + b_{n-2}\dot{z}_k(0) + \cdots + b_1 z_k^{(n-2)}(0) + b_0 z_k^{(n-1)}(0) = 0. \tag{12}$$

If we denote by $M(p)$ the polynomial

$$b_0 p^{n-1} + b_1 p^{n-2} + \cdots + b_{n-2}p + b_{n-1}$$

we may write (12) in the form

$$M(p)z_k|_{t=0} = 0.$$

By virtue of formulas (5) and (7), we thus obtain

$$M(\lambda_k) = 0,$$

which is impossible, since the degree of $M(p)$ does not exceed $n - 1$, so that it cannot have n distinct roots $\lambda_1, \ldots, \lambda_k, \ldots, \lambda_n$. This contradiction indicates that the determinant of (10) cannot be zero; consequently, the constants $c^1, c^2, \ldots, c^n$ can be determined (and uniquely, moreover) so that the solutions $z_*(t)$ and $z(t)$ satisfy identical initial conditions. For this choice (and only for this choice) of constants the solution (8) coincides with the given solution $z_*(t)$. Thus Theorem 4 is proved.

If the coefficients of the polynomial $L(p)$ in equation (6) are real, then we have the problem of separating the real solutions from the set (8) of all complex solutions. The solution of this problem rests upon proposition (D), in the formulation and proof of which we shall use vector notation. We shall recall that notation here.

(C) We shall call the sequence consisting of n numbers,

$$\mathbf{u} = (u^1, u^2, \ldots, u^n),$$

a *vector* of an n-dimensional space. Here $\mathbf{u}$ is the vector and the numbers $u^1, u^2, \ldots, u^n$ are called its *coordinates* (or *components*). We shall always designate vectors by boldface letters. If the coordinates of the vector are real numbers, then the vector is called *real*; if its coordinates are *complex*, then the vector itself is considered complex. The vector $\bar{\mathbf{u}}$, which is the *complex conjugate* of the vector $\mathbf{u}$, is defined by the equality

$$\bar{\mathbf{u}} = (\overline{u^1}, \overline{u^2}, \ldots, \overline{u^n}).$$

It is clear that vector $\mathbf{u}$ is real if and only if

$$\bar{\mathbf{u}} = \mathbf{u}.$$

The *product* of the vector $\mathbf{u} = (u^1, u^2, \ldots, u^n)$ with a real or complex number α is defined by the formula

$$\alpha\mathbf{u} = \mathbf{u}\alpha = (\alpha u^1, \alpha u^2, \ldots, \alpha u^n).$$

The *sum* of the vectors

$$\mathbf{u} = (u^1, u^2, \ldots, u^n) \quad \text{and} \quad \mathbf{v} = (v^1, v^2, \ldots, v^n)$$

is defined by the formula

$$\mathbf{u} + \mathbf{v} = (u^1 + v^1, u^2 + v^2, \ldots, u^n + v^n).$$

The vector 0, all of whose coordinates are zero, is called the *null* vector.

Let

$$\mathbf{u}_1, \mathbf{u}_2, \ldots, \mathbf{u}_r$$

be a finite system of vectors. The relation

$$\alpha^1\mathbf{u}_1 + \alpha^2\mathbf{u}_2 + \cdots + \alpha^r\mathbf{u}_r = 0,$$

where not all of $\alpha^1, \alpha^2, \ldots, \alpha^r$ are zero, is called a *linear dependence* among the vectors $\mathbf{u}_1, \mathbf{u}_2, \ldots, \mathbf{u}_r$. If the vectors are not linearly dependent, then they are called *linearly independent.* Let

$$\mathbf{u}_j = (u_j^1, u_j^2, \ldots, u_j^n), \qquad j = 1, \ldots, r.$$

The numbers u_j^i form a matrix (u_j^i); $i = 1, \ldots, n$; $j = 1, \ldots, r$. If we assume that the upper index i designates the number of the row and the lower index j the number of the column, then the matrix (u_j^i) has height n and width r. Thus we have a correspondence between vector $\mathbf{u}_j$ in the matrix (u_j^i) and the jth column, which consists of the coordinates of this vector. Hence, it is clear that to a linear dependence among the vectors $\mathbf{u}_1, \mathbf{u}_2, \ldots, \mathbf{u}_r$ corresponds a linear dependence between the columns of the matrix (u_j^i). Whenever $r = n$, the matrix (u_j^i) is a square matrix, and the vectors $\mathbf{u}_1, \mathbf{u}_2, \ldots, \mathbf{u}_n$ are linearly independent if and only if the determinant $|u_j^i|$ of this matrix is not zero.

(D) Let

$$\mathbf{z}_1, \mathbf{z}_2, \ldots, \mathbf{z}_n \tag{13}$$

be a system of n linearly independent complex vectors in an n-dimensional space. Let us assume that (13) also contains the conjugate of every vector in (13). Under these assumptions, the vector $\mathbf{z}$, defined by the formula

$$\mathbf{z} = c^1\mathbf{z}_1 + \cdots + c^n\mathbf{z}_n, \tag{14}$$

is real if and only if the coefficients of all pairs of conjugate vectors are conjugate and the coefficients of all real vectors are real.

Let us prove this. To be definite, we shall assume that the relations

$$\bar{\mathbf{z}}_1 = \mathbf{z}_2, \ldots, \bar{\mathbf{z}}_{2k-1} = \mathbf{z}_{2k},$$
$$\bar{\mathbf{z}}_j = \mathbf{z}_j; \qquad j = 2k + 1, \ldots, n$$

are satisfied. Then according to formula (14) the vector $\mathbf{z}$ has the form

$$\mathbf{z} = c^1\mathbf{z}_1 + c^2\mathbf{z}_2 + \cdots + c^{2k-1}\mathbf{z}_{2k-1} + c^{2k}\mathbf{z}_{2k} + c^{2k+1}\mathbf{z}_{2k+1} + \cdots + c^n\mathbf{z}_n, \tag{15}$$

while the vector $\bar{\mathbf{z}}$ has the form

$$\bar{\mathbf{z}} = \bar{c}^2\mathbf{z}_1 + \bar{c}^1\mathbf{z}_2 + \cdots + \bar{c}^{2k}\mathbf{z}_{2k-1} + \bar{c}^{2k-1}\mathbf{z}_{2k} + \bar{c}^{2k+1}\mathbf{z}_{2k+1} + \cdots + \bar{c}^n\mathbf{z}_n. \tag{16}$$

If

$$c^1 = \bar{c}^2, \quad \ldots, \quad c^{2k-1} = \bar{c}^{2k}, \qquad c^{2k+1} = \bar{c}^{2k+1}, \quad \ldots, \quad c^n = \bar{c}^n, \quad (17)$$

then it follows from (15) and (16) that $\bar{\mathbf{z}} = \mathbf{z}$, i.e., the vector $\mathbf{z}$ is real. If, conversely, we assume that $\mathbf{z}$ is real, i.e., that $\bar{\mathbf{z}} = \mathbf{z}$, then (15) and (16) give [by the linear independence of the vectors (13)] the system of relations (17). Thus, proposition (D) is proved.

Proposition (E) below presents a method of separating real solutions from the set of all complex solutions of equation (6) in the case that the coefficients of the polynomial $L(p)$ are real.

(E) Let us assume that the coefficients of $L(p)$ are real; then to every complex root λ of $L(p)$ corresponds a conjugate root $\bar{\lambda}$. The solutions $e^{\lambda t}$ and $e^{\bar{\lambda} t}$ of equation (6) are conjugate to each other [see §5, (D)]. If the root λ is real, then the solution $e^{\lambda t}$ is real. Thus, corresponding to every solution of the fundamental system (7), there is also a complex conjugate solution. For the solution (8) of (6) to be real, it is necessary and sufficient that the coefficients of pairs of complex conjugate solutions be conjugate and the coefficients of real solutions be real.

For proof let us denote by $\mathbf{z}_k$ the vector with coordinates

$$\{z_k(0), \dot{z}_k(0), \ldots, z_k^{(n-2)}(0), z_k^{(n-1)}(0)\}$$

and by $\mathbf{z}$ the vector with coordinates $\{z_0, \dot{z}_0, \ldots, z_0^{(n-1)}\}$. Then (10) takes the form

$$c^1\mathbf{z}_1 + c^2\mathbf{z}_2 + \cdots + c^n\mathbf{z}_n = \mathbf{z}.$$

The vectors $\mathbf{z}_1, \mathbf{z}_2, \ldots, \mathbf{z}_n$ are linearly independent, since the determinant of (11) is not zero. Thus, the necessity of the condition presented in (E) follows directly from (D). On the other hand, if this condition is fulfilled, then the solution (8) is real. Indeed, if λ_1 and λ_2 are two complex conjugate roots, and c^1 and c^2 are two complex conjugate constants, then $c^1e^{\lambda_1 t}$ and $c^2e^{\lambda_2 t}$ are complex conjugate functions and, consequently, their sum is real. Thus proposition (E) is proved.

Examples

1. We shall find all complex solutions of the equation

$$z^{(3)} - 3\ddot{z} + 9\dot{z} + 13z = 0,$$

which can be written in the form (6), where

$$L(p) = p^3 - 3p^2 + 9p + 13.$$

By direct inspection, we see that $p = -1$ is a root of the characteristic polynomial $L(p)$. Factoring $L(p)$ by $p + 1$, we obtain

$$L(p) = (p + 1)(p^2 - 4p + 13),$$

from which we obtain two more roots, $2 \pm 3i$. Thus the roots of $L(p)$ are the numbers

$$\lambda_1 = 2 + 3i, \qquad \lambda_2 = 2 - 3i, \qquad \lambda_3 = -1.$$

By Theorem 4 the general complex solution of the equation under consideration has the form

$$z = c^1 e^{(2+3i)t} + c^2 e^{(2-3i)t} + c^3 e^{-t}.$$

In Examples 2 and 3 below we give two general rules for determining the real solutions. These rules follow directly from proposition (E).

2. We shall assume that the fundamental system of solutions (7) satisfies the conditions

$$\overline{z_1} = z_2, \quad \ldots, \quad \overline{z}_{2k-1} = z_{2k}, \quad \overline{z}_{2k+1} = z_{2k+1}, \quad \ldots, \quad \overline{z_n} = z_n, \tag{18}$$

and set

$$z_1 = x_1 + iy_1, \quad \ldots, \quad z_{2k-1} = x_k + iy_k,$$

where $x_1, \ldots, x_k, y_1, \ldots, y_k$ are real functions. We shall further assume that the numbers $c^1, c^2, \ldots, c^n$ satisfy (17), and set

$$c^1 = \tfrac{1}{2}(a^1 - ib^1), \quad \ldots, \quad c^{2k-1} = \tfrac{1}{2}(a^k - ib^k),$$

where $a^1, \ldots, a^k, b^1, \ldots, b^k$ are real numbers. With this notation the general real solution of (6) may be written in the form

$$z = a^1 x_1 + b^1 y_1 + \cdots + a^k x_k + b^k y_k + c^{2k+1} z_{2k+1} + \cdots + c^n z_n,$$

where

$$a^1, b^1, \ldots, a^k, b^k, c^{2k+1}, \ldots, c^n$$

are arbitrary real numbers. Thus, if for every pair of conjugate complex solutions we substitute their real and imaginary parts in the fundamental system (7), we shall obtain a *fundamental system of real solutions.*

3. Again we shall assume that the solutions (7) satisfy (18); let us set

$$\lambda_1 = \mu_1 + i\nu_1, \quad \ldots, \quad \lambda_{2k-1} = \mu_k + i\nu_k.$$

Under the hypothesis that $c^1, c^2, \ldots, c^n$ satisfy (17), we may set

$$c^1 = \tfrac{1}{2}\rho_1 e^{i\alpha_1}, \quad \ldots, \quad c^{2k-1} = \tfrac{1}{2}\rho_k e^{i\alpha_k}.$$

In this notation every real solution z may be written in the form:

$$z = \rho_1 e^{\mu_1 t} \cos(\nu_1 t + \alpha_1) + \cdots + \rho_k e^{\mu_k t} \cos(\nu_k t + \alpha_k) + c^{2k+1} e^{\lambda_{2k+1} t} + \cdots + c^n e^{\lambda_n t}.$$

Here $\rho_1, \ldots, \rho_k, \alpha_1, \ldots, \alpha_k, c^{2k+1}, \ldots, c^n$ are arbitrary real constants. From the last expression it is evident that every imaginary part $\nu_j \neq 0$ of the root λ_j gives the solution an oscillatory character with a frequency ν_j, while every real part μ_j of the root λ_j causes the solution either to grow (if $\mu_j > 0$) or become smaller (if $\mu_j < 0$).

4. By using the results of Examples 2 and 3, we can write all the real solutions of the equation investigated in Example 1 in the following two forms:

$$z = a^1 e^{2t} \cos 3t + b^1 e^{2t} \sin 3t + c^3 e^{-t},$$
$$z = \rho_1 e^{2t} \cos(3t + \alpha_1) + c^3 e^{-t}.$$

8. The linear homogeneous equation with constant coefficients. Case of multiple roots. If the characteristic polynomial

$$L(p) = p^n + a_1 p^{n-1} + \cdots + a_{n-1} p + a_n$$

of the equation

$$L(p)z = 0 \tag{1}$$

[see §7, (A)] has multiple roots, then by functions of the form $e^{\lambda t}$ it is not possible to find n distinct solutions of equation (1). To find solutions of another form in this case, we can use the following heuristic reasoning. Let λ_1 and λ_2 be two distinct real roots of the characteristic polynomial $L(p)$; then the function $(e^{\lambda_1 t} - e^{\lambda_2 t})/(\lambda_1 - \lambda_2)$ is a solution of (1). If we now assume that with a change in the coefficients of $L(p)$ the number λ_2 tends to λ_1, then this solution tends (in the limit) to the function $te^{\lambda_1 t}$, which may naturally be assumed to be a solution of (1) whenever λ_1 is a double root of the polynomial $L(p)$. Similarly, we arrive at the conjecture that, if λ is a k-tuple root of the characteristic polynomial $L(p)$, then all the functions

$$e^{\lambda t}, te^{\lambda t}, \ldots, t^{k-1} e^{\lambda t}$$

are solutions of (1). Extending this conjecture to the case of complex multiple roots, we come to the question of the validity of the following theorem (which is a generalization of Theorem 4):

THEOREM 5. Let

$$L(p)z = 0 \tag{2}$$

be a linear nth-order homogeneous equation with constant coefficients.

Further, let $\lambda_1, \ldots, \lambda_m$ be the set of all mutually distinct roots of the characteristic polynomial $L(p)$ of equation (2), the root λ_j having the multiplicity k_j, so that $k_1 + k_2 + \cdots + k_m = n$. If we set

$$\begin{aligned} z_1 &= e^{\lambda_1 t}, & z_2 &= te^{\lambda_1 t}, & \ldots, \quad z_{k_1} &= t^{k_1-1}e^{\lambda_1 t}; \\ z_{k_1+1} &= e^{\lambda_2 t}, & z_{k_1+2} &= te^{\lambda_2 t}, & \ldots, \quad z_{k_1+k_2} &= t^{k_2-1}e^{\lambda_2 t}; \\ & & & & \vdots \\ & & & & \ldots, \quad z_n &= t^{k_m-1}e^{\lambda_m t}, \end{aligned} \tag{3}$$

then all the functions (3) are solutions of (2), so that for any complex constants $c^1, c^2, \ldots, c^n$, the function

$$z = c^1 z_1 + \cdots + c^n z_n \tag{4}$$

is also a solution of (2). This solution is the general solution in the sense that every solution of (2) can be obtained from (4) by a proper choice of the constants $c^1, \ldots, c^n$, where the constants $c^1, \ldots, c^n$ are defined uniquely for every given solution z.

Thus the functions (3) constitute the so-called *fundamental system of solutions* [see §18, (C)] of equation (2). We note that the functions (3) are defined on the entire axis $-\infty < t < +\infty$.

We shall preface the proof of Theorem 5 with a proof of the so-called *shift formula.*

(A) Let $L(p)$ be an arbitrary polynomial, λ any complex number, and $f(t)$ a function which can be differentiated an arbitrary and sufficient number of times. Then the following important formula is valid:

$$L(p)\big(e^{\lambda t}f(t)\big) = e^{\lambda t} \cdot L(p + \lambda)f(t). \tag{5}$$

We shall prove formula (5), but first we shall verify it for the case $L(p) \equiv p$. We have

$$p\big(e^{\lambda t}f(t)\big) = \lambda e^{\lambda t}f(t) + e^{\lambda t}\dot{f}(t) = e^{\lambda t}(p + \lambda)f(t).$$

Now it is easy to verify formula (5) for any first-degree polynomial $L(p) = ap + b$. We have

$$\begin{aligned} (ap + b)\big(e^{\lambda t}f(t)\big) &= ap\big(e^{\lambda t}f(t)\big) + be^{\lambda t}f(t) \\ &= ae^{\lambda t}(p + \lambda)f(t) + be^{\lambda t}f(t) = e^{\lambda t}[a(p + \lambda) + b]f(t). \end{aligned}$$

In the general case, we shall carry out the proof of formula (5) inductively with respect to the degree n of polynomial $L(p)$. As we have seen, the formula is valid for $n = 1$. Let us assume that it is also valid for a polynomial of degree $n - 1$, $n \geq 2$, and prove it for a polynomial $L(p)$ of degree n. To do this we shall factor the nth-degree polynomial $L(p)$ into two factors, $L(p) = L_1(p) \cdot L_2(p)$, where $L_1(p)$ is of the first degree and $L_2(p)$ is of degree $n - 1$. Since formula (5) is correct for each of the polynomials $L_1(p)$ and $L_2(p)$, we have [see §7, (A)]

$$\begin{aligned} L(p)(e^{\lambda t}f(t)) &= L_1(p)[L_2(p)(e^{\lambda t}f(t))] = L_1(p)(e^{\lambda t}L_2(p + \lambda)f(t)) \\ &= e^{\lambda t}L_1(p + \lambda)L_2(p + \lambda)f(t) = e^{\lambda t}L(p + \lambda)f(t). \end{aligned}$$

Thus formula (5) is proved.

We shall now prove proposition (B), which includes Theorem 5 almost entirely.

(B) Let $L(p)$ be an arbitrary polynomial in the symbol p, and let the function $\omega_r(t)$ of the real variable t be defined by the formula

$$\omega_r(t) = L(p)t^r e^{\lambda t},$$

where λ is a complex number. We find that, if λ is a k-tuple root of $L(p)$, then the functions $\omega_0(t), \ldots, \omega_{k-1}(t)$ are identically zero. On the other hand, we find that, if the functions $\omega_0(t), \ldots, \omega_{k-1}(t)$ equal zero for even one value $t = t_0$, i.e., if the equalities

$$\omega_0(t_0) = \omega_1(t_0) = \cdots = \omega_{k-1}(t_0) = 0 \tag{6}$$

are valid, then λ is a root of $L(p)$ of multiplicity not less than k.

Let us prove proposition (B). By the shift formula [see (5)] we have

$$\omega_r(t) = e^{\lambda t}L(p + \lambda)t^r. \tag{7}$$

Let us assume first that λ is a k-tuple root of $L(p)$, i.e., that

$$L(p) = M(p)(p - \lambda)^k.$$

If we replace p by $p + \lambda$ in this identity, we obtain

$$L(p + \lambda) = M(p + \lambda)p^k. \tag{8}$$

From formulas (7) and (8) we obtain

$$\omega_r(t) = e^{\lambda t}M(p + \lambda)(p^k t^r) = 0 \quad \text{for} \quad r = 0, 1, \ldots, k - 1,$$

since $p^k t^r = 0$ for $r < k$. Thus the first part of proposition (B) is proved.

Let us now assume that (6) holds. If we expand $L(p + \lambda)$ into powers of p, we obtain

$$L(p + \lambda) = b_0 + b_1 p + \cdots + b_{n-1}p^{n-1} + b_n p^n. \tag{9}$$

From (7) and (9) we obtain

$$\omega_0(t_0) = e^{\lambda t_0} b_0,$$

which by virtue of (6) gives

$$b_0 = 0.$$

Let us now assume that the equalities

$$b_0 = b_1 = \cdots = b_{r-1} = 0, \qquad r \leq k - 1, \tag{10}$$

are valid and prove that $b_r = 0$. By formulas (7), (9), and (10) we have

$$\omega_r(t_0) = e^{\lambda t_0} \cdot r! b_r.$$

It follows from this and formula (6) that

$$b_r = 0.$$

Thus $b_0 = b_1 = \cdots = b_{k-1} = 0$, and the polynomial $L(p + \lambda)$ has the form

$$L(p + \lambda) = b_k p^k + \cdots + b_n p^n = (b_k + \cdots + b_n p^{n-k}) p^k = M_1(p) p^k.$$

Substituting $p - \lambda$ for p in this identity we obtain

$$L(p) = M_1(p - \lambda) \cdot (p - \lambda)^k,$$

which shows that λ is a root of the polynomial $L(p)$ of multiplicity no less than k. Thus proposition (B) is proved.

Proof of Theorem 5. From the first part of proposition (B) it follows directly that the functions (3), defined in the formulation of Theorem 5, are solutions of (2). We shall prove that they constitute the fundamental system of solutions. To prove this it is sufficient to show that by a proper choice of the constants $c^1, \ldots, c^n$ we can obtain from (4) an arbitrary solution z_* of equation (2). We shall prove this (here, unlike our proof of Theorem 4, we shall not use Theorem 3).

Let z_* be an arbitrary solution of equation (2) defined on the interval $r_1 < t < r_2$, and let t_0 be some point of this interval. Let us set

$$z_*(t_0) = z_0, \qquad \dot{z}_*(t_0) = \dot{z}_0, \quad \ldots, \quad z_*^{(n-1)}(t_0) = z_0^{(n-1)}.$$

We shall now seek constants $c^1, \ldots, c^n$ such that a solution z of equation (2) defined by (4) will satisfy the same initial conditions as the given solution z_*. Then by the uniqueness theorem we shall have $z = z_*$ (on the interval $r_1 < t < r_2$). To determine the constants $c^1, \ldots, c^n$, we have the system of equations

$$c^1 z_1^{(s)}(t_0) + c^2 z_2^{(s)}(t_0) + \cdots + c^n z_n^{(s)}(t_0) = z_0^{(s)}, \qquad s = 0, 1, \ldots, n-1. \tag{11}$$

In order that system (11) be solvable, it is sufficient that the determinant of the matrix

$$\begin{pmatrix} z_1(t_0) & z_2(t_0) & \ldots & z_n(t_0) \\ \vdots & & & \\ z_1^{(s)}(t_0) & z_2^{(s)}(t_0) & \ldots & z_n^{(s)}(t_0) \\ \vdots & & & \\ z_1^{(n-1)}(t_0) & z_2^{(n-1)}(t_0) & \ldots & z_n^{(n-1)}(t_0) \end{pmatrix} \tag{12}$$

be different from zero. We shall show that this determinant is not equal to zero. For this we shall show that the rows of (12) are linearly independent. Let us assume the contrary and let $b_{n-1}, b_{n-2}, \ldots, b_0$ be constants, not all zero, by which the first, second, ..., rows of the matrix are to be multiplied in order that their sum be zero. By writing the sum of the elements of the jth column, we obtain the equality

$$b_0 z_j^{(n-1)}(t_0) + b_1 z_j^{(n-2)}(t_0) + \cdots + b_{n-2}\dot{z}_j(t_0) + b_{n-1} z_j(t_0) = 0,$$

which can be written in the form

$$M(p) z_j|_{t=t_0} = 0, \tag{13}$$

where $M(p) = b_0 p^{n-1} + b_1 p^{n-2} + \cdots + b_{n-2} p + b_{n-1}$. The equality (13) for $j = 1, \ldots, k_1$ shows that λ_1 is a root of multiplicity at least k_1 of the polynomial $M(p)$ [see proposition (B)]. In exactly the same way the equality obtained for $j = k_1 + 1, \ldots, k_1 + k_2$ shows that λ_2 is a root of multiplicity at least k_2 of the polynomial $M(p)$. The set of all equalities (13) leads to the conclusion that (taking multiplicities into account) the polynomial $M(p)$ has not less than n roots, but this is impossible since its degree is no higher than $n - 1$. Thus the assumption that the determinant of (12) is equal to zero has led us to a contradiction; this means that (11) is solvable (and, moreover, uniquely) in terms of the unknowns

$$c^1, c^2, \ldots, c^n.$$

Thus Theorem 5 is fully proved.

We shall note one obvious corollary of Theorem 5.

(C) Every solution $z(t)$ of equation (2) can be written in the form

$$z(t) = f_1(t)e^{\lambda_1 t} + f_2(t)e^{\lambda_2 t} + \cdots + f_m(t)e^{\lambda_m t},$$

where $f_j(t)$ is a polynomial of degree not exceeding the number $k_j - 1$, $j = 1, \ldots, m$. Here the polynomials $f_1(t), \ldots, f_m(t)$ are defined uniquely by the solution $z(t)$, since their coefficients are the integration constants $c^1, c^2, \ldots, c^n$ which by Theorem 5 are defined uniquely by the solution $z(t)$.

If the coefficients of equation (2) are real, then the problem facing us is that of separating real solutions from the set of complex solutions of (2).

(D) We shall assume that the coefficients of the characteristic polynomial $L(p)$ of equation (2) are real. Let λ be a certain root of $L(p)$ of multiplicity k; then, for $r = 0, 1, \ldots, k - 1$, the function $t^r e^{\lambda t}$ is a solution of (2). If the root λ is real, then the function $t^r e^{\lambda t}$ is real; if λ is complex, then in addition to the solution $t^r e^{\lambda t}$ there is also the complex conjugate solution $t^r e^{\overline{\lambda t}}$, since $\bar{\lambda}$ is also a root of multiplicity k of $L(p)$. Thus whenever the fundamental system (3) admits a complex solution it admits the complex conjugate as a solution. In order that the solution (4) be real, it is necessary and sufficient that the coefficients of real solutions be real, and the coefficients of pairs of conjugate complex solutions be complex conjugate.

The proof of proposition (D) is carried out in exactly the same way as the proof of proposition (E) of §7 on the basis of proposition (D) of §7.

Examples

1. We shall solve the equation

$$z^{(5)} + 3z^{(4)} + 3z''' + z'' = 0.$$

This equation can be written in the form (2), where the characteristic polynomial $L(p)$ has the form

$$p^5 + 3p^4 + 3p^3 + p^2 = p^2(p + 1)^3.$$

The numbers

$$\lambda_1 = 0,$$
$$\lambda_2 = -1$$

are roots of multiplicity $k_1 = 2$, $k_2 = 3$, respectively, of this polynomial. Therefore by Theorem 5 the fundamental system of solutions of this equation has the form

$$z_1 = 1, \qquad z_2 = t, \qquad z_3 = e^{-t}, \qquad z_4 = te^{-t}, \qquad z_5 = t^2e^{-t}.$$

The general solution is given by the formula

$$z = (c^1 + c^2t) + (c^3 + c^4t + c^5t^2)e^{-t}.$$

2. Let us solve the equation

$$z^{(4)} + 2z'' + z = 0.$$

The characteristic polynomial is $L(p) = (p^2 + 1)^2$; the numbers $\lambda_1 = i$, $\lambda_2 = -i$ are its two roots, and its general solution may be written in the form

$$z = (c^1 + c^2t)e^{it} + (c^3 + c^4t)e^{-it}.$$

The following two examples give the general rules for distinguishing real solutions stemming directly from proposition (D). Examples 3 and 4 are completely analogous to Examples 2 and 3 of §7.

3. In Example 2 of §7 the concrete form of the solution was not taken into account, and it was only assumed that the fundamental system of solutions consisted of mutually conjugate solutions and real solutions. By the same reasoning, therefore, in the case of multiple roots we have the following general rule: In the fundamental system (3) it is necessary that each pair of complex conjugate solutions be replaced by the real and imaginary parts of one of these solutions. The system of functions so obtained is the fundamental system of real solutions.

4. Let

$$t^r e^{\lambda t}, \qquad t^r e^{\bar{\lambda} t}$$

be two complex conjugate solutions of (3). In the case of a real solution z, that part of the sum (4) corresponding to these solutions may be written in the form

$$\hat{z} = ct^r e^{(\mu + i\nu)t} + \bar{c}t^r e^{(\mu - i\nu)t}.$$

If we set

$$c = \tfrac{1}{2}\rho e^{i\alpha},$$

we shall have

$$\hat{z} = \rho t^r e^{\mu t} \cos(\nu t + \alpha). \tag{14}$$

In this way it is possible to replace every pair of complex conjugate solutions appearing in (4) by a real function of the form (14) containing two arbitrary real constants ρ and α. Here again, as in Example 3 of §7, it is evident that if a root λ has a nonzero imaginary part ν, the solution has an oscillatory character, and if the real part μ of λ is nonzero, then the solution either increases (for $\mu > 0$) or decreases (for $\mu < 0$). Finally, if the root λ is multiple, there appears an additional term t^r, which causes a further increase of the solution; however, as $t \to \infty$ in the case $\mu < 0$, the increase of the solution caused by factor t^r is considerably less than the decrease caused by factor $e^{\mu t}$, so that for $\mu < 0$ (and for any order of multiplicity of the root) the solution tends to zero as t increases.

5. Using the results of Examples 3 and 4 we can write all real solutions of the equation considered in Example 2 in the following two forms:

$$z = (a^1 + a^2t)\cos t + (b^1 + b^2t)\sin t,$$
$$z = \rho_1 \cos (t + \alpha_1) + \rho_2 t \cos (t + \alpha_2).$$

9. Stable polynomials. Let

$$L(p)z = 0 \tag{1}$$

be a linear homogeneous equation with constant coefficients. The question of how the solutions of this equation behave as $t \to +\infty$ (whether they tend to zero, remain bounded, or increase without limit) plays a very important role in a whole series of applications of the theory of ordinary differential equations. In Examples 3 of §7 and 4 of §8 it has already been noted that this question of the behavior of solutions of equation (1) is related to the nature of the real parts of the roots of polynomial $L(p)$. We shall now formulate this problem more precisely.

(A) The polynomial $L(p)$ is called *stable* if all its roots have negative real parts or, in geometrical terms, are located on the left-hand side of the imaginary axis in the complex plane. Let

$$\lambda_j = \mu_j + i\nu_j, \qquad j = 1, \ldots, m$$

denote all the roots of the polynomial $L(p)$. If this polynomial is stable, there exists a positive number α such that

$$\mu_j < -\alpha, \qquad j = 1, \ldots, m. \tag{2}$$

We shall show that in this case for every solution $\varphi(t)$ of equation (1) a positive number M can be found such that

$$|\varphi(t)| < Me^{-\alpha t} \quad \text{for} \quad t \geq 0. \tag{3}$$

This formula not only shows that every solution of equation (1) tends to zero as $t \to \infty$, but also gives us an estimate of the rate of convergence to zero.

We shall prove (3) first for an arbitrary fundamental solution z_s, $s = 1, \ldots, n$, of equation 1 [see (3), §8]. We have

$$z_s = t^r e^{\lambda_j t}, \qquad \text{whence} \qquad \left|\frac{z_s}{e^{-\alpha t}}\right| = t^r e^{(\mu_j + \alpha)t}.$$

Since the number $\mu_j + \alpha$ is negative because of (2), the function $t^r e^{(\mu_j+\alpha)t}$ tends to zero as $t \to \infty$ and therefore is bounded for $t \geq 0$. Thus we have

$$\left|\frac{z_s}{e^{-\alpha t}}\right| < M_s \qquad \text{for} \qquad t \geq 0,$$

or, what is the same thing,

$$|z_s| < M_s e^{-\alpha t} \qquad \text{for} \qquad t \geq 0.$$

If now

$$\varphi(t) = c^1 z_1 + c^2 z_2 + \cdots + c^n z_n$$

is an arbitrary solution of equation (1), then for $t > 0$ we have

$$|\varphi(t)| \leq (|c^1| \cdot M_1 + |c^2| \cdot M_2 + \cdots + |c^n| \cdot M_n) e^{-\alpha t} = M e^{-\alpha t}.$$

Thus the inequality (3) is proved. It should be noted that if even one of the roots λ_j of $L(p)$ has a positive real part $\mu_j > 0$, then there exists a solution $e^{\lambda_j t}$ of equation (1) which increases without bound as $t \to \infty$.

Many studies by mathematicians have been devoted to an applicable formulation of stability conditions of polynomials. For second-degree polynomials the stability condition is derived directly from the solution of a quadratic equation [see (B)]. The stability problem for polynomials of an arbitrary degree n was solved in several different forms by the mathematicians Rauss and Hurwitz. The Rauss-Hurwitz conditions, however, are inconvenient for actual computation, and, therefore, the work on new formulations of stability conditions continues. Here, a proof of the Rauss-Hurwitz criterion for $n = 3$ will be presented, and the stability condition for an arbitrary degree n will be given in the Hurwitz form without proof.

(B) A second-degree polynomial $L(p) = p^2 + ap + b$ with real coefficients a and b is stable if and only if its coefficients are positive.

This assertion is easy to verify with the aid of the formula for the solution of a quadratic equation.

(C) If the polynomial $L(p) = p^n + a_1 p^{n-1} + \cdots + a_n$ with real coefficients is stable, then all its coefficients are positive.

For the proof, we shall factor $L(p)$ into its first- and second-degree factors, i.e., into factors of the form $p + c$ and $p^2 + ap + b$. Since $L(p)$ is stable, each of its factors in this form is also stable. For the stability of the factor $p + c$ it is necessary that the number c be positive, and for the stability of the factor $p^2 + ap + b$ it is necessary that both a and b be positive. Since the coefficients of the factors are positive, it follows that the coefficients of the product are also positive.

The following theorem gives a stability criterion for third-degree polynomials.

THEOREM 6. The polynomial

$$L(p) = a_0 p^3 + a_1 p^2 + a_2 p + a_3, \qquad a_0 > 0,$$

with real coefficients is stable if and only if the numbers a_1, a_2, a_3 are positive and, in addition, the inequality

$$a_1 a_2 > a_0 a_3$$

is satisfied.

Proof. For the proof we shall examine the polynomial

$$L(p) = p^3 + ap^2 + bp + c; \tag{4}$$

the case of the general polynomial $L(p)$ can be easily reduced to this. By proposition (C) it is sufficient for us to prove that the polynomial (4) with *positive* coefficients a, b, c is stable if and only if the inequality

$$ab > c \tag{5}$$

is valid. In our proof we shall use the fact that the roots of the polynomial are continuous functions of its coefficients.

First of all we shall determine those conditions for which the polynomial (4) has purely imaginary roots, in particular, the root $p = 0$, which must also be considered purely imaginary since it is located on the imaginary axis. We have

$$L(p) = (p + a)(p^2 + b) - ab + c. \tag{6}$$

If $L(p)$ has the root 0, then $c = 0$, and this case is excluded by hypothesis, since $c > 0$. Let us assume that the number $i\omega$, where $\omega \neq 0$, is a root of $L(p)$. If it is further assumed that $-\omega^2 + b$ is not zero, then the number $(i\omega + a)(-\omega^2 + b)$ has a nonzero imaginary part and cannot cancel the real number $-ab + c$. Thus the number $i\omega$ can be a root of the polynomial $L(p)$ only when $-\omega^2 + b = 0$; in this case we have the equality

$$L(i\omega) = -ab + c = 0.$$

Conversely, if $ab = c$, then by formula (6) the polynomial $L(p)$ has purely imaginary roots $p = \pm i\sqrt{b}$. Thus $L(p)$ (with positive coefficients) has purely imaginary roots if and only if $ab = c$. In particular, by a continuous change of the positive coefficients a, b, c, a root of polynomial $L(p)$ can intersect the imaginary axis only if the equality $ab = c$ is satisfied.

Let us assume that (5) is not satisfied. Then either $ab = c$ or $ab < c$. In the first case, the polynomial $L(p)$ has purely imaginary roots and, consequently, is nonstable. We shall show that in the second case, i.e., when the inequality

$$ab < c, \tag{7}$$

is satisfied, polynomial $L(p)$ is also nonstable. We shall vary the coefficients

a and b continuously, keeping them positive in such a way that they tend to zero without violating inequality (7). During this change no root will pass from one side of the imaginary axis to the other, and consequently the property of stability or nonstability of the polynomial remains invariant. When $a = b = 0$, we obtain the polynomial $p^3 + c$, whose roots $\sqrt[3]{c}\,(\cos(\pi/3) \pm i \sin(\pi/3))$ are located on the right-hand side of the imaginary axis. Since the roots depend continuously on the coefficients, the nonstability (that is, the presence of roots to the right of the imaginary axis) is also retained for sufficiently small positive a and b.

Let us now assume that the inequality (5) is satisfied and show that $L(p)$ is stable. For this, we shall vary the coefficient c in such a way that it tends to zero through positive values without violating (5). For $c = 0$, we obtain the polynomial

$$L(p) = p(p^2 + ap + b),$$

which has one zero root and two roots with negative real parts. For a small positive c these two roots will not change very much, so that their product will remain positive and the zero root will assume a small positive or negative value. Since the product of all three roots equals the negative number $-c$, that root close to zero will be negative. Thus Theorem 6 is proved.

In order to formulate the necessary and sufficient stability conditions for an arbitrary polynomial with real coefficients, we shall first fix our terminology. Let

$$P = \begin{pmatrix} p_{11} & p_{12} & \cdots & p_{1n} \\ p_{21} & p_{22} & \cdots & p_{2n} \\ \vdots & & & \\ p_{n1} & p_{n2} & \cdots & p_{nn} \end{pmatrix}$$

be an arbitrary square matrix of order n. We shall call the determinant of the matrix

$$\begin{pmatrix} p_{11} & p_{12} & \cdots & p_{1k} \\ p_{21} & p_{22} & \cdots & p_{2k} \\ \vdots & & & \\ p_{k1} & p_{k2} & \cdots & p_{kk} \end{pmatrix}$$

its principal kth minor, and denote it by $\Delta_k(P)$. Thus the determinant $\Delta_k(P)$ consists of those elements of the matrix P appearing in the first k rows and columns.

THEOREM 7. Let

$$a_0p^n + a_1p^{n-1} + \cdots + a_n, \qquad a_0 > 0, \tag{8}$$

be an arbitrary polynomial of degree n with real coefficients. In order to determine the stability of (8), the nth-order matrix

$$Q = \left\{\begin{matrix} a_1 & a_3 & a_5 & \dots & 0 \\ a_0 & a_2 & a_4 & \dots & 0 \\ 0 & a_1 & a_3 & \dots & \dots \\ \vdots & & & & \\ 0 & \dots & \dots & a_{n-2} & a_n \end{matrix}\right\}$$

is formed. Then, the polynomial (8) is stable if and only if all principal minors $\Delta_k(Q)$, $k = 1, \dots, n$, of Q are positive.*

For further clarity, we shall describe the form of the matrix Q. The kth column of Q has the form

$$\dots a_{k+2} \ a_{k+1} \ a_k a_{k-1} a_{k-2} \dots ,$$

where the element a_k lies on the principal diagonal; here, any element a_{k+j}, whose index $k + j$ is negative or larger than n, is taken to be zero.

Examples

1. We shall derive Theorem 6 from Theorem 7. In the case $n = 3$, the matrix Q has the form

$$\begin{pmatrix} a_1 & a_3 & 0 \\ a_0 & a_2 & 0 \\ 0 & a_1 & a_3 \end{pmatrix}.$$

Its three principal minors have the values

$$\Delta_1(Q) = a_1, \qquad \Delta_2(Q) = a_1a_2 - a_0a_3, \qquad \Delta_3(Q) = a_3 \cdot \Delta_2(Q).$$

The conditions that these values and the coefficient a_0 be positive are

$$a_0 > 0, \qquad a_1 > 0, \qquad a_3 > 0, \qquad a_1a_2 > a_0a_3.$$

From this set of conditions it clearly follows that coefficient a_2 is positive. Thus in the case $n = 3$ Theorem 7 reduces to Theorem 6.

---

* Theorem 7 will not be proved completely in this book. For a proof of this theorem, together with a comprehensive discussion of this question, see Fuchs, B.A., and Levin, V.I., *Functions of a Complex Variable*, Vol. II, pp. 264 ff., Reading, Mass.: Addison-Wesley, 1961.

2. In the case $n = 4$, the matrix Q has the form

$$Q = \begin{Bmatrix} a_1 & a_3 & 0 & 0 \\ a_0 & a_2 & a_4 & 0 \\ 0 & a_1 & a_3 & 0 \\ 0 & a_0 & a_2 & a_4 \end{Bmatrix}.$$

Its principal minors have the following values:

$$\begin{aligned} \Delta_1(Q) &= a_1; \\ \Delta_2(Q) &= a_1a_2 - a_0a_3; \\ \Delta_3(Q) &= a_3\,\Delta_2(Q) - a_1^2a_4; \\ \Delta_4(Q) &= a_4\,\Delta_3(Q). \end{aligned}$$

The conditions that these minors be positive, together with the condition $a_0 > 0$, are obviously

$$a_0 > 0, \qquad a_1 > 0, \qquad a_2 > 0, \qquad a_3 > 0, \qquad a_4 > 0,$$
$$\Delta_3(Q) = a_1a_2a_3 - a_0a_3^2 - a_1^2a_4 > 0.$$

10. The linear nonhomogeneous equation with constant coefficients. Here we shall give the solution of a linear equation with constant coefficients containing a free term of a special form which is the so-called *quasipolynomial.*

(A) We define a *quasipolynomial* to be any function $F(t)$ which can be written in the form

$$F(t) = f_1(t)e^{\lambda_1 t} + f_2(t)e^{\lambda_2 t} + \cdots + f_m(t)e^{\lambda_m t}, \tag{1}$$

where $\lambda_1, \lambda_2, \ldots, \lambda_n$ are complex numbers, and $f_1(t), f_2(t), \ldots, f_m(t)$ are polynomials in t. From proposition (C) of §8 it follows that every solution of a linear homogeneous equation with constant coefficients is a quasipolynomial. Conversely, it can be proved that every quasipolynomial is a solution of some linear homogeneous equation with constant coefficients. If any two numbers of the sequence $\lambda_1, \lambda_2, \ldots, \lambda_m$ coincide, for example if $\lambda_1 = \lambda_2$, then the terms of the sum (1) corresponding to these numbers can be combined and replaced by the term $(f_1(t) + f_2(t))e^{\lambda_1 t}$. Thus the expression (1) can always be reduced to a form in which the numbers $\lambda_1, \lambda_2, \ldots, \lambda_m$ are mutually distinct. Let us note that the sum and product of two arbitrary quasipolynomials are also quasipolynomials; further, if we apply an arbitrary operator $L(p)$ to an arbitrary quasipolynomial, then we shall again obtain a quasipolynomial.

Thus in the present section we shall study the equation

$$L(p)z = F(t), \tag{2}$$

where $F(t)$ is a certain quasipolynomial. Along with equation (2) we shall study the corresponding homogeneous equation

$$L(p)u = 0. \tag{3}$$

The following proposition follows directly from (B) of §6.

(B) If $\hat{z}$ is some solution of equation (2), then an arbitrary solution z of this equation can be written in the form

$$z = \hat{z} + u,$$

where u is a certain solution of equation (3).

Since we already know how to find an arbitrary solution of a homogeneous equation, the problem reduces to finding one solution, or, as it is called, a *particular solution*, of equation (2) in the case when $F(t)$ is a quasipolynomial. Furthermore, since every quasipolynomial may be written in the form (1) by virtue of (C) of §6, the problem becomes one of determining a particular solution of equation (2) in the case when $F(t) = f(t)e^{\lambda t}$, where $f(t)$ is a polynomial. For this case, the answer is given by the following theorem.

THEOREM 8. Consider the nonhomogeneous equation

$$L(p)z = f(t)e^{\lambda t}, \tag{4}$$

in which $f(t)$ is a polynomial of degree r in t, and λ is a complex number. Let $k = 0$ if $L(\lambda) \neq 0$, and let k be the multiplicity of the root λ if $L(\lambda) = 0$. Then there exists a particular solution of equation (4) of the form

$$z = t^k g(t)e^{\lambda t}, \tag{5}$$

where $g(t)$ is an rth-degree polynomial in t. The coefficients of $g(t)$ can be found by the method of undetermined coefficients.

Proof. We set

$$f(t) = a_0 t^r + f^*(t) \tag{6}$$

and look for a polynomial $g(t)$ in the form

$$g(t) = b_0 t^r + g^*(t), \tag{7}$$

where the polynomials $f^*(t)$ and $g^*(t)$ are of degree $r - 1$. Further, we may choose the number k in such a way that

$$L(p) = M(p)(p - \lambda)^k, \tag{8}$$

where $M(\lambda) \neq 0$. In order that the function (5) be a solution of (4), the condition [see §8, (A)]

$$L(p)e^{\lambda t}t^k g(t) = e^{\lambda t}L(p + \lambda)t^k g(t) = e^{\lambda t}f(t)$$

must be satisfied, i.e., $g(t)$ must satisfy the condition

$$L(p + \lambda)t^k g(t) = f(t). \tag{9}$$

The polynomial $M(p + \lambda)$ has as its free term the number $M(\lambda) \neq 0$, and hence may be written in the form

$$M(p + \lambda) = M(\lambda) + M^*(p)p, \qquad M(\lambda) \neq 0. \tag{10}$$

Taking into account (6), (7), (8), and (10), we can write condition (9), as it is applied to $g(t)$, in the form

$$b_0M(\lambda)p^k t^{k+r} + b_0M^*(p)p^{k+1}t^{k+r} + L(p + \lambda)t^k g^*(t) = a_0t^r + f^*(t). \tag{11}$$

By equating the terms containing t^r in (11), we obtain the relation

$$b_0M(\lambda)p^k t^{k+r} = a_0t^r, \tag{12}$$

from which the coefficient b_0 of the unknown polynomial $g(t)$ is uniquely determined since $M(\lambda) \neq 0$. We shall now assume that b_0 is already chosen, so that (12) is satisfied; then (11) takes the form

$$L(p + \lambda)t^k g^*(t) = f^*(t) - b_0M^*(p)p^{k+1}t^{k+r}, \tag{13}$$

where the right-hand side consists of a known polynomial of degree $r - 1$, while the left-hand side consists of an unknown polynomial $g^*(t)$ of degree $r - 1$. Equation (13) differs from (9) only by the fact that the degrees of the polynomials in it have been decreased by one. By repeating for equation (13) the calculations which were carried out earlier for equation (9), we calculate the coefficient b_1 of the term of highest degree in t [in this case, the $(r - 1)$st degree] of the polynomial $g^*(t)$. By carrying out this process further, we can calculate all the coefficients $b_0, b_1, \ldots, b_r$ of $g(t)$ in such a way that (9) is satisfied, and, in the same way, we can find a solution of (4) in the form (5).

It would be possible to substitute a solution of the form (5) directly into (4) and, by assuming that the coefficients of $g(t)$ are unknown, to obtain for these coefficients a system of linear equations by equating the coefficients of identical terms in the left-hand and right-hand sides of equation (4). The calculations carried out above show that the system of equations obtained for the coefficients of the polynomial $g(t)$ is solvable. Thus theorem 8 is proved.

Note. The system obtained for determining the coefficients of $g(t)$ is a system of linear equations with a triangular matrix: By equating the coefficients of $t^r e^{\lambda t}$, we obtain an equation containing only b_0; by equating the coefficients of $t^{r-1}e^{\lambda t}$, we obtain an equation containing only b_0 and b_1; and so on.

We now establish an important property of quasipolynomials.

(C) If the quasipolynomial

$$F(t) = f_1(t)e^{\lambda_1 t} + f_2(t)e^{\lambda_2 t} + \cdots + f_m(t)e^{\lambda_m t},$$

where $\lambda_1, \lambda_2, \ldots, \lambda_m$ are mutually distinct numbers which are identically zero on a certain interval $r_1 < t < r_2$, then all the polynomials $f_1(t)$, $f_2(t), \ldots, f_m(t)$ are identically zero, so that all coefficients of the quasipolynomial $F(t)$ are zero. From this it follows that if two quasipolynomials $F(t)$ and $F^*(t)$ are identically equal on the interval $r_1 < t < r_2$, then their corresponding coefficients coincide.

We shall prove proposition (C) by induction on the number m, which we shall call here the *order* of $F(t)$. For $m = 1$ proposition (C) is valid, since in this case the equalities $F(t) = f_1(t)e^{\lambda_1 t} = 0$ and $f_1(t) = 0$ are equivalent. We shall now carry out the induction step from $m - 1$ to $m (m \geq 2)$. If $F(t)$ is identically zero on the interval $r_1 < t < r_2$, then this is also true for the quasipolynomial

$$G(t) = p^{l+1}(F(t)e^{-\lambda_m t}),$$

where p is a differentiation operator and l is the degree of $f_m(t)$. By proposition (A) of §8 we have

$$G(t) = g_1(t)\, e^{(\lambda_1-\lambda_m)t} + g_2(t)\, e^{(\lambda_2-\lambda_m)t} + \cdots + g_{m-1}(t)\, e^{(\lambda_{m-1}-\lambda_m)t},$$

where

$$g_i(t) = (p + \lambda_i - \lambda_m)^{l+1}f_i(t), \qquad i = 1, \ldots, m - 1.$$

The quasipolynomial $G(t)$ is of order $m - 1$, and, since it is identically zero on the interval $r_1 < t < r_2$, it follows from the induction hypothesis that all the polynomials $g_1(t), \ldots, g_{m-1}(t)$ are identically zero. Let us suppose that one of the polynomials $f_1(t), \ldots, f_{m-1}(t)$ is not zero, and then show that this assumption leads to a contradiction. Let us assume that $f_1(t)$ is of degree k, i.e., that $f_1(t) = a_0t^k + a_1t^{k-1} + \cdots + a_k$, where $a_0 \neq 0$. By direct inspection we see that

$$g_1(t) = (p + \lambda_1 - \lambda_m)^{l+1} f_1(t) = (\lambda_1 - \lambda_m)^{l+1} a_0 t^k + \cdots,$$

and since $g_1(t)$ is identically zero on the interval $r_1 < t < r_2$, we have

$$(\lambda_1 - \lambda_m)^{l+1} a_0 = 0.$$

Since the numbers λ_1 and λ_m are distinct, it follows that $a_0 = 0$. This contradiction shows that all coefficients of the polynomials $f_1(t), \ldots, f_{m-1}(t)$ are equal to zero, i.e., $F(t) = f_m(t)\, e^{\lambda_m t}$. Hence we conclude that all the coefficients of $f_m(t)$ are also zero.

The case of the identity of two quasipolynomials $F(t)$ and $F^*(t)$ on the interval $r_1 < t < r_2$ is reduced to the case already studied by forming the quasipolynomial $F(t) - F^*(t)$. Thus proposition (C) is proved.

Examples

1. We shall find a particular solution of the equation

$$\ddot{z} + z = t \cos t = \tfrac{1}{2} t e^{it} + \tfrac{1}{2} t e^{-it}. \tag{14}$$

We solve separately the equations

$$\ddot{z} + z = \tfrac{1}{2} t e^{it}, \tag{15}$$

$$\ddot{z} + z = \tfrac{1}{2} t e^{-it}. \tag{16}$$

It is evident that, if z is a solution of (15), then $\bar{z}$ is a solution of (16). Thus it is sufficient to solve only equation (15). Here $r = 1, \lambda = i, k = 1$. Therefore a particular solution must be found in the form

$$t(c^1 + c^2 t) e^{it}.$$

The relation (9) takes the form

$$[(p + i)^2 + 1](c^1 t + c^2 t^2) = \tfrac{1}{2} t,$$

or

$$(p^2 + 2ip)(c^1 t + c^2 t^2) = \tfrac{1}{2} t.$$

This gives

$$2c^2 + 2ic^1 + 4ic^2 t = \tfrac{1}{2} t,$$

whence $c^2 = -\frac{1}{8} i$, $c^1 = ic^2 = \frac{1}{8}$. Thus the particular solution of (15) has the form

$$z = \left(\frac{1}{8} t - \frac{i}{8} t^2\right) e^{it},$$

and the solution of (14) is found to be

$$z + \bar{z} = \frac{1}{8}\,t(e^{it} + e^{-it}) + \frac{1}{8i}\,t^2(e^{it} - e^{-it}) = \frac{t}{4}\cos t + \frac{t^2}{4}\sin t.$$

2. We shall study the function

$$f(t) = \cos 2t \cdot \cos 3t \cdot e^{4t}.$$

Since every factor $\cos 2t$, $\cos 3t$, e^{4t} is a quasipolynomial, their product $f(t)$ is also a quasipolynomial. Let us reduce this quasipolynomial to the form (1):

$$\begin{aligned}\cos 2t \cdot \cos 3t \cdot e^{4t} &= \frac{e^{2it} + e^{-2it}}{2} \cdot \frac{e^{3it} + e^{-3it}}{2}\, e^{4t} \\ &= \tfrac{1}{4}e^{(4+5i)t} + \tfrac{1}{4}e^{(4+i)t} + \tfrac{1}{4}e^{(4-i)t} + \tfrac{1}{4}e^{(4-5i)t}.\end{aligned}$$

The reduction of polynomials to form (1) is useful in the solution of nonhomogeneous equations on the basis of Theorem 8.

11. Method of elimination. Until now we have been occupied with the solution of only one linear equation with constant coefficients. However, a quite general system of linear equations with constant coefficients may, in a certain sense, be reduced to one equation. This reduction is realized by the *method of elimination*, which is similar to that used in the theory of algebraic (not differential) linear equations. We shall explain this method here and then draw certain conclusions.

We shall investigate the system of equations

$$\sum_{s=1}^{n} L_s^j(p)x^s = f^j(t), \qquad j = 1, \ldots, n, \tag{1}$$

where $x^1, \ldots, x^n$ are unknown functions of the independent variable t, and $f^1(t), \ldots, f^n(t)$ are given functions of time t. Each symbol $L_s^j(p)$ represents a polynomial with constant coefficients with respect to a differentiation operator p, so that one term, $L_s^j(p)x^s$, represents a linear combination with constant coefficients in the function x^s and its derivatives. The number of equations in system (1) is equal to the number of unknown functions.

The order of the system (1) with respect to the unknown function x^s is denoted by q_s, so that the general order of (1) is determined by the formula $q = q_1 + q_2 + \cdots + q_n$. In posing the problem of solving (1), we naturally *must assume* that every unknown function x^s has derivatives up to the order q_s inclusive; no assumption on the existence of derivatives of higher orders is contained in the statement of the problem.

In applying the method of elimination to (1), we *assume* that each of the unknown functions x^s, along with each of the functions $f^j(t)$, has an appropriate number of derivatives. In making these assumptions, we are narrowing down, on the one hand, the class of the solutions under consideration (assumption on the adequate differentiability of the unknown functions), and, on the other hand, we are narrowing down the class of the equations under consideration [assumption of adequate differentiability of the functions $f^j(t)$]. The first of these restrictions can be removed by *proving* that if $x^1, \ldots, x^n$ is a solution of (1) and if the right-hand sides of $f^j(t)$ have an appropriate number of derivatives, then each of the functions x^s has an appropriate number of derivatives (see Examples 3 and 4).

Let us proceed to a description of the method of elimination.

(A) Let us study the matrix

$$\begin{pmatrix} L_1^1(p) & \ldots & L_n^1(p) \\ \vdots & & \\ L_1^n(p) & \ldots & L_n^n(p) \end{pmatrix} \tag{2}$$

of the system (1). Every element $L_s^j(p)$ of the matrix (2) is a polynomial in p. Thus it is possible to calculate the determinant $D(p)$ of the matrix (2) and its minors. The cofactor of the element $L_s^j(p)$ of (2) (i.e., the minor of this element taken with the proper sign) will be denoted by $M_j^s(p)$. From any course in higher algebra it is known that the identity

$$\sum_{j=1}^{n} M_j^i(p) L_s^j(p) = \delta_s^i D(p) \tag{3}$$

is valid, where δ_s^i is the so-called *Kronecker symbol:*

$$\delta_i^i = 1, \qquad \delta_s^i = 0 \qquad \text{for} \qquad i \neq s.$$

Multiplying equation (1) by $M_j^i(p)$ (i.e., carrying out a series of differentiations, multiplications and additions) and then summing up with respect to j, we obtain the equality

$$\sum_{j,s=1}^{n} M_j^i(p) L_s^j(p) x^s = \sum_j M_j^i(p) f^j(t). \tag{4}$$

[In going from (1) to (4), we have used the differentiability assumptions imposed on the functions x^s and $f^j(t)$]. By virtue of (3) the equality (4) can be rewritten in the form

$$D(p)x^i = \sum_j M_j^i(p) f^j(t). \tag{5}$$

The system (5), $i = 1, \ldots, n$, has the property that every unknown function x^i appears in only one equation of (5). We have thus proved that if the system of functions $x^1, \ldots, x^n$ is a solution of (1), then each individual function x^i is a solution of (5).

It should not be supposed, however, that if for every number i a solution x^i of equation (5) is selected in an arbitrary manner and the system of functions $x^1, \ldots, x^n$ is then formed, the system of functions obtained will be a solution of system (1). In order to find the *general solution* $x^1, \ldots, x^n$ of (1), it is necessary to find the general solution x^i of every equation (5), $i = 1, \ldots, n$, then form the system of functions $x^1, \ldots, x^n$, and finally determine those conditions (the relations between the constants of integration) under which this system of functions satisfies the system (1).

We shall now draw certain conclusions from the method of elimination. First we shall formulate the result obtained in proposition (A) for the case of the homogeneous system of equations

$$\sum_{s=1}^{n} L_s^j(p)x^s = 0, \qquad j = 1, \ldots, n. \tag{6}$$

(B) If the system of functions $x^1, \ldots, x^n$ is a solution of the system (6), then each individual function x^i appearing in this solution satisfies the equation

$$D(p)x^i = 0,$$

where $D(p)$ is the determinant of the matrix $\left(L_s^j(p)\right)$ of the system (6). In particular, it follows from this that if the determinant $D(p)$ is a stable polynomial [see §9, (A)], then every solution $x^1, \ldots, x^n$ of (6) satisfies the inequality

$$(x^1)^2 + \cdots + (x^n)^2 < R^2 e^{-2\alpha t} \quad \text{for} \quad t \geq 0, \tag{7}$$

where α is a positive constant depending on the system (6), and R is a constant depending on the solution $x^1, \ldots, x^n$.

The inequality (7) follows directly from inequalities (3) of §9.

We shall now show how the system of equations (6) can be solved by using the elimination method.

(C) Let us assume that the determinant $D(p)$ of (6) does not vanish identically, and let λ be a root of multiplicity k of $D(p)$. We shall seek a solution of (6) in the form

$$x^s = g^s(t)e^{\lambda t}, \qquad s = 1, \ldots, n, \tag{8}$$

where

$$g^1(t), \ldots, g^n(t) \tag{9}$$

is a polynomial of degree $k - 1$. By substituting the functions (8) into

(6), we obtain [see §8, (A)]

$$0 = \sum_{s=1}^{n} L_s^j(p)g^s(t)e^{\lambda t} = \sum_{s=1}^{n} e^{\lambda t}L_s^j(p+\lambda)g^s(t), \qquad j = 1, \ldots, n;$$

after factoring out $e^{\lambda t}$ we have

$$\sum_{s=1}^{n} L_s^j(p+\lambda)g^s(t) = 0, \qquad j = 1, \ldots, n. \tag{10}$$

Thus, the system of functions (8) is a solution of (6) if and only if the polynomials (9) satisfy (10). Each term

$$L_s^j(p+\lambda)g^s(t)$$

of the left-hand side of (10) is a polynomial in t, and its coefficients are linear homogeneous functions of the coefficients of the polynomial $g^s(t)$. Thus the set of conditions (10) is equivalent to a certain system of linear homogeneous equations in the coefficients of the polynomials (9). From the theory of linear homogeneous algebraic equations it follows that, by finding the coefficients of the polynomials (9) from this linear system, we may take a certain number of these coefficients as arbitrary and express the remaining ones in terms of them. (The case where the number of arbitrary coefficients is equal to zero is not excluded *a priori.*) The solutions (8) of the system (6) obtained in this manner will be called the solutions *corresponding* to the root λ. It is clear that these solutions are defined for all values of t, $-\infty < t < \infty$.

THEOREM 9. Let us assume that the determinant $D(p)$ of the system (6) does not vanish identically, and let

$$\lambda_1, \lambda_2, \ldots, \lambda_m$$

be the set of all distinct roots of the polynomial $D(p)$. Then an arbitrary solution $x^1, \ldots, x^n$ of (6) can be written in the form

$$x^s = \overset{s}{x}_1 + \cdots + \overset{s}{x}_m, \qquad s = 1, \ldots, n, \tag{11}$$

where

$$\overset{1}{x}_i, \overset{2}{x}_i, \ldots, \overset{n}{x}_i$$

is some solution of system (6) which corresponds to the root λ_i [see (C)]. Hence, in particular, it follows that every solution of (6) is defined for all values of t.

Proof. Let us assume that

$$x^1, x^2, \ldots, x^n$$

is a solution of (6) which is defined on the interval $r_1 < t < r_2$; we shall show that on this interval it can be written in the form (11). By proposition (B) every function x^s on the interval $r_1 < t < r_2$ satisfies the equation

$$D(p)x^s = 0,$$

so that on the same interval by proposition (C) of §8 it can be written in the form

$$x^s = \sum_{i=1}^{m} g_i^s(t)e^{\lambda_i t}, \qquad s = 1, \ldots, n, \tag{12}$$

where $g_i^s(t)$ is a polynomial whose degree is less than the multiplicity of the root λ_i. By substituting the system (12) into (6), we obtain [see §8, (A)]

$$0 = \sum_{s=1}^{n}\sum_{i=1}^{m} L_s^j(p)g_i^s(t)e^{\lambda_i t} = \sum_{i=1}^{m}\left(e^{\lambda_i t}\sum_{s=1}^{n} L_s^j(p+\lambda_i)g_i^s(t)\right),$$

$$j = 1, \ldots, n. \tag{13}$$

Since $\sum_{s=1}^{n} L_s^j(p+\lambda_i)g_i^s(t)$ is a polynomial in t and the numbers $\lambda_1, \ldots, \lambda_m$ are mutually distinct, it follows from proposition (C) of §10 and from (13) that

$$\sum_{s=1}^{n} L_s^j(p+\lambda_i)g_i^s(t) = 0; \qquad i = 1, \ldots, m, \quad j = 1, \ldots, n. \tag{14}$$

Multiplying (14) by $e^{\lambda_i t}$, we obtain

$$0 = \sum_{s=1}^{n} e^{\lambda_i t}L_s^j(p+\lambda_i)g_i^s(t) = \sum_{s=1}^{n} L_s^j(p)g_i^s(t)e^{\lambda_i t},$$

which shows that

$$x_i^s = g_i^s(t)e^{\lambda_i t}, \qquad s = 1, \ldots, n,$$

is a solution of the system (6) for each $i = 1, \ldots, n$. Thus Theorem 9 is proved.

EXAMPLES

1. We shall solve by the elimination method the system of equations

$$\dot{x}^1 + x^1 + \dot{x}^2 = 0, \qquad \ddot{x}^1 - x^1 + \ddot{x}^2 + x^2 = 0.$$

We rewrite this system symbolically as

$$(p + 1)x^1 + px^2 = 0,$$
$$(p^2 - 1)x^1 + (p^2 + 1)x^2 = 0.$$

The determinant of the system, as is easily seen, is equal to $p^2 + 2p + 1$; it has a double root $\lambda = -1$. According to Theorem 9, the solution of the system should have the form

$$x^1 = (at + b)e^{-t},$$
$$x^2 = (ct + d)e^{-t}.$$

Substituting these functions into the first equation (and factoring out e^{-t}), we have

$$a + c - ct - d = 0,$$

whence

$$c = 0,$$
$$a = d.$$

The same relations for the coefficients may also be obtained by substitution into the second equation of the system. Thus the general solution of the system under consideration may be written in the form

$$x^1 = (at + b)e^{-t},$$
$$x^2 = ae^{-t},$$

where a and b are arbitrary constants.

2. We shall apply the elimination method to the normal system of linear homogeneous equations with constant coefficients,

$$\dot{x}^j = \sum_{s=1}^{n} a_s^j x^s, \qquad j = 1, \ldots, n, \tag{15}$$

(such a system will be studied more thoroughly in §14). We rewrite the system (15) by using the symbolism

$$px^j = \sum_{s=1}^{n} a_s^j x^s, \qquad j = 1, \ldots, n,$$

or

$$\sum_{s=1}^{n} (a_s^j - p\,\delta_s^j)x^s = 0, \qquad j = 1, \ldots, n, \tag{16}$$

where δ_s^j is the Kronecker symbol. The system (16) is a special case of the general system (6), in which

$$L_s^j(p) = a_s^j - p\delta_s^j,$$

and the determinant $D(p)$, in this case, is the characteristic determinant of the matrix (a_s^j) of the system (15). The solution of (16) must now be obtained by the method of undertermined coefficients described in Theorem 9. In the particular case when the roots $\lambda_1, \ldots, \lambda_n$ of $D(p)$ are simple, there corresponds to every root λ_i the solution

$$x_i^s = g_i^s e^{\lambda_i t}, \qquad s = 1, \ldots, n, \tag{17}$$

where

$$g_i^1, g_i^2, \ldots, g_i^n \tag{18}$$

are polynomials of zero degree, i.e., numbers. Substituting the functions (17) into (15), we obtain for the undetermined coefficients (18) the system of linear equations

$$\lambda_i g_i^j = \sum_{s=1}^n a_s^j g_i^s, \qquad j = 1, \ldots, n,$$

which shows that the numbers (18) are the components of the eigenvector of the matrix (a_s^j) with eigenvalues λ_i. Since all eigenvectors corresponding to the same eigenvalue λ_i are proportional in the case of simple roots, then by denoting the components of any eigenvector with eigenvalues λ_i by

$$h_i^1, h_i^2, \ldots, h_i^n,$$

we obtain

$$g_i^s = c^i h_i^s, \qquad s = 1, \ldots, n.$$

Thus the general solution of (15) in the case of simple roots may be written in the form

$$x^s = \sum_{i=1}^n c^i h_i^s e^{\lambda_i t}, \qquad s = 1, \ldots, n,$$

where $c^1, \ldots, c^n$ are arbitrary constants.

3. We shall study the linear system

$$\sum_{s=1}^n L_s^j(p) x^s = f^j(t), \qquad j = 1, \ldots, n, \tag{19}$$

with constant coefficients [see (1)]. Let q_i be its order with respect to the unknown x^i and let

$$q = q_1 + q_2 + \cdots + q_n$$

be the order of the system (19). Further, let

$$\begin{pmatrix} L_1^1(p) & \dots & L_n^1(p) \\ \vdots & & \\ L_1^n(p) & \dots & L_n^n(p) \end{pmatrix} \tag{20}$$

be the matrix of (19) and $D(p)$ its determinant. We shall show that the degree of the polynomial $D(p)$ does not exceed the number q. If this degree is equal to q, then we shall call the system (19) *normalizable*. In this case it can be solved for the higher derivatives

$$(x^1)^{(q_1)}, \dots, (x^n)^{(q_n)}, \tag{21}$$

and therefore it can be reduced to a normal system [see §4, (B)].

By assumption, the degree of the polynomial $L_s^j(p)$ does not exceed the number q_s, so we can write

$$L_s^j(p) = a_s^j p^{q_s} + \cdots, \tag{22}$$

where the dots denote terms of degree lower than q_s. By calculating the determinant $D(p)$ of (20), while taking into account formula (22), we have at once that

$$D(p) = \Delta \cdot p^q + \cdots,$$

where Δ is the determinant of the matrix (a_s^j). In this formula we have omitted terms of degree lower than q. Thus we have established that the maximum possible degree of the polynomial $D(p)$ is q, and if this degree is equal to q, then $\Delta \neq 0$. If we select those terms in (19) with the highest derivatives (21), we arrive at the system

$$\sum_{s=1}^{n} a_s^j (x^s)^{(q_s)} + \cdots = f^j(t), \qquad j = 1, \dots, n. \tag{23}$$

Thus, if (19) can be normalized, then $\Delta \neq 0$ and the system (23) is solvable for highest derivatives (21).

Since the normalizable system (19) reduces to a normal one, then, by what was said in Example 3 of §3, every solution of (19) has an arbitrary prescribed number of derivatives if only the right-hand sides $f^j(t)$ of (19) are differentiable an appropriate number of times.

4. We shall now study the case when the determinant $D(p)$ of the system (19) is not identically zero, but the degree of $D(p)$ is less than the order q of the system (19). We shall show that even in this case every solution of (19) has any prescribed number of derivatives, if only the right-hand sides $f^j(t)$ are sufficiently differentiable.

By hypothesis, the degree of $D(p)$ is less than q, so that the determinant Δ is zero. Thus, the columns of the matrix (a_s^j) are linearly dependent; let $b^1, \ldots, b^n$ be the coefficients expressing this dependence. Some of the numbers $b^1, \ldots, b^n$ may be zero. We shall change the numbering of the functions $x^1, \ldots, x^n$ so that the relations

$$b^1 \neq 0, \quad b^2 \neq 0, \ldots, b^m \neq 0, \quad b^{m+1} = \cdots = b^n = 0, \qquad 1 \leq m \leq n, \tag{24}$$
$$q_1 \geq q_2, \quad q_1 \geq q_3, \ldots, q_1 \geq q_m,$$

are valid. Since by (24) we have $b^1 \neq 0$, we may assume that $b^1 = 1$.

We shall now replace the indeterminates $x^1, \ldots, x^n$ by new indeterminates $y^1, \ldots, y^n$ by setting

$$x^1 = y^1; \quad x^i = y^i + b^i p^{q_1 - q_i} y^1, \quad i = 2, \ldots, m;$$
$$x^i = y^i, \quad i = m+1, \ldots, n. \tag{25}$$

The relations (25) can be solved for the new indeterminates $y^1, \ldots, y^n$, namely,

$$y^1 = x^1; \quad y^i = x^i - b^i p^{q_1 - q_i} x^1, \quad i = 2, \ldots, m;$$
$$y^i = x^i, \quad i = m+1, \ldots, n. \tag{26}$$

Substituting the new indeterminates $y^1, \ldots, y^n$ into (19) in place of $x^1, \ldots, x^n$, we obtain the new system of equations

$$\sum_{s=1}^{n} L_s^{*j}(p) y^s = f^j(t), \quad j = 1, \ldots, n. \tag{27}$$

It is immediately evident that the order q_1^* of the system (27) with respect to the function y^1 is less than q_1, and its orders with respect to the remaining unknowns $y^2, \ldots, y^n$ are correspondingly equal to $q_2, \ldots, q_n$. Thus the order q^* of (27) is less than the order q of (19).

If we consider the transformations (25) and (26) as linear transformations of the variables $y^1, \ldots, y^n$ into the variables $x^1, \ldots, x^n$, and conversely, with coefficients which are polynomials in p, then it is evident that the determinant of each of the linear transformations (25) and (26) is equal to $+1$. From this it follows that the determinant $D^*(p)$ of (27) is equal to the determinant $D(p)$ of (19). Thus the difference between the order and the degree of the determinant in system (27) is less than in system (19); by applying the transformation described a finite number of times, we obtain a normalizable system.

Now let

$$x^i = \varphi^i(t), \quad i = 1, \ldots, n, \tag{28}$$

be some solution of the system (19). Since the order of (19) with respect to the unknown x^i is equal to q_i, the function $\varphi^i(t)$ may be assumed to be differentiable q_i times. By virtue of the transformation (26) the solution

$$y^i = \psi^i(t), \qquad i = 1, \ldots, n, \tag{29}$$

of (27) corresponds to the solution (28) of (19). From (26) it is evident that $\psi^i(t)$ is differentiable q_i times. Consequently, from every solution (28) of (19) we can obtain a certain solution (29) of (27), so that no solution is lost in going from (19) to (27). Since, after a series of transformations, we arrive at a normalizable system whose solutions have any given number of derivatives, it is evident from (25) that a solution (28) of (19) also has any prescribed number of derivatives.

12. The method of complex amplitudes. In many branches of engineering and physics dealing with oscillating processes, *harmonic oscillations* play an important role. Mathematically, a harmonic oscillation is defined by a function

$$r \cos(\omega t + \alpha), \qquad r \geq 0. \tag{1}$$

Here r is the *amplitude* of the oscillation, α is its *initial phase*, and the number ω, determining the frequency of the oscillation, is usually called the *frequency*. Actually, if ν is the number of oscillations per second, then

$$\omega = 2\pi\nu,$$

so that ω is the number of oscillations in 2π seconds, not in one second. We also noticed (see Example 1, §4) that the equation

$$\ddot{x} + \omega^2 x = 0 \tag{2}$$

has as its general solution the harmonic function (1) of frequency ω with an arbitrary amplitude and phase. Equation (2) is called the *equation of the harmonic oscillator.*

In the study of harmonic oscillations it is often necessary to deal with the equation

$$L(p)x = r \cos(\omega t + \alpha), \tag{3}$$

where the right-hand side contains a harmonic function. Equation (3) is easily solved by using the method set forth in Theorem 8, since the harmonic function is a quasipolynomial. In the case when the coefficients of the polynomial $L(p)$ are real, Theorem 8 can be utilized in a somewhat different way. In electrical engineering this method is called the *method of complex amplitudes* and is explained in what follows.

(A) In conjunction with the real harmonic function (1) we shall study its corresponding complex harmonic function

$$\rho e^{i\omega t}, \tag{4}$$

where

$$\rho = re^{i\alpha}. \tag{5}$$

The function (4) has the property that its real part coincides with the function (1):

$$\rho e^{i\omega t} = re^{i(\omega t+\alpha)} = r \cos (\omega t + \alpha) + ir \sin (\omega t - \alpha).$$

The complex number (5) is called the *complex amplitude* of the complex harmonic function (4); it combines the amplitude r and the initial phase α. Let us note that

$$r = |\rho|.$$

In the case that the coefficients of $L(p)$ are real, the equation

$$L(p)z = \rho e^{i\omega t} \tag{6}$$

is solved as a preliminary to the solution of equation (3). It is immediately evident that, if $z = x + iy$ is a solution of (6), then x is a solution of (3). Assuming that $i\omega$ is not a root of $L(p)$,

$$L(i\omega) \neq 0, \tag{7}$$

we seek (see Theorem 8) a solution of equation (6) in the form of a complex harmonic function $z = \sigma e^{i\omega t}$ with complex amplitude $\sigma = se^{i\beta}$. By substituting $z = \sigma e^{i\omega t}$ into (6) we obtain

$$\sigma = \frac{\rho}{L(i\omega)} \tag{8}$$

[see §7, (B)]. Thus a solution of (3) is given by

$$x = s \cos (\omega t + \beta). \tag{9}$$

The amplitude s and initial phase β of this solution are determined from the formula

$$se^{i\beta} = \frac{re^{i\alpha}}{L(i\omega)}$$

[see (8)]. In particular, $s = |\sigma| = r/|L(i\omega)|$. If the polynomial $L(p)$ is stable, then (7) is obviously satisfied. In this case the general solution of equation (3) has the form

$$x = u + s \cos (\omega t + \beta), \tag{10}$$

where u is the solution of the homogeneous equation $L(p)u = 0$. The solution u of this homogeneous equation tends to zero as $t \to \infty$, and therefore any solution of (3) tends to the solution (9). The solution (9) is called *stable* and it corresponds to a *steady-state* process, while the solution (10) describes a *transient* process. The stable solution (9) is the unique periodic solution among all solutions (10).

In applying the method of complex amplitudes, we usually do not study the solutions of the real equation (3); instead, we proceed directly from the complex equation (6).

We shall now present the method of complex amplitudes as it is applied to a system of equations. The problem is to find a particular solution of the system of equations

$$\sum_{s=1}^{n} L_s^j(p)x^s = r^j \cos(\omega t + \alpha^j), \qquad j = 1, \ldots, n, \tag{11}$$

with real coefficients, whose right-hand sides are harmonic oscillations of the same frequency ω.

(B) We shall assume that the determinant $D(p)$ of the system (11) [see §11, (A)] does not vanish at $p = i\omega$. To find the solution of (11), we shall first find the solution of the system of equations

$$\sum_{k=1}^{n} L_k^j(p)z^k = \rho^j e^{i\omega t}, \qquad j = 1, \ldots, n, \tag{12}$$

where

$$\rho^j = r^j e^{j\alpha^j}.$$

Since the coefficients of all the polynomials $L_k^j(p)$ are real, we shall obtain the following solution of (11) from any solution $z^1, \ldots, z^n$ of (12):

$$x^k = \operatorname{Re} z^k, \qquad k = 1, \ldots, n.$$

We find the solution of the system (12) in the form

$$z^k = \sigma^k e^{i\omega t}, \qquad k = 1, \ldots, n. \tag{13}$$

By substituting the functions (13) into (12) and factoring out $e^{i\omega t}$, we obtain the system of equations

$$\sum_{k=1}^{n} L_k^j(i\omega)\sigma^k = \rho^j,$$

which can be solved uniquely for the unknowns σ^k, because its determinant $D(i\omega)$ is assumed to be different from zero. We shall find solutions of this system and assume that

$$\sigma^k = s^k e^{i\beta^k},$$

so that by (13) we obtain the solution

$$x^k = s^k \cos(\omega t + \beta^k), \qquad k = 1, \ldots, n, \tag{14}$$

of (11). If the determinant $D(p)$ of (11) is a stable polynomial, then the inequality $D(i\omega) \neq 0$ is satisfied, and in addition every solution of (11) differs from the solution (14) by a term which tends to zero as $t \to \infty$ [see §11, (B)]. Thus in the case of a stable polynomial $D(p)$, the solution (14) of (11) not only consists of the particular solutions, but is itself a *stable solution*.

Example

We shall solve the equation

$$\ddot{x} + \omega_1^2 x = r \cos(\omega t + \alpha) \tag{15}$$

of the harmonic oscillator which is subject to an external harmonic force. Instead of (15) we shall study the corresponding complex equation

$$\ddot{z} + \omega_1^2 z = re^{i(\omega t + \alpha)}. \tag{16}$$

If $\omega \neq \omega_1$, then (16) has a solution of the form $z = \sigma e^{i\omega t}$, so that by (8)

$$\sigma = \frac{re^{i\alpha}}{\omega_1^2 - \omega^2}.$$

Thus (15) has the solution

$$x = \frac{r}{|\omega_1^2 - \omega^2|} \cos(\omega t + \beta), \tag{17}$$

where $\beta = \alpha$ for $\omega_1 > \omega$ and $\beta = \alpha + \pi$ for $\omega_1 < \omega$. Formula (17) gives the *forced oscillations* of the oscillator under a harmonic external force. Here it is important to note the phenomenon of *resonance*, which incorporates the fact that the amplitude

$$\frac{r}{|\omega_1^2 - \omega^2|}$$

of a forced oscillation *increases* as the difference $|\omega_1 - \omega|$ decreases. It is also interesting to note that the phase β of the oscillation (17) coincides with the phase α of the external force for $\omega_1 > \omega$ and is opposite to it for $\omega_1 < \omega$. The general solution of (15) may be written in the form

$$x = r_1 \cos(\omega_1 t + \alpha_1) + \frac{r}{|\omega_1^2 - \omega^2|} \cos(\omega t + \beta),$$

where $u = r_1 \cos(\omega_1 t + \alpha_1)$ is a solution of the corresponding homogeneous equation. The term u is called the *natural oscillation* of the oscillator.

If $\omega_1 = \omega$, then the formula (17) loses its meaning. In this case the solution of (16) must be sought in the form

$$z = \rho t e^{i\omega t},$$

where ρ is a complex number (see Theorem 8). By formula (9) of §10, we have

$$[(p + i\omega)^2 + \omega^2]\rho t = re^{i\alpha},$$

whence

$$\rho = \frac{re^{i\alpha}}{2i\omega}.$$

Thus for $\omega_1 = \omega$ the particular solution of equation (16) has the form

$$z = \frac{rte^{i(\omega t+\alpha)}}{2i\omega} = \frac{rte^{i[\omega t+\alpha-(\pi/2)]}}{2\omega},$$

and the solution of equation (15) proves to be

$$x = \frac{rt}{2\omega}\cos\left(\omega t + \alpha - \frac{\pi}{2}\right) = \frac{rt}{2\omega}\sin(\omega t + \alpha).$$

Thus for $\omega = \omega_1$, the phenomenon of resonance involves the fact that the amplitude $rt/2\omega$ becomes variable and increases without limit with time. In a real apparatus this phenomenon cannot be observed due to the presence of "friction."

13. Electrical circuits. The theory of ordinary differential equations finds applications in various fields of engineering; it is applied in electrical engineering and in particular in radio engineering. With some idealization, the performance of a radio device can be mathematically described by a system of ordinary differential equations; the values of currents flowing through various elements of the device or the voltage drops between individual junctions of the device are the unknown functions of time in this system. Radio devices afford a very rich source of material illustrating the application of the theory of ordinary differential equations—much richer than, for example, the problems of mechanics. This richness is characterized in particular by the fact that a system of ordinary differential equations arising from a certain engineering problem often can be *simulated* by an electrical device, i.e., an electrical device may be designed whose performance is described by the *same* system of equations as the engineering object in which we are interested. An electrical device so designed can help solve the system of equations, since in observing its performance we are at the same time observing the behavior of the unknown functions which satisfy the system of equations. The physical laws governing the performance of electrical devices are formulated so simply that they can be easily communicated even to a beginner in physics. Here, in a somewhat dogmatic form, the simplest laws of electrical engineering are given along with several examples of the application of differential equations to the study of the performance of electrical devices.

Resistors, inductors (self-induction), and *capacitors* (condensers) are some of the most important components from which electrical devices are designed. Each of these components is a *two-terminal element*, i.e., has two contacts which in assembling an electrical device are connected to the terminals of other components. During the operation of an electrical device, an electric current passes through the two-terminal element installed in the device, and the electrical state of the two-terminal element at each instant t is characterized by two values: the *current* $I_{ab}(t)$ which flows from the pole a to the pole b of the two-terminal element ab, and the *voltage drop* $U_{ab}(t)$ from pole a to pole b. The current $I_{ab}(t)$ can take positive as well as negative values; if the current flows from pole a to pole b (keeping in mind the so-called *technical* direction of the current), then the number $I_{ab}(t)$ is positive, and in the opposite case it is negative. The voltage drop $U_{ab}(t)$ from pole a to pole b is the difference $V_a(t) - V_b(t)$ of the potentials at the pole a and b. Thus both values, $I_{ab}(t)$ and $U_{ab}(t)$, which describe the state of a two-terminal element ab at instant t, depend on which pole is put in the first place and which in the second. When the order of the poles changes, each of the values $I_{ab}(t)$ and $U_{ab}(t)$ obviously changes its sign, so that we have the relations

$$I_{ba}(t) = -I_{ab}(t), \tag{1}$$

$$U_{ba}(t) = -U_{ab}(t). \tag{2}$$

For every two-terminal element ab, the functions $I_{ab}(t)$ and $U_{ab}(t)$ of time t are not independent, but are related to each other by physical laws governing the performance of a two-terminal element. For resistance, inductance, and capacitance, these laws are given by the following proposition:

(A) For a two-terminal element ab representing resistance, the relationship (Ohm's law)

$$U_{ab}(t) = R_{ab}I_{ab}(t) \tag{3}$$

is valid; here R_{ab} is a positive coefficient called the *resistance*, which for different two-terminal elements can take different values but which is constant for a given element; here we always have

$$R_{ba} = R_{ab}. \tag{4}$$

For a two-terminal element ab which represents inductance, the relation

$$U_{ab}(t) = L_{ab}\frac{d}{dt}I_{ab}(t) \tag{5}$$

is valid; here L_{ab} is a positive coefficient called the *inductance* which can take different values for different two-terminal elements but which is constant for each given element. Here

$$L_{ba} = L_{ab}. \tag{6}$$

For a two-terminal element ab which is the capacitance (the condenser) the relationship

$$I_{ab}(t) = C_{ab}\frac{d}{dt}U_{ab}(t) \tag{7}$$

is valid, where C_{ab} is a positive coefficient called the *capacitance*, which can take different values for different two-terminal elements but has only one value for a given element; here we have

$$C_{ab} = C_{ba}. \tag{8}$$

If we integrate the relation (7), we obtain

$$U_{ab}(t) = U_{ab}(t_0) + \frac{1}{C_{ab}}\int_{t_0}^{t} I_{ab}(t)\,dt. \tag{9}$$

The function

$$Q_{ab}(t) = C_{ab}U_{ab}(t)$$

represents a physical quantity describing the state of the condenser at a given instant and is called the *charge* of the condenser ab. The relation (9) is often written in the form

$$U_{ab}(t) = \frac{1}{C_{ab}}\int I_{ab}(t)\,dt,$$

where $\int I_{ab}(t)\,dt$ represents the charge of the condenser.

Relation (4) follows from (1), (2), and (3):

$$\begin{aligned} R_{ab}I_{ab}(t) &= U_{ab}(t) = -U_{ba}(t) = -R_{ba}I_{ba}(t) \\ &= R_{ba}(-I_{ba}(t)) = R_{ba}I_{ab}(t). \end{aligned}$$

The relations (6) and (8) are established in a similar way.

An important role in the performance of electrical devices is given to the phenomenon of *mutual induction* between two inductances.

(B) Two inductances a_1b_1 and a_2b_2 with values $L_{a_1b_1} = L_1$ and $L_{a_2b_2} = L_2$ can be in a state of mutual induction which is described by the *coefficient of mutual induction*, $M = M_{a_1b_1,a_2b_2}$. In this case the voltage drop $U_{a_1b_1}(t) = U_1(t)$ in a two-terminal element a_1b_1 is related not only to the current $I_{a_1b_1}(t) = I_1(t)$, but also to the current $I_{a_2b_2}(t) = I_2(t)$. In the same way, the voltage $U_{a_2b_2}(t) = U_2(t)$ in a two-terminal element a_2b_2 is related not only to the current $I_2(t)$ but also to the current $I_1(t)$. The exact relations are given by the formulas

$$U_1(t) = L_1 \frac{d}{dt} I_1(t) + M \frac{d}{dt} I_2(t), \tag{10}$$

$$U_2(t) = L_2 \frac{d}{dt} I_2(t) + M \frac{d}{dt} I_1(t). \tag{11}$$

In this case the equalities

$$M_{a_1b_1,a_2b_2} = M_{a_2b_2,a_1b_1} = -M_{a_1b_1,b_2a_2}$$

are valid for the coefficient of mutual induction $M_{a_1b_1,a_2b_2}$, as well as the inequality

$$M^2 \leq L_1L_2.$$

The greater the "interaction" of two inductances, the more closely the coefficient of mutual induction M approximates in magnitude the quantity $\sqrt{L_1L_2}$.

The two-terminal elements described in proposition (A) are called *passive* because they cannot themselves initiate an electrical phenomenon in a device. The *active* two-terminal elements which serve as the direct cause of electrical currents in a device are *voltage sources* and *current sources*.

(C) In the two-terminal element ab which is a voltage source, the relationship

$$U_{ab}(t) = U(t) \tag{12}$$

is valid, where $U(t)$ is a given function of time t describing a voltage source. Equation (12) may be considered as relating the functions $U_{ab}(t)$ and $I_{ab}(t)$, except that this relation is such that the function $I_{ab}(t)$ does not appear in it. For a current source ab, the relationship

$$I_{ab}(t) = I(t)$$

is similarly valid, where $I(t)$ is a given function of t describing the current source. The voltage sources and current sources most often studied are those for which $U(t)$ and $I(t)$ are either constants or periodic functions of the form

$$r \cos (\omega t + \alpha).$$

These are the principal, and at the same time the simplest, components from which electrical devices are assembled. The devices themselves are called *electrical networks*, and the components from which they are assembled are called their *elements*. It should be noted that elements exist which are different from those described above; in particular, there exist *multiterminal* elements. An example of a three-terminal element is the *electron tube* (triode), whose performance will be discussed later (see §29).

Kirchhoff's laws. We shall now proceed to formulate *Kirchhoff's laws*, which govern the performance of electrical networks.

(D) An *electrical network* is defined as a finite set of elements (in particular, two-terminal elements of the form described above) with poles that are connected in so-called "junctions" of the network so that at each junction two or more poles of the various elements of the network are joined. *Kirchhoff's first law* asserts that *the sum of all currents entering each junction of a network from all elements connected to this junction is equal to zero. Kirchhoff's second law* follows from the proposition that at each junction a of a network there is an electric potential $V_a(t)$, and the voltage drop $U_{ab}(t)$ from junction a to junction b is the difference between the potentials in junctions a and b, so that $U_{ab}(t) = V_a(t) - V_b(t)$. From this assumption it follows that, *if $a, b, c, \ldots, h, k$ is a certain sequence of junctions in an electrical network, then the relation*

$$U_{ab}(t) + U_{bc}(t) + \cdots + U_{hk}(t) + U_{ka}(t) = 0$$

is valid. This relation is Kirchhoff's second law. It is stated as follows: *The sum of voltage drops around a closed circuit of a network is equal to zero.* We do not assume in these formulations of Kirchhoff's laws that all the elements have two terminals. We shall now define more clearly Kirchhoff's laws as applied to networks composed of only two-terminal elements.

(E) Let S be a certain electrical network composed of two-terminal elements. Kirchhoff's first law states that, if a is an arbitrary junction of a network S, and if $b_1a, b_2a, \ldots, b_qa$ is the set of all two-terminal elements connected to the junction a (Fig. 12), then

$$I_{b_1a}(t) + I_{b_2a}(t) + \cdots + I_{b_qa}(t) = 0.$$

Kirchhoff's second law states that if $ab, bc, \ldots, hk, ka$ is a sequence of two-terminal elements in a network S [each successive two-terminal element starts at that junction at which the preceding one ends (Fig. 13)], then

$$U_{ab}(t) + U_{bc}(t) + \cdots + U_{hk}(\mathrm{t}) + U_{ka}(t) = 0.$$

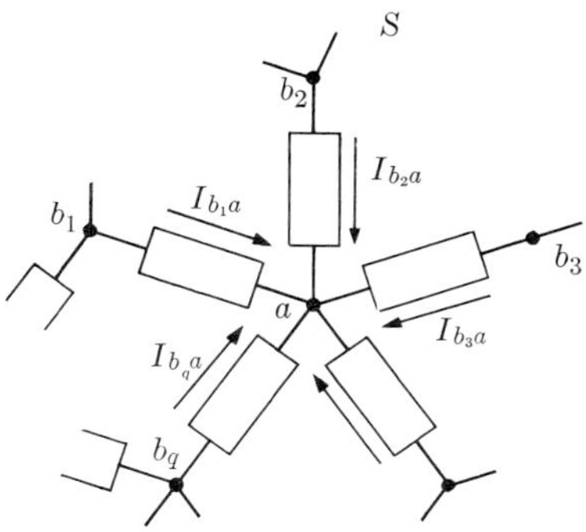

FIGURE 12

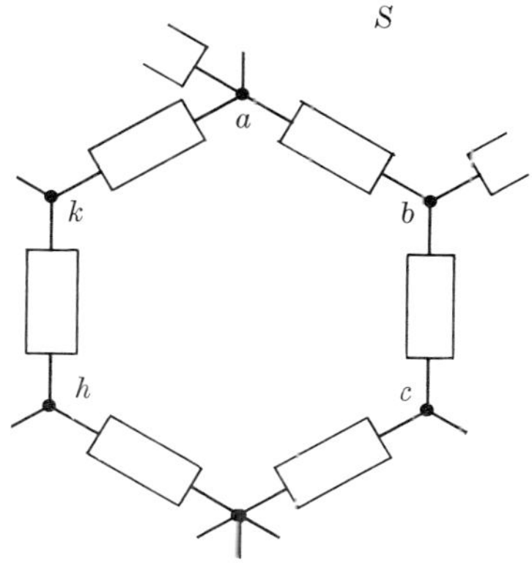

FIGURE 13

To calculate the performance of an electrical network consisting of two-terminal elements we must find the current and voltage in each two-terminal element in the network; thus if the network consists of n two-terminal elements, we face the problem of finding $2n$ functions of time. The law governing the performance of each two-terminal element gives one relation between the functions sought so that by this law we obtain n relations for $2n$ unknown functions. The remaining n relations are given by Kirchhoff's laws. (It can be proved that Kirchhoff's laws give exactly n independent relations, but we shall omit this proof.) As a result of using all the relations, we obtain a system of $2n$ equations for $2n$ unknown functions. These equations are partly differential and partly finite (algebraic). Kirchhoff's laws give finite equations which must first be used for eliminating part of the unknown functions. For this elimination one of the following two methods is usually employed. The first method relies on the fact that it is possible to take currents for basic unknown functions and express the voltages in terms of them. In this case it is first necessary to use Kirchhoff's first law: to express all currents in terms of the minimal number of independent currents (by this law). Such independent currents are called *loop currents*. After this, Kirchhoff's second law is to be used by replacing every voltage by its expression in terms of the corresponding current. This method is called the *method of loop currents*. In the second method the voltages on the two-terminal elements are taken as the basic unknown functions, and the currents are expressed in terms of these voltages (with the aid of laws governing the performance of each two-terminal element). In this case it is first necessary to express all voltages in terms of the minimal number of independent voltages by means of Kirchhoff's second law. Independent voltages are called *nodal* voltages. Next it is necessary to use Kirchhoff's first law by replacing every current

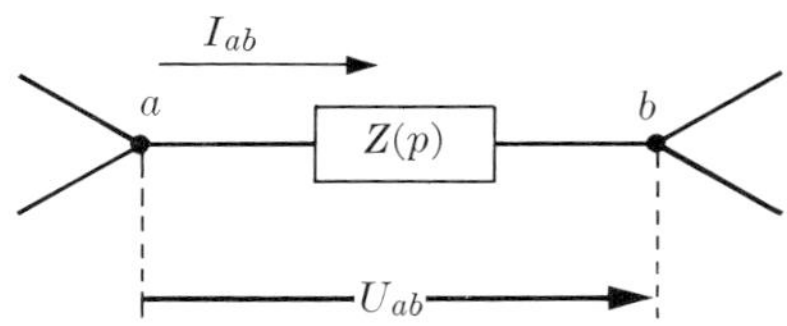

FIGURE 14

by its expression in terms of the corresponding voltage. This method is called the *method of nodal voltages.*

Operational impedance of a two-terminal element. Before proceeding to the analysis of examples of electrical network calculation, we shall write relations (3), (5), (9), (10), and (11), i.e., the laws governing the performance of two-terminal elements, in a symbolic notation.

(F) Let ab be a two-terminal element representing resistance, inductance, or capacitance. Let us set

$$U_{ab}(t) = U(t), \qquad I_{ab}(t) = I(t),$$
$$R_{ab} = R, \qquad L_{ab} = L, \qquad C_{ab} = C.$$

If in addition to the symbolic notation previously used [see §7, (A)], we introduce the natural notation $(1/p)f(t) = \int f(\tau)\,d\tau$, then (3), (5), and (9) can be expressed by one formula

$$U(t) = Z(p)I(t) \tag{13}$$

(Fig. 14), where, respectively, $Z(p) = R$, $Z(p) = Lp$, $Z(p) = 1/Cp$. The function $Z(p)$ is called the *impedance* of the two-terminal element ab *in operator form*, or the *operational impedance.* For the capacitance it is not a polynomial, but a rational function $1/Cp$:

$$U(t) = \frac{1}{Cp}\, I(t). \tag{14}$$

The relation (14) after multiplication by Cp acquires the usual form $I(t) = CpU(t)$, which is merely a polynomial in p. If we set

$$G(p) = \frac{1}{Z(p)},$$

then relation (13) takes the form

$$I(t) = G(p)U(t).$$

The function $G(p)$ is called the *admittance* of two-terminal element *ab in*

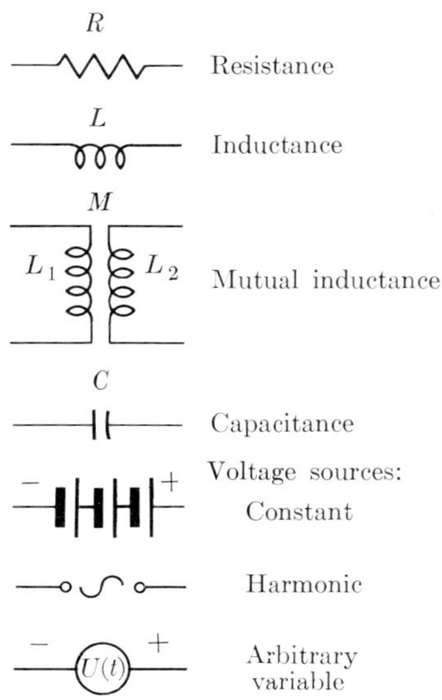

FIG. 15. Standard symbols of certain elements of electrical networks.

operator form and has the corresponding form:

$$G(p) = \frac{1}{R}, \qquad G(p) = \frac{1}{Lp}, \qquad G(p) = Cp.$$

The relations (10) and (11) in operational notation have the form

$$U_1(t) = L_1 pI_1(t) + MpI_2(t),$$
$$U_2(t) = L_2 pI_2(t) + MpI_1(t).$$

We shall proceed to the analysis of some examples. As a visual aid, electrical networks are represented graphically by a point for every junction; and for every two-terminal element, a straight line segment or a curve for the connection to the corresponding junctions; on every such segment the corresponding two-terminal element is represented by the conventional designation (Fig. 15).

EXAMPLES

1. (*Oscillatory loop.*) Let S be an electrical network with four junctions a, b, c, and d consisting of four two-terminal elements ab, bc, cd, and da (Fig. 16). The element ab is the inductance L, bc is the resistance R, cd is the capacitance C, and, finally, the element ad is a voltage source $U_{ad}(t) = U(t)$. For the calculations we shall use the method of loop

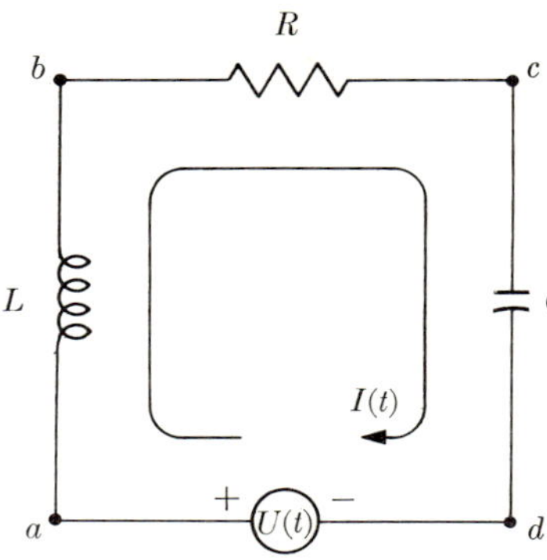

FIGURE 16

currents. By applying Kirchhoff's first law to junction b, we obtain $I_{ab}(t) + I_{cb}(t) = 0$, or $I_{ab}(t) = I_{bc}(t)$. This is always the case when exactly two two-terminal elements are connected to one junction. Thus we have

$$I_{ab}(t) = I_{bc}(t) = I_{cd}(t) = I_{da}(t) = I(t).$$

Here $I(t)$ is the *loop current*. Further, by writing for every two-terminal element the law governing its performance, we obtain

$$\begin{aligned} U_{ab}(t) &= LpI(t), \qquad & U_{bc}(t) &= RI(t), \\ U_{cd}(t) &= \frac{1}{Cp}\, I(t), \qquad & U_{da}(t) &= -U(t). \end{aligned} \tag{15}$$

Kirchhoff's second law gives

$$U_{ab}(t) + U_{bc}(t) + U_{cd}(t) + U_{da}(t) = 0. \tag{16}$$

From (15) and (16) we obtain

$$\left(Lp + R + \frac{1}{Cp}\right) I(t) = U(t). \tag{17}$$

Both sides of (17) can be multiplied by p (which involves differentiation term by term), whence we obtain

$$\left(Lp^2 + Rp + \frac{1}{C}\right) I(t) = pU(t). \tag{18}$$

This is the differential equation of the network under examination.

If the two-terminal element ad is removed from the network, we shall obtain a so-called *open* circuit, consisting of three passive two-terminal

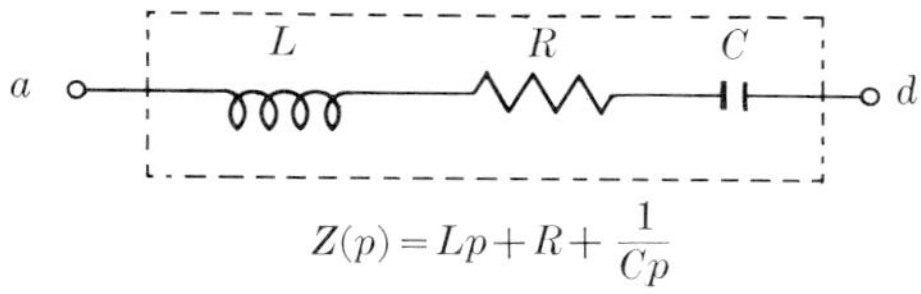

FIGURE 17

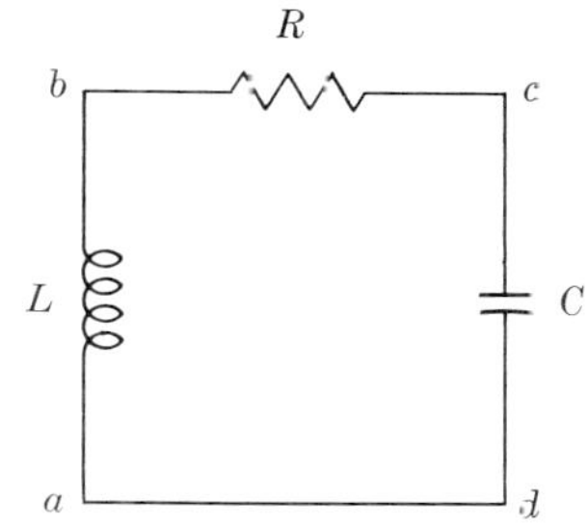

FIGURE 18

elements ab, bc, cd. This network (as a whole) can be considered as a two-terminal element with poles a and d (Fig. 17). The law governing the performance of such a two-terminal element is given by relation (17), which is analogous to (13). Here the function $Z(p) = Lp + R + 1/Cp$ is an *impedance in operator form* and its inverse,

$$G(p) = \frac{Cp}{LCp^2 + RCp + 1},$$

is an *admittance in operator form.*

If we set $U(t) = 0$, then this will be equivalent to the assumption that in our network the active two-terminal element ad is absent, and the network consists of three passive two-terminal elements ab, bc, cd, where the junctions a and d coincide (Fig. 18). The equation describing the performance of this passive electrical network S^* has the form

$$\left(Lp^2 + Rp + \frac{1}{C}\right) I(t) = 0. \tag{19}$$

As previously noted, electrical phenomena do not arise by themselves in a passive electrical circuit, and this is reflected by the fact that the particular solution of equation (19) is the function $I(t) \equiv 0$. It is possible, however, to study the performance of the electrical network S^* by first assuming that it already has a current, and then determining how this current will vary with time. Let λ_1 and λ_2 be roots of the polynomial

$$Lp^2 + Rp + \frac{1}{C}\cdot \tag{20}$$

Since the numbers L, R, C are positive [see (A)], the real parts of the roots λ_1 and λ_2 are negative, so that the electrical process in the network S^* will be damped with time [see §9, (A)]. This damping can occur, however, in various ways: if λ_1 and λ_2 are complex, then every nonzero solution of (19)

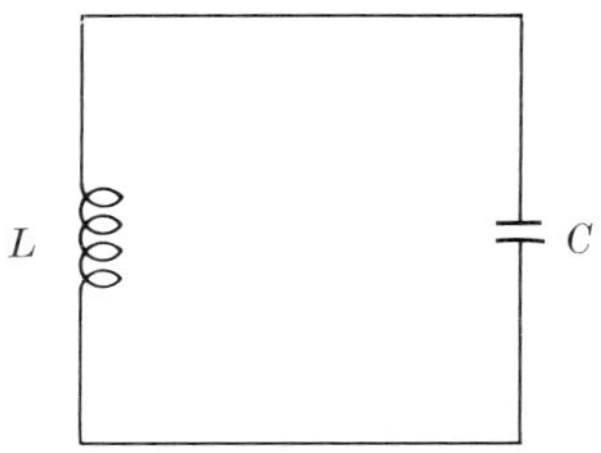

FIGURE 19

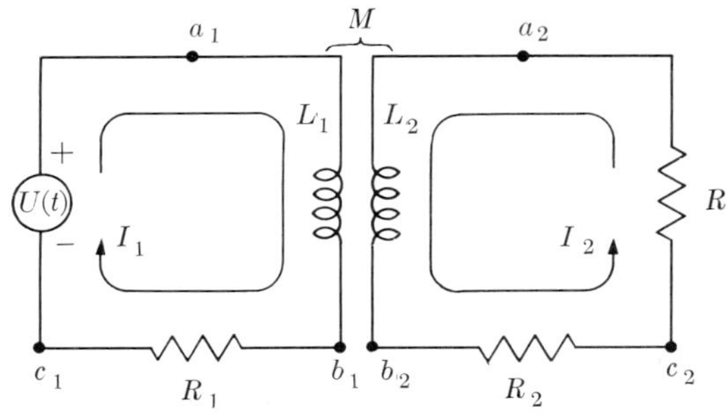

FIGURE 20

has an oscillatory character (see Example 3, §7); however, if λ_1 and λ_2 are real, then the damping occurs aperiodically, that is, any solution of (19), starting at a certain instant, becomes monotonic. The question of whether the roots λ_1 and λ_2 are complex or real is determined by the sign of the number

$$\Delta = \left(\frac{R}{2L}\right)^2 - \frac{1}{LC};$$

if $\Delta < 0$, then the solutions of (19) are oscillatory; if $\Delta > 0$, they are aperiodic.

Of particular interest is the oscillatory loop S^* where the resistance R is completely absent. In this case our circuit consists only of two passive elements ab and cd, and $b = c$, $a = d$ (Fig. 19). Under this assumption the equation of the electrical network has the form

$$\left(p^2 + \frac{1}{LC}\right) I(t) = 0.$$

The general solution of this equation may be written in the form

$$I(t) = s \cos (\omega_1 t + \beta_1),$$

where $\omega_1 = 1/\sqrt{LC}$. Thus, in the absence of resistance in a passive oscillatory circuit, there occur nondamping oscillations with the frequency

$$\omega_1 = \frac{1}{\sqrt{LC}}.$$

In the general case the value $1/\sqrt{LC}$ is called the *natural frequency* of the oscillatory loop S.

We shall now return to the study of the oscillatory loop S and examine the case of a harmonic voltage source $U(t)$.

Since the roots of polynomial (20) have negative real parts, it is possible to examine the steady-state process in the network S. We shall seek a solution by the method of complex amplitudes (see §12). Let $U(t) = re^{i\omega t}$ be the complex harmonic oscillation with real amplitude $r > 0$. Then the right-hand side of equation (18) has the form

$$pU(t) = p(re^{i\omega t}) = ir\omega e^{i\omega t}.$$

We have

$$I_{\text{steady state}}(t) = \sigma e^{i\omega t},$$

where the complex amplitude σ of the current $I_{\text{steady state}}(t)$ is defined by the formula

$$\sigma = \frac{ir\omega}{iR\omega + (-L\omega^2 + 1/C)} = \frac{r}{R + i\,(L\omega - 1/C\omega)}$$

[see §12, (A)]. Hence for the real amplitude we obtain

$$s = |\sigma| = \frac{r}{\sqrt{R^2 + (L\omega - 1/C\omega)^2}}.$$

From this formula it is clear that for a given amplitude r of the voltage source, the current amplitude s is maximum at the natural frequency $\omega = \omega_1 = 1/\sqrt{LC}$ of the loop S. For this frequency the amplitudes s and r are connected by the relation $s = r/R$, i.e., at this frequency the loop behaves as though only resistance were present. For the remaining frequencies the current amplitude s has a value smaller than r/R. This phenomenon is called *resonance* (compare the example in §12). The oscillatory loop L, R, C *resonates* at its own natural frequency $1/\sqrt{LC}$.

2. (*Transformer.*) A transformer consists of two windings, *primary* and *secondary*, placed on a single core. A source of variable voltage is connected to the primary winding, and to the secondary winding a load, for example, external resistance. Each winding has inductance and resistance (internal). Between the windings there exists a mutual induction. Thus the transformer may be considered an electrical network consisting of two separate loops inductively connected. The first loop consists of three two-terminal elements: a_1b_1 is the inductance L_1, b_1c_1 is the internal resistance R_1, and a_1c_1 is the voltage source $U_{a_1c_1} = U(t)$. The second loop also consists of three two-terminal elements: a_2b_2 is the inductance L_2, b_2c_2 is the internal resistance R_2, and c_2a_2 is the load resistance R. In addition, there exists a mutual induction $M_{a_1b_1,a_2b_2} = M$ (Fig. 20). By Kirchhoff's first law we have

$$I_{a_1b_1} = I_{b_1c_1} = I_{c_1a_1} = I_1; \qquad I_{a_2b_2} = I_{b_2c_2} = I_{c_2a_2} = I_2.$$

Thus we have two loop currents, I_1, I_2. By applying Kirchhoff's second law we obtain

$$L_1pI_1 + MpI_2 + R_1I_1 - U(t) = 0,$$
$$L_2pI_2 + MpI_1 + R_2I_2 + RI_2 = 0,$$

or

$$(L_1p + R_1)I_1 + MpI_2 = U(t).$$
$$MpI_1 + (L_2p + R_2 + R)I_2 = 0.$$

The determinant $D(p)$ of this system has the form

$$D(p) = (L_1L_2 - M^2)p^2 + (L_1R_2 + L_1R + L_2R_1)p + R_1(R_2 + R).$$

By virtue of proposition (B) of §9 this polynomial is stable since $L_1L_2 - M^2 > 0$. We shall examine the performance of a transformer in the case where the voltage $U(t)$ varies harmonically, and we shall seek the steady-state solution by the method of §12, (B). Let us set

$$U(t) = u_1e^{i\omega t},$$

where u_1 is a positive real number (the amplitude of the voltage applied to the primary winding). We shall seek the unknown functions I_1 and I_2 in the form

$$I_1 = \sigma_1e^{i\omega t}, \qquad I_2 = \sigma_2e^{i\omega t},$$

where σ_1 and σ_2 are the complex amplitudes of the currents.

The so-called *ideal transformer*, i.e., a transformer in which the values of R_1, R_2 and $L_1L_2 - M^2$ are negligible, is of the greatest theoretical interest. By not using these values in the equations for the values σ_1 and σ_2, we obtain

$$L_1 \cdot i\omega\sigma_1 + M \cdot i\omega\sigma_2 = u_1,$$
$$M \cdot i\omega\sigma_1 + (L_2 \cdot i\omega + R)\sigma_2 = 0.$$

Since $M \approx \sqrt{L_1L_2}$, by subtracting from the second equation the first equation multiplied by $\sqrt{L_2/L_1}$, we obtain

$$R\sigma_2 = -\sqrt{\frac{L_2}{L_1}}\, u_1.$$

Thus the amplitude $u_2 = R|\sigma_2|$ of the voltage drop across the load resistance R is

$$u_2 = \sqrt{\frac{L_2}{L_1}}\, u_1;$$

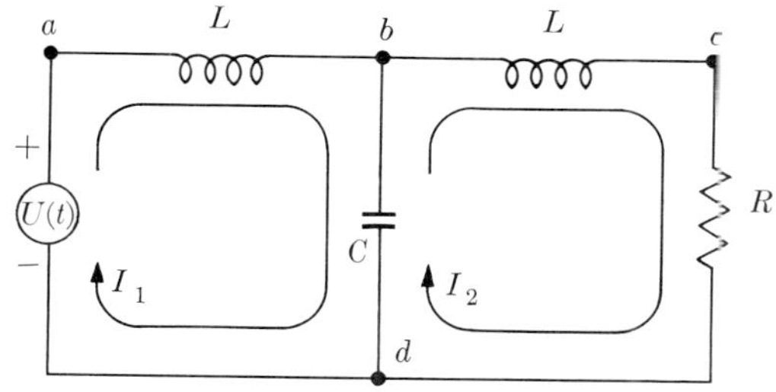

FIGURE 21

the value $\sqrt{L_2/L_1}$ is called the *transformer ratio*. Thus if $L_2 > L_1$, we have a transformer which increases the voltage,

$$\frac{u_2}{u_1} > 1;$$

if $L_2 < L_1$, we have a transformer which decreases the voltage,

$$\frac{u_2}{u_1} < 1.$$

3. (*Electrical filter.*) We shall study an electrical network with four junctions a, b, c, d and five two-terminal elements (Fig. 21), where

ab is the inductance L,
bc is the inductance of the same quantity L,
bd is the capacitance C,
ad is the voltage source $U_{ad}(t) = U(t)$,
cd is the load resistance R.

Let us set

$$I_{ab} = I_1, \qquad I_{bc} = I_2.$$

Then by Kirchhoff's first law we have

$$I_{bd} = I_1 - I_2, \qquad I_{cd} = I_2.$$

By Kirchhoff's second law we have

$$U_{ab} + U_{bc} + U_{cd} + U_{da} = 0,$$
$$U_{bc} + U_{cd} + U_{db} = 0,$$

or

$$LpI_1 + LpI_2 + RI_2 - U(t) = 0,$$

$$LpI_2 + RI_2 + \frac{1}{Cp}(I_2 - I_1) = 0.$$

Multiplying the second equation by p, we obtain the following system:

$$LpI_1 + (Lp + R)I_2 = U(t), \qquad -\frac{1}{C}\,I_1 + \left(Lp^2 + Rp + \frac{1}{C}\right) I_2 = 0.$$

The determinant of this system has the form

$$D(p) = L^2p^3 + LRp^2 + \frac{2Lp}{C} + \frac{R}{C}\cdot$$

By Theorem 6, the polynomial $D(p)$ is stable. We shall now assume that $U = be^{i\omega t}$, where b is the real amplitude of the voltage (see §12).

We shall seek unknown functions I_1 and I_2 in the form

$$I_1 = a_1e^{i\omega t}, \qquad I_2 = a_2e^{i\omega t},$$

where a_1 and a_2 are the complex amplitudes of the currents, i.e., we restrict ourselves to finding a steady-state process.

Determining the voltage drop $U_{cd} = RI_2$ across the load, we have

$$a_2 = \frac{b/C}{(R/C - LR\omega^2) + i\omega\,(2L/C - L^2\omega^2)}\,,$$

from which we determine the amplitude $a = |a_2|R$ of the voltage U_{cd}:

$$a = |a_2|R = \frac{bR/C}{\sqrt{(R/C - LR\omega^2)^2 + \omega^2\,(2L/C - L^2\omega^2)^2}}\cdot$$

For small values of the frequency ω we have $a/b \approx 1$; in other words, voltages of small frequency are easily transmitted through a filter with almost no change of amplitude. For large values of the frequency we have

$$a/b \approx R/CL^2\omega^3,$$

so that high-frequency voltages almost fail to pass through, i.e., they are "filtered."

14. The normal linear homogeneous system with constant coefficients. Here the system of equations

$$\dot{x}^i = \sum_{j=1}^{n} a^i_j x^j, \qquad i = 1, \ldots, n, \tag{1}$$

with constant coefficients is solved. This system can be solved by the method of elimination (see §11, particularly Example 2). Here it is solved by reduction of the matrix $A = (a_j^i)$ to the Jordan form. In the case when all eigenvalues of A are distinct, the task of reducing it to Jordan form, i.e., to diagonal form, is quite elementary. In the general case, however, the problem of reducing A to the Jordan form is one of the most complicated in linear algebra. In what follows we will use only those results of the present section which are based on the reduction of A to a diagonal form in the case of distinct eigenvalues. Results involving the reduction of A to the Jordan form in the general case will not be applied later. They are presented in propositions (C), (D), and (G) and in Theorem 28, but these propositions and Theorem 28 may be skipped without impairing our comprehension of subsequent material.

Usually the reduction of A to the Jordan form for a solution of system (1) is effected by means of a linear transformation of the indeterminates $x^1, \ldots, x^n$. This method is presented at the end of the present section under the title "Transformation of variables." A second method, also based on the reduction of A to the Jordan form, is presented in the first part of this section.

In this section we shall make no distinction between a matrix A and the corresponding transformation $\mathbf{A}$ in the vector space $\mathbf{x} = (x^1, \ldots, x^n)$, because the basis will not change. The only exception is the proof of proposition (F).

The case of simple roots of the characteristic equation. (A) The system of differential equations (1) may be rewritten in vector form

$$\dot{\mathbf{x}} = A\mathbf{x}. \tag{2}$$

Here $A = (a_j^i)$, and in place of the system of indeterminates $x^1, \ldots, x^n$, we introduce the indeterminate vector

$$\mathbf{x} = (x^1, \ldots, x^n);$$

by the *derivative* $\dot{\mathbf{x}}$ of $\mathbf{x}$ we mean the vector $(\dot{x}^1, \ldots, \dot{x}^n)$. If $\mathbf{h}$ is an eigenvector of A with eigenvalues λ, i.e., if

$$A\mathbf{h} = \lambda\mathbf{h}$$

[see §32, (B)], then the vector function $\mathbf{x}$ defined by the relation

$$\mathbf{x} = \mathbf{h}e^{\lambda t},$$

is a solution of (2).

This last proposition may be verified by substituting $\mathbf{x} = \mathbf{h}e^{\lambda t}$ into (2).

THEOREM 10. Let

$$\dot{\mathbf{x}} = A\mathbf{x} \tag{3}$$

be a system of differential equations [see (A)] such that the eigenvalues

$\lambda_1, \ldots, \lambda_n$ of the matrix A are mutually distinct, and let

$$\mathbf{h}_1, \ldots, \mathbf{h}_n$$

be the corresponding eigenvectors of this matrix. If we set

$$\mathbf{x}_i = \mathbf{h}_i e^{\lambda_i t}, \qquad i = 1, \ldots, n, \tag{4}$$

then the vector function

$$\mathbf{x} = c^1\mathbf{x}_1 + \cdots + c^n\mathbf{x}_n, \tag{5}$$

where $c^1, \ldots, c^n$ are constants, is a solution of equation (3), and any solution of equation (3) is given by (5).

Proof. By proposition (A) every function $\mathbf{x}_i$ is a solution of (3), and therefore, by proposition (A) of §6, formula (5) always gives a solution of equation (3). We shall now prove that every solution of equation (3) can be written in the form (5). Let $\varphi(t)$ be an arbitrary solution of equation (3). By Theorem 3 the solution $\varphi(t)$ can be assumed to be defined on the entire line $-\infty < t < \infty$. Thus this solution is also defined at $t = 0$. Let us set $\varphi(0) = \mathbf{x}_0$. Let

$$\mathbf{x}_0 = c^1\mathbf{h}_1 + \cdots + c^n\mathbf{h}_n$$

be an expansion of the vector $\mathbf{x}_0$ in terms of the basis vectors $\mathbf{h}_1, \ldots, \mathbf{h}_n$; the vectors $\mathbf{h}_1, \ldots, \mathbf{h}_n$ form a basis by virtue of proposition (C) of §32. Then the solution $\mathbf{x}$, defined by formula (5), evidently satisfies the initial conditions

$$\mathbf{x}(0) = \mathbf{x}_0.$$

The same initial conditions $\varphi(0) = \mathbf{x}_0$ are also satisfied by the solution $\varphi(t)$; thus, by the uniqueness theorem (see Theorem 2), $\mathbf{x} = \varphi(t)$, so that Theorem 10 is proved.

In the case that the matrix (a_j^i) defining equations (3) is real, we have the problem of separating the real solutions from all the solutions (5).

(B) We shall assume that the matrix (a_j^i) defining equation (3) is real, and we shall select vectors $\mathbf{h}_1, \ldots, \mathbf{h}_n$ in such a way that real vectors correspond to real eigenvalues and complex conjugate vectors to complex conjugate eigenvalues. Then in the system of solutions (4) to each pair of eigenvalues will correspond a real solution, and to every two complex conjugate eigenvalues will correspond complex conjugate solutions. Therefore, the solution (5) is real if and only if the constants attached to real solutions are real and the constants corresponding to complex conjugate solutions are conjugate.

The validity of proposition (B) follows directly from proposition (D) of §7.

The general case. We turn now to the solution of the system (1) in the general case, where the matrix (a_j^i) may have multiple eigenvalues. The analysis of this case rests upon an important and difficult-to-prove algebraic theorem concerning the reduction of a matrix to the Jordan form (see §34).

(C) Let us write system (1) in vector form

$$\dot{\mathbf{x}} = A\mathbf{x} \tag{6}$$

and let

$$\mathbf{h}_1, \ldots, \mathbf{h}_k$$

be a certain basis set with eigenvalue λ [see §34, (A)] with respect to the matrix A, so that the relations

$$A\mathbf{h}_1 = \lambda\mathbf{h}_1, \qquad A\mathbf{h}_2 = \lambda\mathbf{h}_2 + \mathbf{h}_1, \quad \ldots, \quad A\mathbf{h}_k = \lambda\mathbf{h}_k + \mathbf{h}_{k-1}$$

are satisfied. Let us introduce a sequence of vector functions, assuming that

$$\boldsymbol{\omega}_r(t) = \frac{t^{r-1}}{(r-1)!}\mathbf{h}_1 + \frac{t^{r-2}}{(r-2)!}\mathbf{h}_2 + \cdots + \mathbf{h}_r, \qquad r = 1, \ldots, k. \tag{7}$$

We then find that the vector functions

$$\mathbf{x}_r = \boldsymbol{\omega}_r(t)e^{\lambda t}, \qquad r = 1, \ldots, k, \tag{8}$$

are solutions of equation (6), and in addition

$$\mathbf{x}_r(0) = \mathbf{h}_r. \tag{9}$$

Thus to every basis set of k vectors corresponds a system of k solutions.

To prove that the vector functions (8) are solutions of (6), we shall state the following two identities concerning the vector functions (7):

$$\begin{aligned} \dot{\boldsymbol{\omega}}_r(t) &= \boldsymbol{\omega}_{r-1}(t), & r &= 1, \ldots, k; \\ A\boldsymbol{\omega}_r(t) &= \lambda\boldsymbol{\omega}_r(t) + \boldsymbol{\omega}_{r-1}(t), & r &= 1, \ldots, k. \end{aligned}$$

In these relations $\boldsymbol{\omega}_0(t)$ is taken to equal 0. Both may be directly verified by elementary calculations. With the aid of these identities, it is immediately seen that the functions (8) are solutions of (6). Actually, we have

$$\begin{aligned} \dot{\mathbf{x}}_r(t) &= \dot{\boldsymbol{\omega}}_r(t)e^{\lambda t} + \lambda\boldsymbol{\omega}_r(t)e^{\lambda t} = \left(\boldsymbol{\omega}_{r-1}(t) + \lambda\boldsymbol{\omega}_r(t)\right)e^{\lambda t} \\ &= A\boldsymbol{\omega}_r(t)e^{\lambda t} = A\mathbf{x}_r(t). \end{aligned}$$

We shall now turn to the formulation and proof of a theorem which gives the solution of (1) in the general case.

THEOREM 11. *Let*

$$\dot{\mathbf{x}} = A\mathbf{x} \tag{10}$$

be the vector form of the system (1). By Theorem 28 (see §34) there exists a basis $\mathbf{h}_1, \ldots, \mathbf{h}_n$, consisting of basis sets relative to the matrix A. Specifically, we shall assume that $\mathbf{h}_1, \ldots, \mathbf{h}_{k_1}$, is a basis set with the eigenvalue λ_1, that $\mathbf{h}_{k_1+1}, \ldots, \mathbf{h}_{k_1+k_2}$ is a basis set with the eigenvalue λ_2, etc. By proposition (C), to each basis set corresponds a system of solutions, so that we can write the following solutions of equation (10):

$$\begin{aligned} \mathbf{x}_1 &= \mathbf{h}_1 e^{\lambda_1 t}, \ldots, \mathbf{x}_{k_1} = \left(\frac{t^{k_1-1}}{(k_1-1)!}\mathbf{h}_1 + \cdots + \mathbf{h}_{k_1}\right) e^{\lambda_1 t}, \\ \mathbf{x}_{k_1+1} &= \mathbf{h}_{k_1+1} e^{\lambda_2 t}, \\ &\vdots \\ \mathbf{x}_{k_1+k_2} &= \left(\frac{t^{k_2-1}}{(k_2-1)!}\mathbf{h}_{k_1+1} + \cdots + \mathbf{h}_{k_1+k_2}\right) e^{\lambda_2 t}, \text{ etc.} \end{aligned} \tag{11}$$

Hence, the formula

$$\mathbf{x} = c^1\mathbf{x}_1 + \cdots + c^n\mathbf{x}_n, \tag{12}$$

where $c^1, \ldots, c^n$ are constants, always gives a solution of equation (1), and every solution of (10) may be written in terms of (12).

Thus the functions $\mathbf{x}_1, \ldots, \mathbf{x}_n$ constitute the so-called *fundamental system of solutions of equation* (10).

Proof. Since the functions $\mathbf{x}_1, \ldots, \mathbf{x}_n$ are solutions of (10) [see (C)], then, by proposition (A) of §6, formula (12) always gives a solution of (10). We shall show that every solution of (10) may, by suitable choice of constants $c^1, \ldots, c^n$, be written in form (12). Let $\boldsymbol{\varphi}(t)$ be an arbitrary solution of (10). By Theorem 3, the solution $\boldsymbol{\varphi}(t)$ can be assumed to be defined on the entire line $-\infty < t < \infty$, so that the vector $\boldsymbol{\varphi}(0) = \mathbf{x}_0$ is defined. Let us expand this vector in terms of the basis $\mathbf{h}_1, \ldots, \mathbf{h}_n$:

$$\mathbf{x}_0 = c^1\mathbf{h}_1 + \cdots + c^n\mathbf{h}_n.$$

If the constants $c^1, \ldots, c^n$ determined are now substituted into (12), we obtain a solution $\mathbf{x}(t)$, which satisfies the initial conditions

$$\mathbf{x}(0) = c^1\mathbf{x}_1(0) + \cdots + c^n\mathbf{x}_n(0) = c^1\mathbf{h}_1 + \cdots + c^n\mathbf{h}_n = \mathbf{x}_0$$

[see (9)]. Thus the solutions $\boldsymbol{\varphi}(t)$ and $\mathbf{x}(t)$ have common initial values and therefore coincide. Theorem 11 is thus proved.

It now remains to separate from the solutions given by (12) the real solutions in the case when the matrix (a_j^i) is real. This is done in exactly the same manner as in the case of simple roots of the characteristic equation.

(D) Let us assume that the matrix (a^i_j) which defines equation (10) is real. In this case we choose the basis $\mathbf{h}_1, \ldots, \mathbf{h}_n$ by the method specified in Theorem 28 (see §34) for the case that the matrix (a^i_j) is real. With this choice of basis, there will be on the one hand, among the solutions (11) constructed in Theorem 11, real solutions and, on the other hand, pairs of complex conjugate solutions. Thus the solution (12) is real if and only if the constants corresponding to real solutions are real and the constants corresponding to complex conjugate solutions are complex conjugates. Proposition (D) now follows directly from (D) of §7.

In conclusion it should be noted that the above results of the present section are closely connected with the results of §7 and §8, where we studied one nth-order homogeneous equation with constant coefficients. By the method of §4 such an equation can be written in the form of a normal system of n equations. Thus the results of §7 and §8 can be derived from the results of the present section. In addition it is found that the characteristic polynomial of the normal system obtained coincides with the characteristic polynomial of the initial equation.

Transformation of variables. (E) Let us replace the indeterminates

$$x^1, \ldots, x^n \tag{13}$$

in the system (1), which is defined by the constant matrix $A = (a^i_j)$, by new indeterminates

$$y^1, \ldots, y^n, \tag{14}$$

by setting

$$y^j = \sum_{i=1}^{n} s^j_i x^i, \qquad f = 1, \ldots, n, \tag{15}$$

where s^j_i are constant coefficients with a nonsingular matrix $S = (s^j_i)$. In terms of the new indeterminates our system may be written in the form

$$\dot{y}^i = \sum_{j=1}^{n} b^i_j y^j, \qquad i = 1, \ldots, n, \tag{16}$$

where the matrix $B = (b^i_j)$ is obtained from A by the formula

$$B = SAS^{-1}. \tag{17}$$

We shall now prove this. Differentiating relation (15) with respect to t, we obtain

$$\dot{y}^j = \sum_{i=1}^{n} s^j_i \dot{x}^i, \qquad j = 1, \ldots, n. \tag{18}$$

Thus, the components of the vector $\dot{\mathbf{x}} = (\dot{x}^1, \ldots, \dot{x}^n)$ are transformed in the same way as the components of the vector $\mathbf{x} = (x^1, \ldots, x^n)$. Formula (17) follows directly from this transformation [see §32, (A)]. We shall carry out anew, however, the proof given in §32, (A). The relations (15) and (18) may be rewritten in the matrix form

$$\mathbf{y} = S\mathbf{x}, \qquad \dot{\mathbf{y}} = S\dot{\mathbf{x}},$$

so that we have

$$\dot{\mathbf{y}} = S\dot{\mathbf{x}} = SA\mathbf{x} = SA \cdot S^{-1}\mathbf{y},$$

which proves formula (17).

By a proper choice of the matrix S we can obtain the simplest form of B. Since the transformation (17) of the matrix A into the matrix B is effected by means of the matrix S, we can attain the Jordan form of the matrix B [see §34, (B)].

(F) If each eigenvalue

$$\lambda_1, \ldots, \lambda_n$$

of the matrix A of the system (1) is distinct, then the linear transformation (15) can be chosen in such a way that (16) takes the form

$$\dot{y}^k = \lambda_k y^k, \qquad k = 1, \ldots, n. \tag{19}$$

If, in addition, the matrix A is real so that together with every complex eigenvalue λ_k in the sequence $\lambda_1, \ldots, \lambda_n$ there appears the complex conjugate $\lambda_l = \bar{\lambda}_k$, and if the variables (13) are real, then the transformation (15) can be chosen so that to every real eigenvalue λ_j corresponds a real variable y^j, and to the pair of complex conjugate eigenvalues λ_k and λ_l correspond the complex conjugate variables y^k and $y^l = \bar{y}^k$. Thus to the conjugate eigenvalues correspond the conjugate equations

$$\dot{y}^k = \lambda_k y^k, \qquad \dot{\bar{y}}^k = \bar{\lambda}_k \bar{y}^k. \tag{20}$$

Let

$$\lambda_k = \mu_k + i\nu_k, \qquad y^k = \xi^k + i\eta^k, \tag{21}$$

where μ_k, ν_k, ξ^k, η^k are real numbers. Then the pair of conjugate equations (20) can be replaced by a pair of real equations

$$\dot{\xi}^k = \mu_k \xi^k - \nu_k \eta^k, \qquad \dot{\eta}_k = \nu_k \xi^k + \mu_k \eta^k. \tag{22}$$

Carrying out a similar substitution for every pair of complex conjugate eigenvalues, we shall be able to substitute for the system of real variables (13) a new system of real variables, whose equation has partly the form (19) (for real λ_j) and partly the form (22) (for pairs of complex conjugate eigenvalues).

To prove proposition (F) in the space R of vectors $\mathbf{x} = (x^1, \ldots, x^n)$, we make the matrix A correspond with the linear transformation $\mathbf{A}$ (see §32), and we denote by $\mathbf{h}_k$ the eigenvector of the transformation $\mathbf{A}$ with eigenvalue λ_k. As a basis in R we shall now take the vectors

$$\mathbf{h}_1, \ldots, \mathbf{h}_n, \tag{23}$$

and we denote the coordinates of the vector $\mathbf{x}$ corresponding to this basis by $y^1, \ldots, y^n$. Thus we shall obtain the linear transformation (15). In the new system of coordinates the matrix B corresponds to the transformation $\mathbf{A}$. Finally, since the matrix B has obviously a diagonal form with the numbers $\lambda_1, \ldots, \lambda_n$ in the diagonal, the system (16) takes the form (19).

Now if the matrix A is real, then to each real eigenvalue λ_j will correspond the real vector $\mathbf{h}_j$, and to a pair of complex conjugate eigenvalues λ_k and $\lambda_l = \bar{\lambda}_k$, will correspond a pair of complex conjugate eigenvectors $\mathbf{h}_k$ and $\mathbf{h}_l = \bar{\mathbf{h}}_k$. An arbitrary vector $\mathbf{x}$ in the new coordinate system may be written in the form

$$\mathbf{x} = \sum_{j=1}^{n} y^j \mathbf{h}_j. \tag{24}$$

If it is real then the coefficients of real vectors must be real and the coefficients of complex conjugate vectors must be complex conjugate [see §7, (D)]. Thus a real variable y^j corresponds to every real eigenvalue λ_j, and the complex conjugate values y^k and $y^l = \overline{y^k}$ correspond to the pair of complex conjugate eigenvalues λ_k and $\lambda_l = \bar{\lambda}_k$.

We shall write equation (20), after substituting in it the values λ_k and y^k from (21), for the transition from a pair of complex conjugate equations (20) to a pair of real equations (22). We obtain

$$\dot{\xi}^k + i\dot{\eta}^k = (\mu_k + i\nu_k)(\xi^k + i\eta^k) = \mu_k \xi^k - \nu_k \eta^k + i(\nu_k \xi^k + \mu_k \eta^k).$$

By equating separately the real and imaginary parts of this relation, we obtain the system of equations (22), and proposition (F) is proved.

The system of equations (19) has the obvious solution

$$y^k = c_k e^{\lambda_k t}, \qquad k = 1, \ldots, n,$$

but in order to obtain a solution of the original system (3) it is necessary to go from the indeterminates (14) to the indeterminates (13), and for this transition we must know the eigenvectors (23) of the matrix A [see (24)]. Thus proposition (F) is equivalent to Theorem 10.

For the solution of (1) in the general case, the reduction of the matrix A to the Jordan form can be used. The proposition (G) for this case is equivalent to Theorem 11.

(G) Let the matrix A of the system (1) be arbitrary. We shall select the transformation (15) in such a way that the matrix B has the Jordan form (see §34). Let λ be one of the eigenvalues of matrix A and let k be the dimension of one of the Jordan cells of B with eigenvalue λ. We shall assume that this cell occupies the first k rows. The system of equations corresponding to this cell has the form

$$\begin{aligned} \dot{y}^1 &= \lambda y^1 + y^2, \\ \dot{y}^2 &= \lambda y^2 + y^3, \\ &\vdots \\ \dot{y}^{k-1} &= \lambda y^{k-1} + y^k, \\ \dot{y}^k &= \lambda y^k. \end{aligned}$$

To every other Jordan cell of B corresponds a similar system of equations which is easy to solve.

Examples

1. The application of the method presented in this section to the solution of system (1) requires the determination of the basis $\mathbf{h}_1, \ldots, \mathbf{h}_n$ of the vector space which consists of basis sets (see Theorem 11). This determination itself presents a certain algebraic problem. By using the results of this section, we shall show how the system (1) can be solved by the *method of undetermined coefficients* without finding the basis which is composed of basis sets. Let λ be a certain eigenvalue of the matrix (a_j^i). To this eigenvalue, generally speaking, correspond several basis sets contained in the basis $\mathbf{h}_1, \ldots, \mathbf{h}_n$; let k be the longest of the basis sets corresponding to the eigenvalue λ. By Theorem 11, each of the solutions corresponding to the eigenvalue λ can be written in the form

$$x^i = f^i(t)e^{\lambda t}, \qquad i = 1, \ldots, n, \tag{25}$$

where $f^i(t)$ is a polynomial of degree $\leq k - 1$. Thus, substituting into system (1) a solution in the form (25) and assuming that the coefficients of the polynomials $f^i(t)$, $i = 1, \ldots, n$, are unknown constants, we can, by using the method of undetermined coefficients, find all solutions of (1) corresponding to the eigenvalue λ. In order to solve the system (1) by this method it is not necessary to know the basis sets corresponding to the eigenvalue λ; it is necessary to know only the *lengths* of these basis sets. Determination of the lengths is a simpler algebraic problem than reduction to the Jordan form; it is solved by the *theory of elementary divisors of matrices*, which pertains to linear algebra. The theory of elementary divisors is not used in this book.

2. We shall now show how to solve system (1) by the method of elimination presented in §11. To apply the method of elimination, we shall write (1) in the form

$$\sum_{j=1}^{n} L_j^i(p)x^j = 0,$$

where

$$L_j^i(p) = a_j^i - p\,\delta_j^i.$$

The determinant $D(p)$ of the matrix $(L_j^i(p))$ in this case is a characteristic polynomial of the matrix (a_j^i). Let λ be a certain root of $D(p)$ or, what is the same thing, an eigenvalue of the matrix (a_j^i). The multiplicity of λ will be denoted here by l. By proposition (C) of §11 every solution of (1) corresponding to the root λ is to be found in the form

$$x^i = g^i(t)e^{\lambda t}, \qquad i = 1, \ldots, n,$$

where the degree of the polynomial $g^i(t)$ does not exceed $l - 1$. If we compare the method presented in this example with that in Example 1, we see that the whole difference lies in the determination of the maximal degree of the polynomials. The method of Example 1 gives a more accurate determination of the degree of the polynomials, since the number k, generally speaking, is smaller than the number l. Indeed, the sum of *all* lengths of basis sets corresponding to λ is equal to l. Thus the equality $k = l$ can hold only when there is only one basis set corresponding to the eigenvalue λ.

15. Autonomous systems of differential equations and their phase spaces. We shall give here a geometrical interpretation of an *autonomous* system of equations in the form of the *phase space* of this system. This interpretation differs essentially from the geometrical interpretation of the system of equations in §1 and should, therefore, more correctly be called a *kinematic interpretation,* since to every solution of its system of equations corresponds not a curve in a space, but the motion of a point along the curve. The kinematic interpretation (the phase space) is in certain respects more expressive than the geometrical (the system of integral curves).

Autonomous systems. A system of ordinary differential equations is called *autonomous* if it does not explicitly contain the independent variable t (or, as we shall call it, the *time*). This means that the law of variation of the unknown functions which are described by the system of equations does not change with time, as is usually the case with physical laws. It is very easy to prove that if

$$x^i = \varphi^i(t), \qquad i = 1, \ldots, n,$$

is a solution of a certain autonomous system of equations, then

$$x^i = \overset{i}{\varphi}{}_*(t) = \varphi^i(t + c), \qquad i = 1, \ldots, n,$$

where c is a constant, is also a solution of the same autonomous system of equations. We shall carry out the proof of this fact by an example of a *normal* autonomous system of equations.

(A) Let

$$\dot{x}^i = f^i(x^1, \ldots, x^n), \qquad i = 1, \ldots, n, \tag{1}$$

be an autonomous normal system of nth-order equations and

$$\dot{\mathbf{x}} = f(\mathbf{x})$$

its vector notation. The autonomy of system (1) consists of the fact that the functions $f^i(x^1, \ldots, x^n)$, $i = 1, \ldots, n$, are functions of the variables $x^1, \ldots, x^n$ and *do not depend* on the time t. We shall assume that the functions $f^i(x^1, \ldots, x^n)$ are defined in a certain domain Δ of the n-dimensional space where $x^1, \ldots, x^n$ are the coordinates of a point. We shall assume further that $f^i(x^1, \ldots, x^n)$ and their first-order partial derivatives are continuous in the domain Δ. Thus, if

$$x^i = \varphi^i(t), \qquad i = 1, \ldots, n, \tag{2}$$

is a solution of (1), then

$$x^i = \overset{i}{\varphi}{}_*(t) = \varphi^i(t + c), \qquad i = 1, \ldots, n, \tag{3}$$

is also a solution of (1). It is evident that if the solution (2) has as a maximal interval of existence the interval $m_1 < t < m_2$, then the solution (3) has the maximal interval

$$m_1 - c < t < m_2 - c.$$

From the differentiation formula for a composite function, we have the relation

$$\dot{\overset{i}{\varphi}}{}_*(t) = \dot{\varphi}^i(t + c), \qquad i = 1, \ldots, n. \tag{4}$$

Indeed,

$$\begin{aligned} \dot{\overset{i}{\varphi}}{}_*(t) &= \frac{d}{dt}\overset{i}{\varphi}{}_*(t) = \frac{d}{dt}\varphi^i(t + c) = \frac{d}{d(t + c)}\varphi^i(t + c) \cdot \frac{d(t + c)}{dt} \\ &= \dot{\varphi}^i(t + c) \cdot 1 = \dot{\varphi}^i(t + c). \end{aligned}$$

We shall now prove that (3) is a solution of the system (1). Since (2) is a solution, we have the identities

$$\dot{\varphi}^i(t) = f^i(\varphi^1(t), \ldots, \varphi^n(t)), \qquad i = 1, \ldots, n.$$

Replacing t in these identities by $t + c$ we obtain

$$\dot{\varphi}^i(t + c) = f^i(\varphi^1(t + c), \ldots, \varphi^n(t + c)), \qquad i = 1, \ldots, n.$$

Combining this with (4) and (3), we have

$$\dot{\varphi}_*^i(t) = \dot{\varphi}^i(t + c) = f^i(\varphi^1(t + c), \ldots, \varphi^n(t + c)) = f^i(\varphi_*^1(t), \ldots, \varphi_*^n(t)).$$

We shall now turn to the kinematic interpretation of the solutions of system (1). Formally we shall speak about an interpretation in n-dimensional space, but for the sake of clarity it is reasonable to imagine the case of a plane ($n = 2$).

(B) To every solution

$$x^i = \varphi^i(t), \qquad i = 1, \ldots, n, \tag{5}$$

of the autonomous system (1) we make correspond the motion of a point in n-dimensional space defined by equations (5), where $x^1, \ldots, x^n$ are the coordinates of the point in space and t is the time. In the course of its motion the point describes a curve known as the *trajectory* of the motion. If we associate with the solution (5) not the process of motion, but the trajectory of the motion of the point, then we shall obtain a less complete picture of the solution, since it is also desirable to indicate the direction of the motion on the trajectory. Thus, if there is another solution

$$x^i = \psi^i(t), \qquad i = 1, \ldots, n, \tag{6}$$

in addition to (5), then the trajectories corresponding to these solutions either do not intersect in the space or else they coincide. That is, if the trajectories have even one common point, i.e.,

$$\varphi^i(t_1) = \psi^i(t_2), \qquad i = 1, \ldots, n, \tag{7}$$

then

$$\psi^i(t) = \varphi^i(t + c), \qquad \text{where} \qquad c = t_1 - t_2. \tag{8}$$

These last equalities show that the trajectories described by the first and second solutions coincide, but the first solution describes the same trajectory as the second with the time "delay" c. If the point corresponding to the first solution has reached a certain position on the trajectory at instant $t + c$, then the point corresponding to the second solution has already been in this position at the instant t.

In order to derive (8) from (7), we shall examine the solution

$$\varphi_*^i(t) = \varphi^i(t + c) \tag{9}$$

along with (5) [see (A)]. Equations (7), for $c = t_1 - t_2$, yields the equality

$$\varphi_*^i(t_2) = \varphi^i(t_2 + c) = \varphi^i(t_1) = \psi^i(t_2), \qquad i = 1, \ldots, n.$$

Thus, the solutions (6) and (9) of the system (1) have common initial conditions (namely, their values at the instant t_2), and therefore by the uniqueness theorem they must coincide, so that we have

$$\psi^i(t) = \varphi_*^i(t) = \varphi^i(t + c), \qquad i = 1, \ldots, n.$$

States of equilibrium and closed trajectories. We pose the question of whether a trajectory representing a solution of the system can intersect itself.

(C) Let

$$x^i = \varphi^i(t), \qquad i = 1, \ldots, n, \tag{10}$$

be a certain solution of the system (1) defined on a maximal interval $m_1 < t < m_2$. We shall assume that the equalities

$$\varphi^i(t_1) = \varphi^i(t_2), \qquad i = 1, \ldots, n, \qquad t_1 \neq t_2, \tag{11}$$

are valid, where t_1 and t_2, of course, belong to the interval $m_1 < t < m_2$. It then turns out that $m_1 = -\infty$, $m_2 = +\infty$ [i.e., the maximum interval of existence for the solution (10) is the entire line] and that the following two mutually exclusive cases are possible.

1. For all values of t the equality

$$\varphi^i(t) = a^i, \qquad i = 1, \ldots, n,$$

is valid, where $(a^1, a^2, \ldots, a^n)$ is a point of the domain Δ which does not depend on t. Thus in this case the point $\big(\varphi^1(t), \ldots, \varphi^n(t)\big)$ actually does not move as t varies but remains fixed. In this case the solution (10) itself with the point $(a^1, \ldots, a^n)$ is called a *state of equilibrium* of the system (1).

2. There exists a positive number T such that for arbitrary t, the equalities

$$\varphi^i(t + T) = \varphi^i(t), \qquad i = 1, \ldots, n,$$

are valid, but for $|\tau_1 - \tau_2| < T$ and for at least one $i = 1, \ldots, n$, the inequality

$$\varphi^i(\tau_1) \neq \varphi^i(\tau_2)$$

is valid. In this case the solution (10) is called *periodic* with period T, and the trajectory described by (10) is called a *closed trajectory* or a *cycle*.

First of all, we shall show that the maximal interval of existence of solution (10) is the entire straight line. As was noted in proposition (B), the identities

$$\varphi^i(t + c) = \varphi^i(t), \qquad i = 1, \ldots, n, \qquad c = t_1 - t_2, \tag{12}$$

follow from equality (11). Since by this equality the interval $m_1 - c < t < m_2 - c$ coincides with the interval $m_1 < t < m_2$, we have $m_1 = -\infty, m_2 = +\infty$.

Every number c for which (12) is satisfied will be called a *period* of the solution (10); the set of all periods of the solution (1) is designated by F, which is a certain set of numbers. We shall establish some of its properties. Substituting $t - c$ for t in (12), we obtain $\varphi^i(t) = \varphi^i(t - c)$. Thus, if c is a period, then $-c$ is also a period. Let us assume that c_1 and c_2 are periods, i.e., that

$$\varphi^i(t + c_1) = \varphi^i(t), \qquad \varphi^i(t + c_2) = \varphi^i(t); \qquad i = 1, \ldots, n.$$

Then

$$\varphi^i((t + c_2) + c_1) = \varphi^i(t + c_2) = \varphi^i(t), \qquad i = 1, \ldots, n.$$

Thus if c_1 and c_2 are periods, then $c_1 + c_2$ is also a period. Let us assume that $c_1, c_2, \ldots, c_m, \ldots$ is a sequence of periods which converges to a certain number c_0; then we have

$$\varphi^i(t + c_m) = \varphi^i(t); \qquad i = 1, \ldots, n, \qquad m = 1, 2, \ldots.$$

Since the functions $\varphi^i(t)$ are continuous, then for $m \to \infty$ we have

$$\varphi^i(t + c_0) = \varphi^i(t),$$

i.e., we see that c_0 is also a period, because the set F is *closed*.

Since the number c in (12) is distinct from zero ($t_1 \neq t_2$), the set F contains numbers distinct from zero. From the properties of the set F which have been established it is easily seen that there exist only two possibilities: (1) the set F coincides with the set of all real numbers; (2) in the set F there is a minimal positive number T, such that F consists of all integer multiples of the number T. Let us prove that there are actually only these two possibilities. Since the set F contains the number $-c$ whenever it contains the number c, and since in F there are numbers different from zero, there are positive numbers in F.

Let us assume that there is no least positive number in F, i.e., that for every positive number ϵ there is a positive period $c < \epsilon$. From the above properties of the set F it follows, since c is a period, that all numbers mc, where m is an integer, are also periods. Since $c < \epsilon$, then for an arbitrary real number c_0 it is possible to find an integer m such that $|c_0 - mc| < \epsilon$. Thus an arbitrary number c_0 is a limit point of the set F, and therefore, in view of the fact that set F is closed, this set coincides with the set of all real numbers.

Let us now assume that F is not the set of all real numbers. By what has been proved, there then exists in F a least positive number T. Let c be an arbitrary period. Then it is possible to select an integer m such that $|c - mT| < T$. Let us assume that $c \neq mT$; then $|c - mT|$ is a period distinct from zero; but this is impossible since $|c - mT| < T$, which contradicts the minimal character of the number T. It is thus proved that every number c from F can be written in the form $c = mT$, where m is an integer.

Now it is easy to verify that, if F is the set of all real numbers, then case (1) occurs, and if F is not the set of all real numbers, then case (2) occurs. Thus proposition (C) is proved.

Proposition (C) can be formulated briefly by saying that there exist three kinds of trajectories: (1) those of the state of equilibrium; (2) periodic trajectories (cycles); and (3) nonintersecting trajectories. It is natural to take the last case as the "most general."

From Theorem 2 it follows that a trajectory representing a solution of the system passes through every point of the domain of definition of the system (1). Thus, the entire domain Δ is filled with trajectories, and in accordance with (B) these trajectories do not intersect each other in pairs. Those trajectories which do not intersect are of particular interest; they represent either states of equilibrium or cycles, and are quite important.

This is the kinematic interpretation of solutions of an autonomous system of equations. The system of equations itself also admits a geometric interpretation.

Phase spaces. (D) Since the autonomous system of equations (1) is defined in the domain Δ, each point $(x_0^1, \ldots, x_0^n)$ of the domain Δ corresponds to a sequence of n numbers, namely the sequence

$$f^1(x_0^1, \ldots, x_0^n), \ldots, f^n(x_0^1, \ldots, x_0^n).$$

These numbers can be thought of as components of a vector $\mathbf{f}(x_0^1, \ldots, x_0^n)$ in an n-dimensional space emanating from the point $(x_0^1, \ldots, x_0^n)$. Thus the autonomous system gives rise to a geometric picture, a *vector field* defined in the domain Δ. The vector $\mathbf{f}(x_0^1, \ldots, x_0^n)$ is defined at every point $(x_0^1, \ldots, x_0^n)$ of Δ, starting from this point. The connection between the geometrical interpretation of the solutions and the geometrical interpretation of the system of equations itself is given by the following. Let $(x_0^1, \ldots, x_0^n)$ be an arbitrary point of Δ. In the geometrical interpretation of the system of equations the vector $\mathbf{f}(x_0^1, \ldots, x_0^n)$ corresponds to the point from which it starts. Further, by Theorem 2 there exists a solution $x^i = \varphi^i(t)$ of (1) which satisfies the initial conditions

$$\varphi^i(t_0) = x_0^i, \qquad i = 1, \ldots, n.$$

According to the kinematic interpretation, the solution $x^i = \varphi^i(t)$ corresponds in the space to the motion of a point which describes a trajectory which, at the instant $t = t_0$, passes through the point $(x_0^1, \ldots, x_0^n)$ in the space. Thus the vector velocity of the point which describes the solution $x^i = \varphi^i(t)$ at the instant of its passage through the point $(x_0^1, \ldots, x^n)$ coincides with the vector $\mathbf{f}(x_0^1, \ldots, x_0^n)$. It is just this coincidence which is expressed by the system of equations (1) for

$$x^i = x_0^i, \qquad i = 1, \ldots, n; \quad t = t_0.$$

The n-dimensional space, in which solutions of autonomous system (1) are interpreted in the form of trajectories and the autonomous system (1) itself in the form of a vector field, is called the *phase space* of the system (1). The trajectories in this space are called the *phase trajectories*, and the vectors $\mathbf{f}(x_0^1, \ldots, x_0^n)$ are called the *phase velocities*. The connection between the two interpretations consists in the fact that the velocity of the motion of a point along a trajectory at each instant coincides with the phase velocity given at that point of the space where the moving point is located at that instant.

Let us now examine states of equilibrium from the point of view of phase velocities.

(E) In order that the point $(a^1, \ldots, a^n)$ of the domain Δ be a state of equilibrium of the system (1), i.e., that $x^i = \varphi^i(t)$ be a solution of the system for which

$$\varphi^i(t) \equiv a^i, \qquad i = 1, \ldots, n, \tag{13}$$

it is necessary and sufficient that the phase velocity $\mathbf{f}(a^1, \ldots, a^n)$ at $(a^1, \ldots, a^n)$ be equal to zero. Thus to find all states of equilibrium of (1) it is necessary to solve the system of equations

$$f^i(a^1, \ldots, a^n) = 0, \qquad i = 1, \ldots, n.$$

This system is not a system of differential equations, but rather a system of *finite* (or algebraic) equations since it does not include derivatives.

To prove proposition (E), we shall assume that $(a^1, \ldots, a^n)$ is a state of equilibrium, i.e., that there exists a solution $x^i = \varphi^i(t)$ for which the relations (13) are satisfied, and we shall substitute this solution into (1). Since the derivative of a constant is zero, the substitution yields

$$f^i(a^1, \ldots, a^n) = \frac{d}{dt}\,\varphi^i(t) = \frac{d}{dt}\,a^i = 0.$$

Thus the phase velocity vector $\mathbf{f}(a^1, \ldots, a^n)$ actually vanishes at the point $(a^1, \ldots, a^n)$. Let us assume that, conversely, the phase velocity vector $\mathbf{f}(a^1, \ldots, a^n)$ vanishes at the point $(a^1, \ldots, a^n)$, i.e., that

$$f^i(a^1, \ldots, a^n) = 0, \qquad i = 1, \ldots, n,$$

and we shall show that in this case the equalities (13) determine a solution of (1). Substitution gives

$$\dot{\varphi}^i(t) = f^i(a^1, \ldots, a^n), \qquad i = 1, \ldots, n;$$

these equalities are satisfied, since on the left we have the derivative of a constant and on the right, zero.

Examples

1. (Compare Example 3 in §2). We shall study the autonomous first-order differential equation

$$\dot{x} = f(x), \tag{14}$$

where $f(x)$ and its derivative are continuous over the entire line P where x is allowed to vary. We shall assume in addition that the zeros of $f(x)$ or, what is the same thing, the states of equilibrium of equation (14) do not have limit points. Under this hypothesis the states of equilibrium divide the straight line P into a system Σ of intervals. Each interval (a, b) of Σ has the property that the function $f(x)$ does not vanish on it and that each of its endpoints, a or b, is either a zero of the function $f(x)$, or is equal to $\pm\infty$. Thus Σ consists of a finite or countable number of finite intervals and not more than two semi-infinite intervals, or consists only of the infinite interval $(-\infty, \infty)$. Let (a, b) be a certain interval of Σ, x_0 a point of this interval, $x = \varphi(t)$ a solution of equation (14) with initial values $0, x_0$, and $m_1 < t < m_2$ the maximal interval of existence of the solution $\varphi(t)$. To be definite we shall assume that $f(x_0) > 0$, so that

$$a < \varphi(t) < b \qquad \text{for} \qquad m_1 < t < m_2, \tag{15}$$

$$\lim_{t \to m_1} \varphi(t) = a, \qquad \lim_{t \to m_2} \varphi(t) = b. \tag{16}$$

Further, if either number a or b is finite, then the number m_1 or m_2, respectively, is infinite. Thus (Fig. 22) every interval (a, b) represents one unique phase trajectory of equation (14).

We shall prove relations (15), (16). From the hypothesis that $f(x_0) > 0$ it follows that the function $f(x)$ is positive on the interval (a, b), and therefore every point of this interval describing a phase trajectory moves from

FIGURE 22

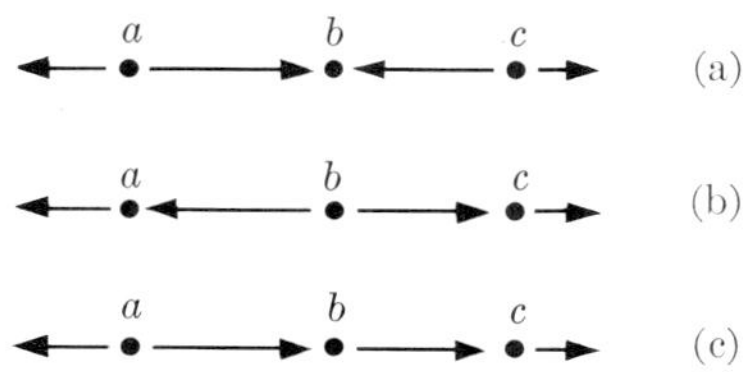

FIGURE 23

left to right. Thus with increasing t the point $\varphi(t)$ can leave the interval (a, b) only by passing over the right-hand endpoint b. Let us assume that this takes place for some $t = t_1$; then at $t = t_1$, we have $\varphi(t_1) = b$, which creates the impossible situation that the two different trajectories $x = \varphi(t)$ and $x = b$ intersect. In exactly the same way we can prove that the point $\varphi(t)$ cannot leave interval (a, b) for decreasing t. Thus (15) is proved.

Let us now assume that $\lim_{t\to m_2} \varphi(t) = c < b$, and let $\psi(t)$ be a solution of (14) with the initial values 0, c. Since $f(c) > 0$, then for some negative value of t_2 we have $\psi(t_2) < c$, but this means that two different trajectories $\varphi(t)$ and $\psi(t)$ intersect, which is impossible. Thus it is proved that $\lim_{t-m_2} \varphi(t) = b$. The relation

$$\lim_{t\to m_1} = \varphi(t) = a$$

is proved in exactly the same way.

Let us assume, finally, that $b < \infty$, and show that under this assumption $m_2 = \infty$. Let us assume the contrary, that is, $m_2 < \infty$. Let us then define a function $\chi(t)$ by setting $\chi(t) = \varphi(t)$ for $m_1 < t < m_2$, and $\chi(t) = b$ for $t \geq m_2$. It is evident that the function $\chi(t)$ is continuous and satisfies equation (14), but this is impossible since two different trajectories $\chi(t)$ and $x = b$ would then intersect. This contradiction shows that $m_2 = +\infty$. In exactly the same way it can be proved that for $a > -\infty$ we have $m_1 = -\infty$.

Let b be an arbitrary state of equilibrium of equation (14) and let (a, b) and (b, c) be the two intervals of Σ adjoining it (on the left and right, respectively). Each of the intervals (a, b) and (b, c) represents one trajectory. If both points describing the trajectories (a, b) and (b, c) *approach* (with increasing t) the state of equilibrium b, then the state of equilibrium b is called *stable* [Fig. 23(a)]. If both points describing the trajectories (a, b) and (b, c) *recede* from the point b, then the state of equilibrium b is called *unstable* [Fig. 23(b)]. If along one of the trajectories the point approaches and along the other it recedes, then the state of equilibrium b is called *semistable* [Fig. 23(c)]. In order that a state of equilibrium b be stable, it is necessary and sufficient that the function $f(x)$ be positive on the interval (a, b) and negative on the interval (b, c). For state of equilibrium b to be unstable, it is necessary and sufficient that the function $f(x)$ be negative on the interval (a, b) and positive on the interval (b, c). For a state of equilibrium b to be semistable, it is necessary and sufficient that the function $f(x)$ have the same sign on both of the intervals (a, b) and (b, c).

Let us assume that $\dot{f}(b) \neq 0$; then the sign of function $f(x)$ in the neighborhood of the point b is the same as the sign of the quantity $\dot{f}(b)(x - b)$. Hence it follows that for $\dot{f}(b) < 0$ the state of equilibrium b of equation (14) is stable and for $\dot{f}(b) > 0$ it is unstable.

2. We shall study the equation

$$\dot{x} = f(x), \tag{17}$$

where $f(x)$ is a periodic function with a continuous first derivative. To be definite we shall assume that its period is equal to 2π. Everything said in Example 1 concerning equation (14) remains valid for equation (17) as well, since equation (17) is a particular case of equation (14). However, in order to take into account the specific character of equation (17) [the periodicity of function $f(x)$], we shall assume that the phase space of equation (17) is not a straight line but a circle K of radius one on which we choose a reference point 0 and a direction of motion (for example, counterclockwise). To every number x we make correspond the point ξ of the circle K by marking counterclockwise from the reference point an arc of length x (Fig. 24). Then to all numbers $x + 2k\pi$, where k is an integer, there corresponds a unique point ξ on the circumference. Since

$$f(x + 2k\pi) = f(x),$$

it is possible to set $f(\xi) = f(x)$, and the function f is then defined on the circumference K. Equation (17) now defines the motion of point ξ along the circumference K. If $x(t)$ is a certain solution of equation (17), then the point $\xi(t)$ corresponding to the number $x(t)$ moves along circumference K. If α is a point on K such that $f(\alpha) = 0$, then there exists a solution $x(t)$ of (17) such that $\xi(t) = \alpha$ and α is a state of equilibrium of (17). Let us assume for simplicity that the state of equilibrium of (17) on K has no limit points; then there is only a finite number of points or none at all (Fig. 25). States of equilibrium divide the circumference into a finite system Σ of intervals. If there are no states of equilibrium at all, then the

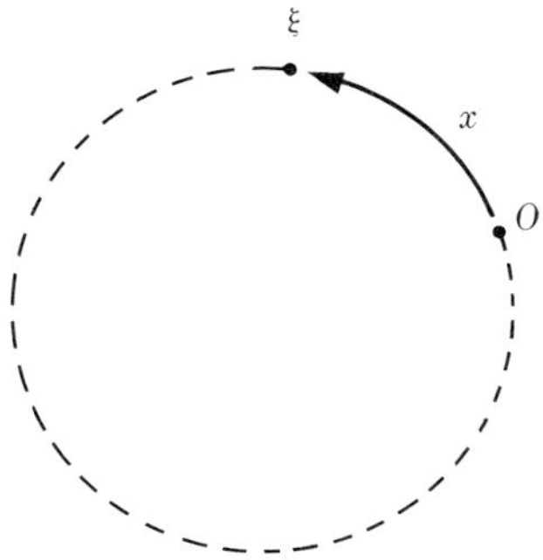

FIGURE 24

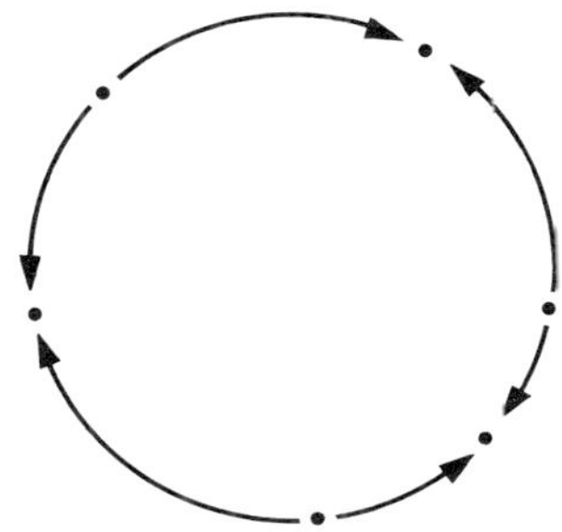

FIGURE 25

system Σ contains only one "interval" (the circumference). If there is only one state of equilibrium α, then the system Σ also contains only one interval, which consists of all points of K with the exception of the point α. In the first case the interval has no endpoints at all; in the second case both its endpoints coincide. Let I be a certain interval of Σ and $x(t)$ a certain solution of (17) with initial values 0, x_0, where ξ_0 is a point of I. The solution $x(t)$ is always defined for all values of t, and the point $\xi(t)$ belongs to the interval I. If the interval I has endpoints (one or two), then the point traverses I in a fixed direction, the solution $\xi(t)$ passing once through each point of I. If the interval I coincides with the entire circumference, then after leaving the position ξ_0, the point will return to it after a certain time T, so that $\xi(0) = \xi(T)$. In this case the motion $\xi(T)$ depends periodically with period T on the number t. The numerical solution $x(t)$ of equation (17) corresponding to the motion $\xi(t)$ satisfies the condition

$$x(t + T) = x(t) \pm 2\pi.$$

From this example it is apparent that it is not always appropriate to consider a euclidean coordinate space as the phase space of a system of equations; sometimes it is necessary to consider a more complex geometrical configuration. In Example 3 below we shall encounter this circumstance in a more complex situation than in this example.

3. We shall investigate the system of equations

$$\dot{x}^i = f^i(x^1, x^2), \qquad i = 1, 2, \tag{18}$$

where the function $f^i(x^1, x^2)$ is periodic of period 2π in each of its arguments:

$$f^i(x^1 + 2k\pi, x^2 + 2l\pi) = f^i(x^1, x^2), \qquad i = 1, 2.$$

As always, we shall assume that the functions $f^i(x^1, x^2)$ are continuous and

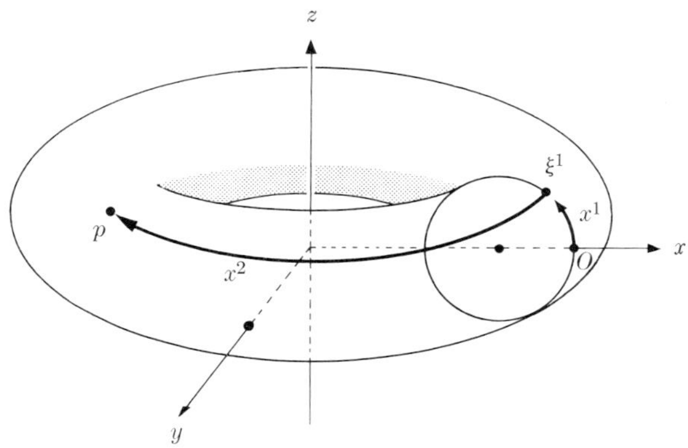

FIGURE 26

have continuous first-order partial derivatives. In view of the periodicity of the functions $f^i(x^1, x^2)$ it is reasonable to assume that the phase space of (18) is not a plane but a more complex geometrical configuration, namely, the *surface of a torus* or, as it is called, a *torus* (Fig. 26). We shall describe this surface.

In a three-dimensional euclidean space with cartesian coordinates x, y, z, we take in the xz-plane the circle K with a radius of one and with its center at the point (2, 0, 0). We take as origin on the circumference the point with coordinates (3, 0, 0). Then to every number x^1 we make correspond the point ξ^1 of K (see Example 2). We now rotate the xz-plane about the z-axis in the (x, y, z) space. The surface P described by the circumference K in this rotation is a torus. Let ξ^1 be some point of K. As a result of rotating the xz-plane through the angle x^2 measured in radians, the point ξ^1 goes over into a certain point p of the torus P (Fig. 26). If the rotation is made not through the angle x^2 but through the angle $x^2 + 2k\pi$, then we arrive at the same point p on the torus P. Thus the point p on the torus P is defined uniquely by two *cyclic coordinates* ξ^1, ξ^2, and to each pair of cyclic coordinates ξ^1, ξ^2 corresponds one well-defined point on the torus. Thus we see that the functions $f^i(x^1, x^2)$ can be considered as defined not on a plane, but on the surface of the torus P:

$$f^i(\xi^1, \xi^2) = f^i(x^1, x^2).$$

Now let $x^1(t)$, $x^2(t)$ be a certain solution of system (18). If we make the correspondence between the numbers $x^1(t)$ and $x^2(t)$ and the cyclic coordinates $\xi^1(t)$ and $\xi^2(t)$, we obtain the point $\xi^1(t)$, $\xi^2(t)$ on the torus P. Thus, every solution $x^1(t)$, $x^2(t)$ of (18) can be represented by the motion of a point on the torus, the law of motion at each instant being defined by that point $\xi^1(t)$, $\xi^2(t)$ of the torus through which the trajectory passes at that instant. This is explained by the fact that functions $f^i(\xi^1, \xi^2)$ are defined on the torus. Thus the entire torus P is found to be covered by trajectories, each two of which either do not intersect, or else coincide. In particular, if a trajectory intersects itself, it is then either closed or it is a state of equilibrium.

The representation of phase trajectories of (18) not on a plane but on the surface of a torus reflects the specific property of the system (18) (periodicity of the functions f^i) and is a convenient means of studying it.

16. The phase plane of a linear homogeneous system with constant coefficients. In this section we shall construct phase trajectories on the phase plane of the system

$$\begin{aligned} \dot{x}^1 &= a_1^1 x^1 + a_2^1 x^2, \\ \dot{x}^2 &= a_1^2 x^1 + a_2^2 x^2, \end{aligned} \tag{1}$$

or, in vector form,

$$\dot{\mathbf{x}} = A(\mathbf{x}), \tag{2}$$

with constant real coefficients a_j^i. Here we shall have to investigate several different cases, since the phase picture of the trajectories of a system depends essentially on the values of the coefficients.

It should be noted that the origin (0, 0) is always a state of equilibrium of the system (1). This state of equilibrium is unique if and only if the determinant of the matrix (a_j^i) is different from zero, or, what is the same thing, if both eigenvalues of this matrix are different from zero.

Let us assume that the eigenvalues of matrix A are real, distinct, and nonzero. Then, according to the results of §14 (Theorem 10), any real solution of (2) can be written in the form

$$\mathbf{x} = c^1 \mathbf{h}_1 e^{\lambda_1 t} + c^2 \mathbf{h}_2 e^{\lambda_2 t}, \tag{3}$$

where $\mathbf{h}_1$ and $\mathbf{h}_2$ are real, linearly independent eigenvectors of the matrix A, λ_1 and λ_2 are its real eigenvalues, and c^1 and c^2 are real constants. We shall expand (3) in terms of the basis vectors $(\mathbf{h}_1, \mathbf{h}_2)$ by setting

$$\mathbf{x} = \xi^1 \mathbf{h}_1 + \xi^2 \mathbf{h}_2, \tag{4}$$

whence we have

$$\xi^1 = c^1 e^{\lambda_1 t}, \qquad \xi^2 = c^2 e^{\lambda_2 t}. \tag{5}$$

Generally speaking, the coordinates ξ^1, ξ^2 on a phase plane P of the system (1) are not rectangular; therefore, we shall make an affine mapping of the phase plane P onto an auxiliary plane P^* in such a way that the vectors $\mathbf{h}_1$, $\mathbf{h}_2$ are transformed into mutually orthogonal unit vectors of

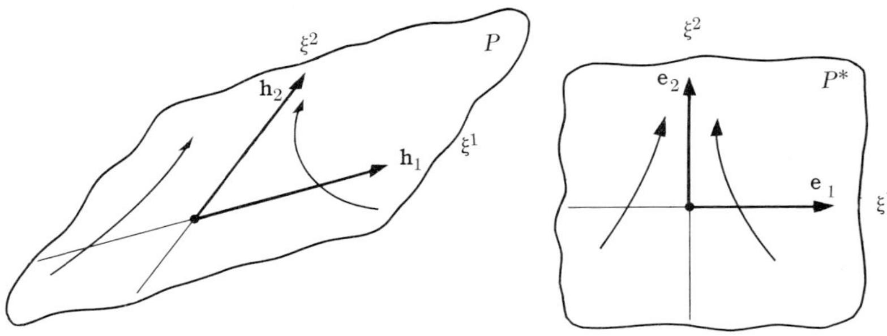

FIGURE 27

the plane directed along the axis of abscissas and the axis of ordinates, respectively (Fig. 27). The point $\mathbf{x} = \xi^1\mathbf{h}_1 + \xi^2\mathbf{h}_2$ of the plane P is transformed by this mapping into a point with rectangular cartesian coordinates ξ^1, ξ^2 in the plane P^*. Thus the trajectory defined by the parametric equations (5) in the plane P will be mapped into a trajectory (which we shall also call *phase trajectory*) defined by the same equations in the rectangular coordinates of the plane P^*. We shall first plot the trajectories defined by the equations (5) in P^*, and then we shall map them back into P.

Together with phase trajectory (5) in P^* there is a trajectory defined by the equations

$$\xi^1 = c^1e^{\lambda_1 t}, \qquad \xi^2 = -c^2e^{\lambda_2 t}, \tag{6}$$

and also a trajectory defined by the equations

$$\xi^1 = -c^1e^{\lambda_1 t}, \qquad \xi^2 = c^2e^{\lambda_2 t}. \tag{7}$$

The trajectory (6) is obtained from trajectory (5) by a mirror reflection in the axis of abscissas, and trajectory (7) is obtained by reflexion in the axis of ordinates. Thus the two mirror images shown leave invariant the picture of the trajectories in P^*. From this it is evident that, if trajectories are drawn in the first quadrant, then it is easy to imagine the entire phase picture in the plane P^*.

We shall note that for $c^1 = c^2 = 0$ we obtain the motion of a point which describes the state of equilibrium $(0, 0)$. For $c^2 = 0$, $c^1 > 0$ we obtain a motion which describes the positive semiaxis of abscissas; for $c^1 = 0$, $c^2 > 0$ we obtain a motion which describes the positive semiaxis of ordinates. If $\lambda_1 < 0$, then the motion which describes the positive semiaxis of abscissas proceeds *toward the origin;* if $\lambda_1 > 0$, then this motion is directed *away from the origin*. In the first case the point approaches the origin as closely as we please; in the second, it recedes without bound toward infinity. The same is true also of the motion which describes the

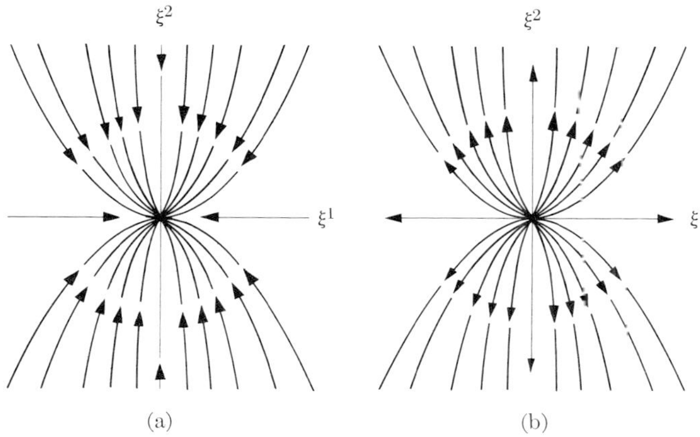

FIGURE 28

positive semiaxis of ordinates. If c_1 and c_2 are positive, then the motion of the point occurs within the first quadrant, without approaching the boundary.

We shall carry out separately a further, more detailed description of the phase plane for several cases, depending on the signs of the numbers λ_1 and λ_2.

(A) *Node.* Let us assume that both numbers, λ_1 and λ_2, are different from zero and have the same sign, and that

$$|\lambda_1| < |\lambda_2|. \tag{8}$$

We shall investigate first the case when

$$\lambda_1 < 0, \qquad \lambda_2 < 0.$$

Under these hypotheses the motion along the positive semiaxis of abscissas is directed toward the origin, just as is the motion along the positive semiaxis of ordinates. Further, the motion along an arbitrary trajectory inside the first quadrant consists of an asymptotic approach of the point toward the origin, the trajectory in this case being tangent to the axis of abscissas at the origin. For $t \to -\infty$, the point moves so that its abscissa and ordinate increase without bound, but the increase of the ordinate is more rapid than that of the abscissa, i.e., the motion goes in the direction of the axis of ordinates. This phase picture is called a *stable node* [Fig. 28, (a)]. If the inequalities

$$\lambda_1 > 0, \qquad \lambda_2 > 0$$

together with inequality (8) are fulfilled, then the trajectories remain as

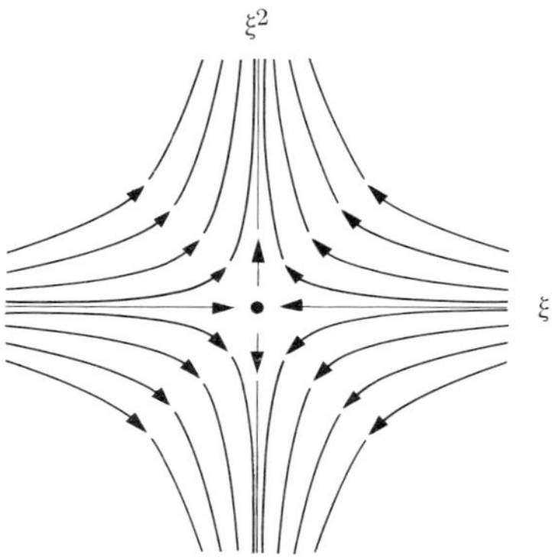

FIGURE 29

before, but the motion along them is in the opposite direction, and we have an *unstable node* [Fig. 28, (b)].

(B) *Saddle point.* Let us assume that the numbers λ_1 and λ_2 have opposite signs. To be definite we assume that

$$\lambda_1 < 0 < \lambda_2.$$

In this case the motion along the positive semiaxis of abscissas is directed toward the origin, and the motion along the positive semiaxis of ordinates is directed away from the origin. The forms of the trajectories lying inside the first quadrant resemble hyperbolas and motions along these trajectories proceed toward the origin along the axis of abscissas and away from the origin along the axis of ordinates. This phase picture is called a *saddle point*, or a *saddle*.

Figures 28(a), 28(b), and 29 give a picture of trajectories on an auxiliary phase plane P^*. The distribution of trajectories on the phase plane P is obtained by means of an affine transformation and depends on the position of the eigenvectors (see, for example, Figs. 30 and 31).

We shall now investigate the case when the eigenvalues of a linear matrix A are complex. In this case they are complex conjugates and can be denoted by $\lambda = \mu + i\nu$ and $\bar{\lambda} = \mu - i\nu$, where $\nu \neq 0$. The eigenvectors of A can be chosen to be conjugate, so that they can be denoted by $\mathbf{h}$ and $\bar{\mathbf{h}}$. Let us set

$$\mathbf{h} = \tfrac{1}{2}(\mathbf{h}_1 - i\mathbf{h}_2),$$

where $\mathbf{h}_1$ and $\mathbf{h}_2$ are real vectors. The vectors $\mathbf{h}_1$ and $\mathbf{h}_2$ are linearly independent, since if they were linearly dependent, then the vectors $\mathbf{h}$ and $\bar{\mathbf{h}}$ would also be linearly dependent. Thus $\mathbf{h}_1$ and $\mathbf{h}_2$ may be assumed to form a basis for the phase plane P of equation (2).

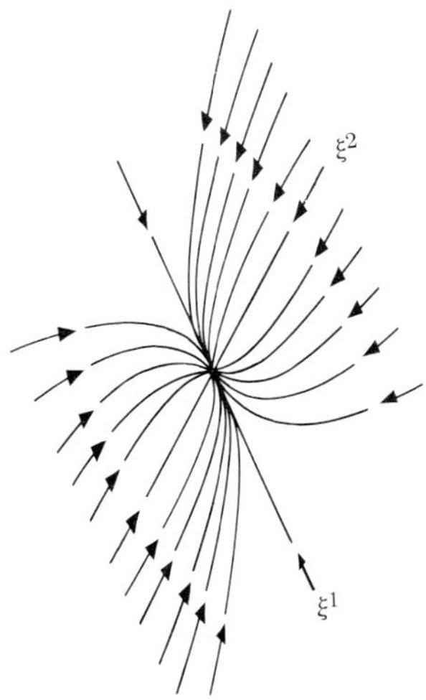

FIGURE 30

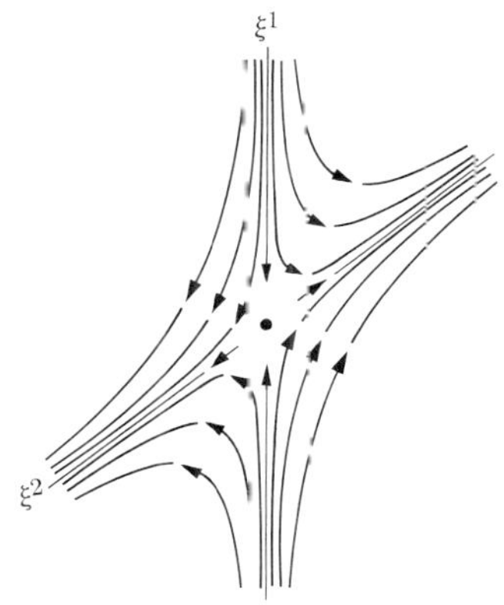

FIGURE 31

An arbitrary real solution of (2) may be written in the form

$$\mathbf{x} = c\mathbf{h}e^{\lambda t} + \overline{c\mathbf{h}}e^{\bar{\lambda} t}, \tag{9}$$

where c is a complex constant. Let

$$\zeta = \xi^1 + i\xi^2 = ce^{\lambda t};$$

then we have

$$\mathbf{x} = \xi^1\mathbf{h}_1 + \xi^2\mathbf{h}_2.$$

Let us map the phase plane P affinely onto the auxiliary plane P^* of the complex variable ζ in such a way that the vector $\mathbf{h}_1$ goes into 1 and the vector $\mathbf{h}_2$ into i; then to the vector $\xi^1\mathbf{h}_1 + \xi^2\mathbf{h}_2$ will correspond the complex number $\zeta = \xi^1 + i\xi^2$. Under this mapping, the phase trajectory (9) will be mapped into a phase trajectory on the plane P^* described by the equation

$$\zeta = ce^{\lambda t}. \tag{10}$$

(C) *Focus and center.* We shall rewrite equation (10) in polar coordinates by setting

$$\zeta = \rho e^{i\varphi}, \qquad c = Re^{i\alpha}.$$

Thus we obtain

$$\rho = Re^{\mu t}, \qquad \varphi = \nu t + \alpha,$$

which is the equation of motion of a point in the phase plane P^*. For

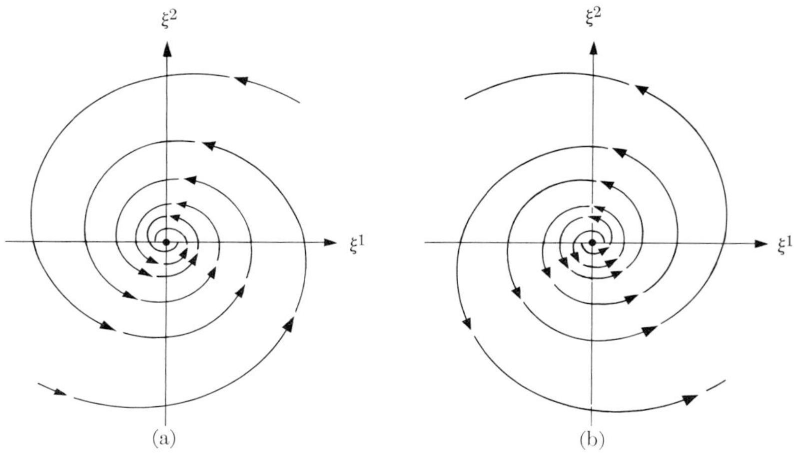

FIGURE 32

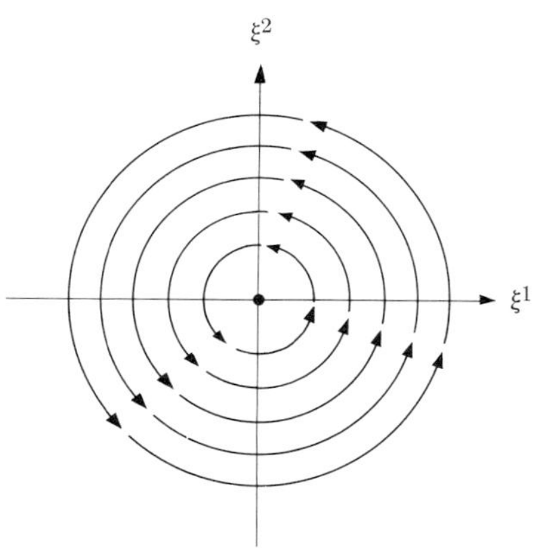

FIGURE 33

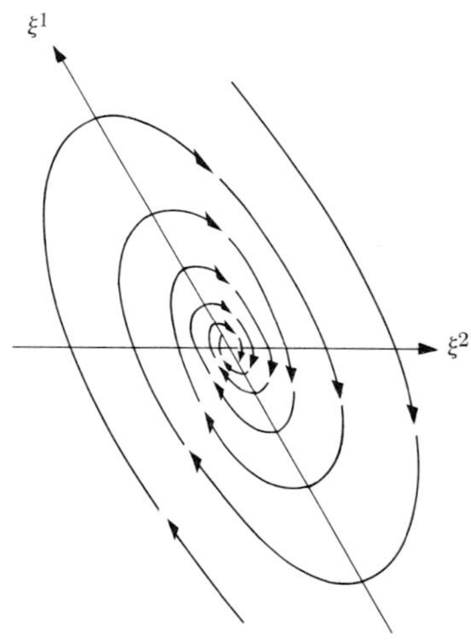

FIGURE 34

$\mu \neq 0$ every trajectory turns out to be a *logarithmic spiral.* The corresponding picture on the plane P is called a *focus.* If $\mu < 0$, then the point approaches the origin asymptotically as t increases, describing a logarithmic spiral. This is a *stable focus* [Fig. 32(a)]. If $\mu > 0$, then the point moves from the origin toward infinity, and we have an *unstable focus* [Fig. 32(b)]. If the number μ is zero, then every phase trajectory except the state of equilibrium $(0, 0)$ is closed, and we have the so-called *center* (Fig. 33).

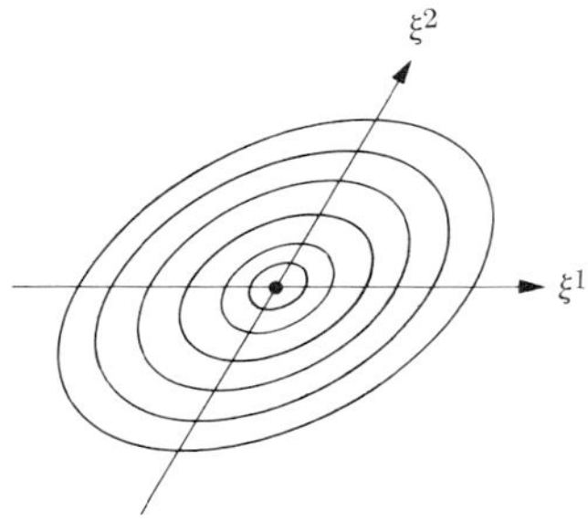

FIGURE 35

Figures 32 and 33 give a picture in an auxiliary phase plane; in the plane P the picture is affinely distorted (see, for example, Figs. 34 and 35).

Above we investigated so-called *nondegenerate cases* where the roots λ_1 and λ_2 are distinct and different from zero. A small variation of the elements of the matrix (a_j^i) does not change the general character of the behavior of phase trajectories in these propositions. The case of a center is an exception: for a small variation of the elements of the matrix (a_j^i) the equality $\mu = 0$ can be violated, and the center will pass into a stable or unstable focus. Because of its importance this degenerate case (of the center) is included in the basic text of the section. The remaining degenerate cases will be investigated in Examples 1 and 3.

EXAMPLES

1. (*Degenerate node.*) If the matrix A of the system (1) has only one eigenvalue λ, then there are possible two essentially different cases; in describing these cases we shall denote by $\mathbf{A}$ the transformation corresponding to the matrix A.

Case I. There exists in the plane P a basis $\mathbf{h}_1$, $\mathbf{h}_2$ consisting of two eigenvectors of the transformation $\mathbf{A}$:

$$\mathbf{A}\mathbf{h}_1 = \lambda\mathbf{h}_1, \qquad \mathbf{A}\mathbf{h}_2 = \lambda\mathbf{h}_2. \tag{11}$$

Case II. There exists in P a basis $\mathbf{h}_1$, $\mathbf{h}_2$ such that

$$\mathbf{A}\mathbf{h}_1 = \lambda\mathbf{h}_1, \qquad \mathbf{A}\mathbf{h}_2 = \lambda\mathbf{h}_2 + \mathbf{h}_1. \tag{12}$$

The existence of a basis of one of the forms (11), (12) follows directly from Theorem 28, but here we shall prove this fact directly. Let $\mathbf{h}_1$ be an eigenvector of the transformation $\mathbf{A}$, and $\mathbf{h}_2$ an arbitrary vector which is not collinear with $\mathbf{h}_1$. Then we have

$$\mathbf{A}\mathbf{h}_1 = \lambda\mathbf{h}_1, \qquad \mathbf{A}\mathbf{h}_2 = \alpha\mathbf{h}_1 + \mu\mathbf{h}_2.$$

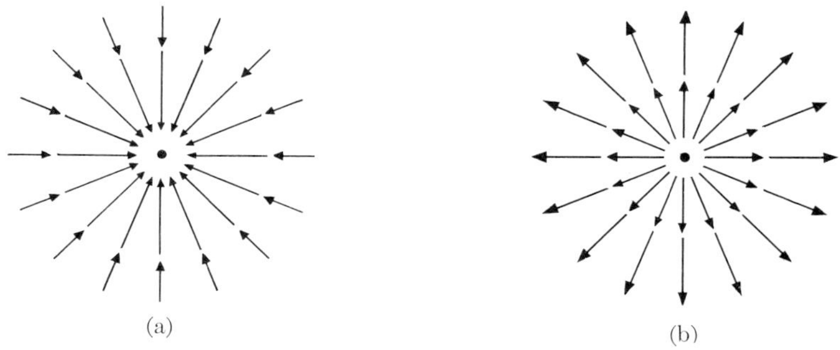

FIGURE 36

From this it is evident that the transformation **A** has, in terms of the basis $\mathbf{h}_1$, $\mathbf{h}_2$, the matrix

$$\begin{pmatrix} \lambda & 0 \\ \alpha & \mu \end{pmatrix},$$

so that λ and μ are its eigenvalues, and therefore $\mu = \lambda$. If $\alpha = 0$, then the relations (11) are satisfied for the basis $\mathbf{h}_1$, $\mathbf{h}_2$. If $\alpha \neq 0$, then by replacing the vector $\mathbf{h}_1$ by the collinear vector $\alpha\mathbf{h}_1$, we obtain a basis which satisfies (12).

By direct inspection we see that in case I the general solution of (2) may be written in the form

$$\mathbf{x} = c^1\mathbf{h}_1e^{\lambda t} + c^2\mathbf{h}_2e^{\lambda t} = \mathbf{x}_0e^{\lambda t}. \tag{13}$$

This solution has the initial value $(0, \mathbf{x}_0)$. For $\lambda \neq 0$ every solution describes a ray emanating from the origin. For $\lambda < 0$ the motion is directed toward the origin [Fig. 36(a)] and for $\lambda > 0$, away from the origin [Fig. 36(b)]; for the case $\lambda = 0$, see Example 3.

By direct inspection it is also seen that in case II an arbitrary solution of (2) has the form

$$\mathbf{x} = c^1\mathbf{h}_1e^{\lambda t} + c^2(\mathbf{h}_1t + \mathbf{h}_2)e^{\lambda t}.$$

If we expand this solution in terms of the basis $\mathbf{h}_1$, $\mathbf{h}_2$,

$$\mathbf{x} = \xi^1\mathbf{h}_1 + \xi^2\mathbf{h}_2,$$

we obtain the equations of the trajectories in P in terms of the basis $\mathbf{h}_1$, $\mathbf{h}_2$:

$$\xi^1 = (c^1 + c^2t)e^{\lambda t}, \qquad \xi^2 = c^2e^{\lambda t}. \tag{14}$$

An affine mapping of the phase plane P, which transforms vectors $\mathbf{h}_1$ and $\mathbf{h}_2$ into unit orthogonal vectors directed along the coordinate axes of the plane P^*, also transforms the trajectories of P into trajectories of P^*, where the trajectories are already given in rectangular coordinates by (14).

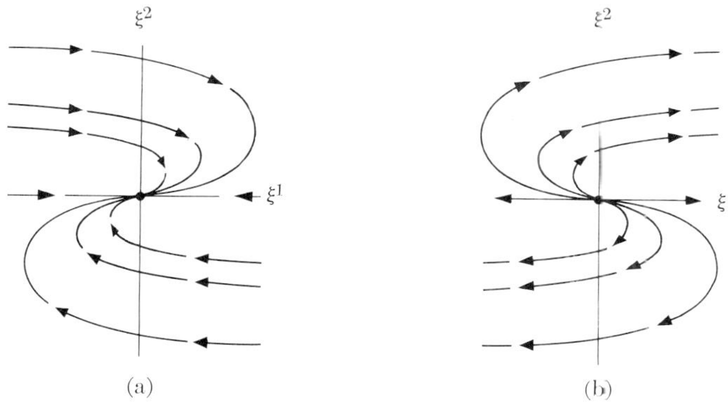

FIGURE 37

We shall investigate the case when $\lambda \neq 0$ (for the case $\lambda = 0$ see Example 3). First let $\lambda < 0$. In this case we shall study the trajectories filling the plane P^*. First of all from equations (14) it is evident that, by changing simultaneously the signs of c^1 and c^2, we shall obtain a reflection of the plane P^* with respect to the origin under which trajectories are transformed into trajectories. Thus it is sufficient to examine those trajectories which fill out the upper half-plane. For $c^2 = 0$, $c^1 \neq 0$ we obtain two trajectories, one for $c^1 > 0$ and the other for $c^1 < 0$. The first one coincides with the positive semiaxis of abscissas and the second, with the negative semiaxis of abscissas; the motion along both is directed toward the origin. We shall consider the trajectory $c^1 = 0$, $c^2 > 0$. We have

$$\xi^1 = c^2 te^{\lambda t}, \qquad \xi^2 = c^2 e^{\lambda t}. \tag{15}$$

For $t = 0$ we obtain the point $(0, c^2)$ on the axis of ordinates. As t increases from zero, the point first moves to the right, then to the left, always descending toward the origin t_0, which it approaches along the trajectory which is tangent to the positive direction of the axis of abscissas. As t decreases from zero to $-\infty$, the point moves to the left and at the same time it ascends less rapidly so that the general tendency of its motion is in the negative direction along the axis of abscissas. If in equations (15) the constant c^2 is assigned all positive values, then the trajectories described in this manner fill out the entire upper half-plane [Fig. 37, (a)]. We have here a *stable degenerate node*. If $\lambda > 0$, then the trajectory is obtained from those described by a mirror reflection of the plane in the axis of ordinates [Fig. 37(b)], and the motion along them proceeds in the opposite direction, i.e., away from the origin of coordinates. This is an *unstable degenerate node*. The phase trajectories in plane P are shown in Fig. 38(a) and 38(b).

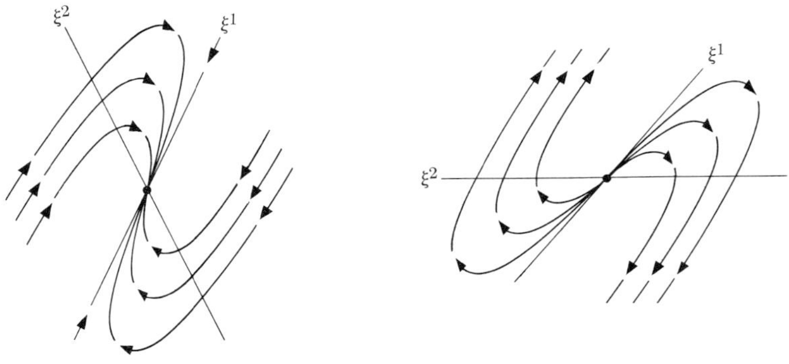

FIGURE 38

2. Let us study the linear homogeneous second-order equation with constant coefficients

$$\ddot{x} + a\dot{x} + bx = 0. \tag{16}$$

Replacing this equation with a normal system by the method presented in §4, we obtain

$$\dot{x} = y, \qquad \dot{y} = -bx - ay. \tag{17}$$

The phase plane of system (17) may be taken as the phase plane of equation (16). By direct inspection it is seen that the characteristic polynomial of (17) coincides with the characteristic polynomial of (16), that is,

$$p^2 + ap + b. \tag{18}$$

Thus, if the roots of (18) are complex, then the phase plane of (16) is a focus or center. We shall study the phase plane in the case of real, distinct, and nonzero roots of (18). Let λ be a root of (18) and $\mathbf{h} = (h^1, h^2)$ its corresponding eigenvector. We have then, taking into account the form of system (17),

$$h^2 = \lambda h^1,$$

so that the characteristic direction corresponding to the eigenvalue λ is determined by a straight line having the equation

$$y = \lambda x;$$

we shall call it the *characteristic line*.

If the roots λ_1 and λ_2 are negative, then we have a *stable node* [see (A)]. In this case both characteristic lines pass through the second and fourth quadrants; the trajectories in the neighborhood of the origin are tangent to whichever of these lines is closer to the axis of abscissas [Fig. 39].

If λ_1 and λ_2 are positive, then we have an *unstable node* [see (A)]. Both characteristic lines pass through the first and third quadrants; in the

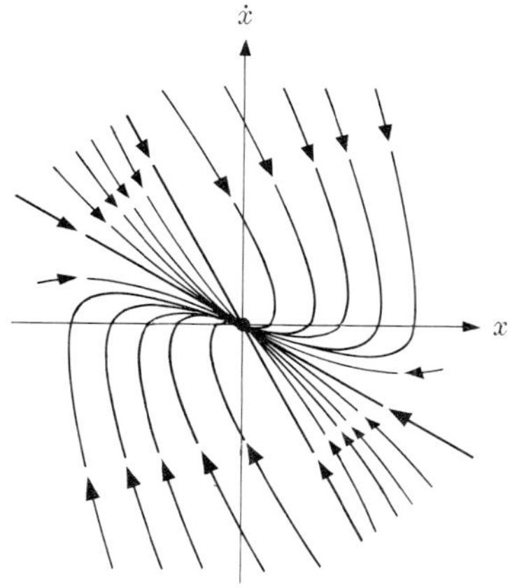

FIGURE 39

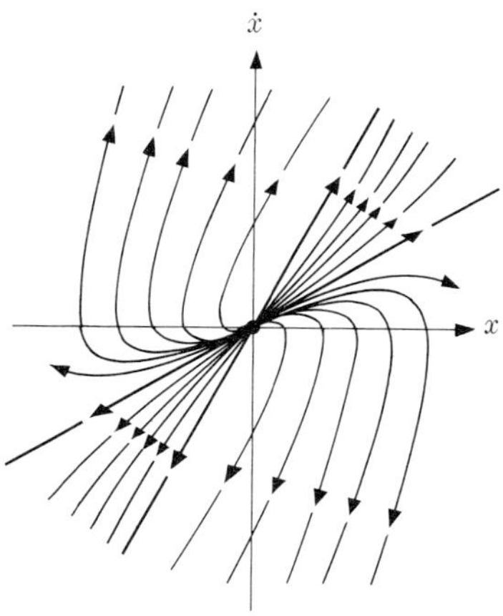

FIGURE 40

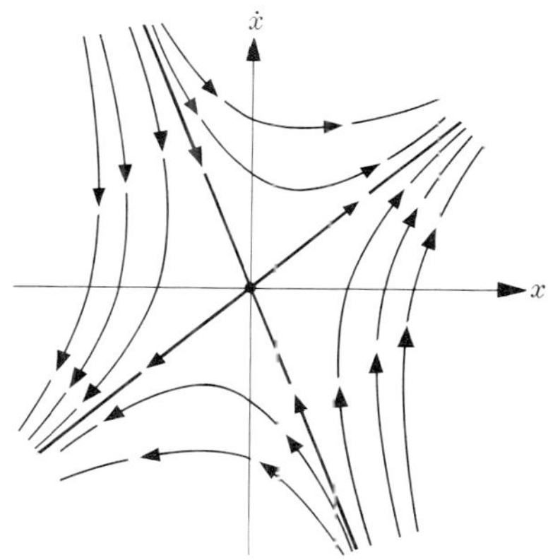

FIGURE 41

neighborhood of the origin the trajectories are tangent to the line closer to the axis of abscissas (Fig. 40).

If λ_1 and λ_2 have different signs, then we have a saddle; one characteristic line passes through the second and fourth quadrants and the other through the first and third quadrants. In the direction of the first of these lines, the trajectories approach the origin, and in the direction of the second, they recede from the origin (Fig. 41).

3. We shall study, finally, the case when at least one of the eigenvalues of the matrix A is zero.

Case I. Only one eigenvalue,

$$\lambda_1 \neq 0, \qquad \lambda_2 = 0,$$

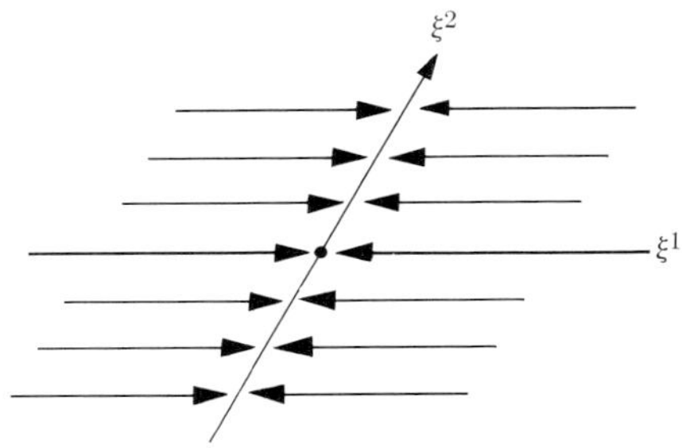

FIGURE 42

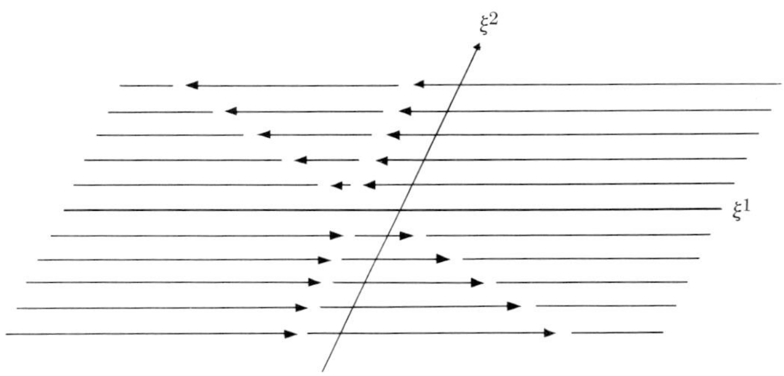

FIGURE 43

is zero. In this case the solution can be written in form (4), where ξ^1 and ξ^2 are defined by the formulas (5). Since $\lambda_2 = 0$, then $\xi^2 = \text{const}$, and the motion takes place along the straight line $\xi^2 = \text{const}$ in the direction of the line $\xi^1 = 0$ or away from it depending on the sign of the number λ_1. All points of the line $\xi^1 = 0$ are states of equilibrium (Fig. 42).

If there exists only one eigenvalue $\lambda_1 = \lambda_2 = 0$, then the two cases in Example 1 can occur.

Case I [see (11), $\lambda = 0$]. The general solution may be written in the form [see (13)]

$$\mathbf{x} = \mathbf{x}_0.$$

This case occurs if all coefficients of (1) are zero; every point of the plane P is a state of equilibrium.

Case II [see (12), $\lambda = 0$]. The general solution may be written in the form [see (14)]

$$\xi^1 = c^1 + c^2 t, \qquad \xi^2 = c^2.$$

The motion takes place uniformly along each of the straight lines $\xi^2 = \text{const}$. All points of the line $\xi^2 = 0$ are states of equilibrium (Fig. 43).

CHAPTER 3

LINEAR EQUATIONS WITH VARIABLE COEFFICIENTS

In this chapter we develop the theory of linear equations first for a normal nth-order system and then for one nth-order equation; almost all results are related to the equation derived from the corresponding results for the normal system. The third section of this chapter is devoted to normal linear homogeneous systems with periodic coefficients. The basic result here is Lyapunov's theorem on transforming a normal system with periodic coefficients into a normal system with constant coefficients by means of a linear periodic transformation of the variables. Later this result will find an important application in stability theory. Its proof is very simple, but is based on a comparatively nonelementary part of the theory of matrix functions. This theory, which is not a part of the theory of ordinary differential equations, is presented for the convenience of the reader in the last chapter of this book, which is a purely algebraic supplement. Thus the third section (§19) of this chapter is nonelementary because it contains matrix calculus. The material of this section is, apparently, presented for the first time in an undergraduate textbook.

17. The normal system of linear equations. We shall study here the normal system of linear equations with variable coefficients

$$\dot{x}^i = \sum_{j=1}^{n} a_j^i(t)x^j + b^i(t), \qquad i = 1, \ldots, n. \tag{1}$$

We recall that if $q_1 < t < q_2$ is the interval where the coefficients $a_j^i(t)$ and the free terms $b^i(t)$ of system (1) exist and are continuous; then, by Theorem 3, the interval $q_1 < t < q_2$ is the maximum interval of existence of every solution of (1). In the sequel we shall assume that every solution to be considered is defined on this interval and that every value t considered belongs to it.

The fundamental system of solutions. First of all we shall study the homogeneous system of equations

$$\dot{x}^i = \sum_{j=1}^{n} a_j^i(t)x^j, \qquad i = 1, \ldots, n, \tag{2}$$

or, in vector form,

$$\dot{\mathbf{x}} = A(t)\mathbf{x}. \tag{3}$$

(A) We shall establish the simplest properties of equation (3). (a) If $\mathbf{x} = \boldsymbol{\varphi}(t)$ is a solution of (3) which vanishes for a certain value t_0, i.e.,

$$\boldsymbol{\varphi}(t_0) = 0, \tag{4}$$

then this solution is identically zero:

$$\boldsymbol{\varphi}(t) \equiv 0, \qquad q_1 < t < q_2.$$

(b) If the vector functions

$$\boldsymbol{\varphi}_1(t), \quad \boldsymbol{\varphi}_2(t), \ldots, \boldsymbol{\varphi}_r(t)$$

are solutions of (3), then the vector function

$$\boldsymbol{\varphi}(t) = c^1\boldsymbol{\varphi}_1(t) + \cdots + c^r\boldsymbol{\varphi}_r(t),$$

where $c^1, \ldots, c^r$ are constants, is also a solution of (3).

Property (b) can be verified directly. Property (a) follows from the fact that the vector $\mathbf{x} = 0$, which is identically zero, is obviously a solution of (3), and therefore the solution $\boldsymbol{\varphi}(t)$, specified in (a) as having common initial conditions (4) with this solution, must coincide with it.

(B) Let

$$\boldsymbol{\varphi}_1(t), \quad \boldsymbol{\varphi}_2(t), \ldots, \boldsymbol{\varphi}_r(t) \tag{5}$$

be a system of solutions of (3). It is called *linearly dependent* if there exist constants $c^1, c^2, \ldots, c^r$, not all simultaneously zero, such that

$$c^1\boldsymbol{\varphi}_1(t) + c^2\boldsymbol{\varphi}_2(t) + \cdots + c^r\boldsymbol{\varphi}_r(t) \equiv 0.$$

Otherwise, the system (5) of solutions of (3) is called *linearly independent.* It follows that if for even one value $t = t_0$ the vectors

$$\boldsymbol{\varphi}_1(t_0), \quad \boldsymbol{\varphi}_2(t_0), \ldots, \boldsymbol{\varphi}_r(t_0) \tag{6}$$

are linearly dependent, then the solutions (5) are linearly dependent. In other words, if the system (5) is linearly independent, then for no value of t can the vectors (5) be linearly dependent.

We shall now prove this. Let us assume that the vectors (6) are linearly dependent, i.e., that

$$c^1\boldsymbol{\varphi}_1(t_0) + \cdots + c^r\boldsymbol{\varphi}_r(t_0) = 0,$$

where not all the numbers $c^1, \ldots, c^r$ are zero. Let us set

$$\boldsymbol{\varphi}(t) = c^1\boldsymbol{\varphi}_1(t) + \cdots + c^r\boldsymbol{\varphi}_r(t).$$

By proposition (A) the vector function $\boldsymbol{\varphi}(t)$ is a solution of (3), and is also identically equal to zero, since it vanishes at the point $t = t_0$.

We now turn to the concept of a *fundamental system of solutions*, which is most important for homogeneous linear systems.

(C) The system of solutions of equation (3)

$$\boldsymbol{\varphi}_1(t), \quad \boldsymbol{\varphi}_2(t), \ldots, \boldsymbol{\varphi}_n(t), \tag{7}$$

where n is the order of the system (2), is called a *fundamental system of solutions* if it is linearly independent [see (B)]. It turns out that (a) a fundamental system of solutions always exists for the equation (3), and (b) if (7) is a fundamental system of solutions of (3), then every solution $\boldsymbol{\varphi}(t)$ of (3) can be represented in the form

$$\boldsymbol{\varphi}(t) = c^1\boldsymbol{\varphi}_1(t) + c^2\boldsymbol{\varphi}_2(t) + \cdots + c^n\boldsymbol{\varphi}_n(t), \tag{8}$$

where $c^1, \ldots, c^n$ are suitably chosen constants. (For a linear system with constant coefficients, the fundamental system of solutions was constructed in §14).

We shall prove first of all that a fundamental system of solutions of (3) exists. Let

$$\mathbf{a}_1, \mathbf{a}_2, \ldots, \mathbf{a}_n$$

be an arbitrary system of constant, linearly independent vectors, where n is the order of (2). We shall define solutions (7) by the initial conditions

$$\boldsymbol{\varphi}_i(t_0) = \mathbf{a}_i, \qquad i = 1, \ldots, n,$$

where t_0 is a certain value of t. Since the vectors $\boldsymbol{\varphi}_1(t_0), \boldsymbol{\varphi}_2(t_0), \ldots \boldsymbol{\varphi}_n(t_0)$ are by assumption linearly independent, then proposition (B) implies that the solutions (7) are also linearly independent, i.e., they form a fundamental system.

We shall show that every solution $\boldsymbol{\varphi}(t)$ can be written in the form (8). Let t_0 be a certain instant of the time t; since the solutions (7) are linearly independent, the vectors $\boldsymbol{\varphi}_1(t_0), \ldots, \boldsymbol{\varphi}_n(t_0)$ are also linearly independent [see (B)], and since their number is equal to the dimension of the vector space considered, they form a basis for it, so that the vector $\boldsymbol{\varphi}(t_0)$ may be written in the form

$$\boldsymbol{\varphi}(t_0) = c^1\boldsymbol{\varphi}_1(t_0) + \cdots + c^n\boldsymbol{\varphi}_n(t_0), \tag{9}$$

where the numbers $c^1, \ldots, c^n$ are suitably chosen. The solutions $\boldsymbol{\varphi}(t)$ and $c^1\boldsymbol{\varphi}_1(t) + \cdots + c^n\boldsymbol{\varphi}_n(t)$ have common initial conditions [see (9)] and therefore coincide, so that (8) is valid.

We shall now proceed to a description in coordinate form of the facts obtained and to the determination of certain other results.

(D) Let

$$\boldsymbol{\varphi}_1(t), \ldots, \boldsymbol{\varphi}_n(t) \tag{10}$$

be a certain system of solutions of equation (3). Let us express the solution $\boldsymbol{\varphi}_k(t)$ in terms of coordinates by setting

$$\boldsymbol{\varphi}_k(t) = (\varphi_k^1(t), \varphi_k^2(t), \ldots, \varphi_k^n(t)).$$

We form the matrix

$$\begin{pmatrix} \varphi_1^1(t) & \cdots & \varphi_k^1(t) & \cdots & \varphi_n^1(t) \\ \varphi_1^2(t) & \cdots & \varphi_k^2(t) & \cdots & \varphi_n^2(t) \\ \vdots & & & & \\ \varphi_1^n(t) & \cdots & \varphi_k^n(t) & \cdots & \varphi_n^n(t) \end{pmatrix}, \tag{11}$$

where the solution $\boldsymbol{\varphi}_k(t)$ of (2), or more precisely, its coordinates, serve as the kth column. We denote the determinant of this matrix by $W(t)$; it is called the *Wronskian determinant* of the system of solutions (10). It is evident that, if the solutions (10) are linearly independent, then the Wronskian $W(t)$ cannot vanish for any value of t; in this case system (10) is a fundamental system of solutions. Furthermore, if (10) is linearly dependent, the Wronskian is identically zero. Whenever the system (10) is fundamental, we shall call the matrix (11) a *fundamental matrix*.

We shall now prove that any given nth-order square matrix, consisting of functions of t and satisfying certain natural conditions, is fundamental for a certain system of equations of the form (2).

(E) We shall assume that the matrix (11) is an arbitrary given matrix of functions of t which are continuously differentiable on the interval $q_1 < t < q_2$, with a determinant which does not vanish on this interval. Then (11) is fundamental for some (unique) system (2) which is defined on the interval $q_1 < t < q_2$.

To prove this, we shall write out the statement that the vector function $\boldsymbol{\varphi}_k(t)$, whose coordinates form the kth column of the matrix (11), constitutes a solution of (3). We have

$$\dot{\varphi}_k^i(t) = \sum_{i=1}^{n} a_j^i(t)\varphi_k^j(t), \qquad i, k = 1, \ldots, n. \tag{12}$$

If in this relation the index i is fixed and only the index k is considered to be variable, then the system of relations obtained can be taken as a system of linear algebraic equations in the unknowns $a_1^i(t), \ldots, a_n^i(t)$. This system can be solved uniquely, since its matrix is obtained by a transposition of the matrix (11) so that its determinant does not vanish. Thus for every fixed i the functions $a_j^i(t)$ may be determined uniquely from (12). They then become continuous functions since the functions $\dot{\varphi}_k^i(t)$ and $\varphi_k^j(t)$ are continuous.

Liouville's formula. To prove proposition (G) we need a formula for the differentiation of a determinant, which we shall derive here.

(F) Let $(\varphi_j^i(t))$ be an nth-order square matrix whose elements are differentiable functions of the variable t, and let $W(t)$ be the determinant of this matrix. The derivative $\dot{W}(t)$ of this determinant may be found from the following formula:

$$\dot{W}(t) = W_1(t) + \cdots + W_n(t), \tag{13}$$

where the term $W_i(t)$ in the ith place on the right-hand side of the equality is defined in the following way. In the matrix $(\varphi_j^i(t))$ all elements of the ith row are differentiated with respect to t, the other rows remaining unchanged and the determinant of the matrix obtained being denoted by $W_i(t)$. It is evident that the role of the rows and columns may be interchanged.

To prove (13), we first consider the determinant U of the nth-order square matrix (u_j^i) as a function of all the elements u_j^i, $i, j = 1, \ldots, n$ of this matrix, and we assume these elements to be independent variables. Let us calculate the partial derivative

$$\frac{\partial U}{\partial u_s^r}$$

of the function U with respect to the variable u_s^r; here r and s are fixed. We denote the cofactor of the element u_j^i in the matrix (u_j^i) by V_i^j, so that

$$U = \sum_{j=1}^{n} u_j^r V_r^j. \tag{14}$$

This formula gives us the expansion of the determinant U in terms of elements of the rth row. The cofactor V_r^j does not depend on the variable u_s^r so that by differentiating (14) with respect to u_s^r, we obtain

$$\frac{\partial U}{\partial u_s^r} = V_r^s. \tag{15}$$

If we set $u_j^i = \varphi_j^i(t)$, then we have $U = W(t)$. Differentiating $W(t)$ as a composite function, we obtain by (15)

$$\dot{W}(t) = \sum_{i,j} \frac{\partial U}{\partial u_j^i} \dot{\varphi}_j^i(t) = \sum_{i,j} \dot{\varphi}_j^i(t) V_i^j = \sum_{i=1}^{n} \left(\sum_{j=1}^{n} \dot{\varphi}_j^i(t) V_i^j \right).$$

Since it is obvious that

$$\sum_{j=1}^{n} \dot{\varphi}_j^i(t) V_i^j = W_i(t),$$

formula (13) is proved.

We now turn to the proof of the so-called *Liouville formula.*

(G) Let $W(t)$ be the Wronskian of a fundamental system of solutions of equations (2); then the formula

$$W(t) = W(t_0) \exp\left(\int_{t_0}^{t} S(\tau)\, d\tau\right) \tag{16}$$

is valid, where

$$S(t) = a_1^1(t) + a_2^2(t) + \cdots + a_n^n(t).$$

To prove (16), we shall introduce the differential equation which the Wronskian satisfies.

We shall calculate the derivative $\dot{W}(t)$ of the Wronskian by means of (13). In order to simplify the calculations, we shall assume the rows of (11) to be vectors; that is, we assume that

$$\mathbf{x}^i(t) = (\varphi_1^i(t), \ldots, \varphi_n^i(t)), \qquad i = 1, \ldots, n.$$

The relation (12) may now be written in the form

$$\dot{\mathbf{x}}^i(t) = a_1^i(t)\mathbf{x}^1(t) + \cdots + a_n^i(t)\mathbf{x}^n(t), \tag{17}$$

which shows that the derivative of the ith row of the matrix (11) is a linear combination of rows of the same matrix. Thus, in calculating the determinant $W_i(t)$, we must replace the ith row of the determinant $W(t)$ by a linear combination (17) of rows of the same determinant. Since adding multiples of the other rows to the given row does not change the value of the determinant, the determinant $W_i(t)$ may be obtained from the determinant $W(t)$ by multiplying its ith row by $a_i^i(t)$, so that we have

$$W_i(t) = a_i^i(t) W(t).$$

Thus by (13) we obtain

$$\dot{W}(t) = S(t) W(t).$$

The unique solution of this equation with the initial condition

$$W(t)|_{t=t_0} = W(t_0)$$

is formula (16). Thus Liouville's formula is proved.

The method of variation of parameters. We turn now to the study of nonhomogeneous systems. Let

$$\dot{\mathbf{y}} = A(t)\mathbf{y} + \mathbf{b}(t) \tag{18}$$

be the vector form of the nonhomogeneous system (1), and let $\mathbf{y} = \boldsymbol{\psi}(t)$ be a certain solution of this equation. Together with equation (18) we shall study the corresponding homogeneous equation (3). From §6 it follows directly that an arbitrary solution of (18) may be written in the form

$$\mathbf{y} = \boldsymbol{\varphi}(t) + \boldsymbol{\psi}(t),$$

where $\boldsymbol{\varphi}(t)$ is an arbitrary solution of (3).

Thus the solution of the nonhomogeneous equation (18) reduces to the solution of a homogeneous equation and to the determination of a particular solution of the nonhomogeneous equation. We shall show how, by knowing the fundamental system of solutions of the homogeneous equation (3), we can find by means of quadratures a particular solution of the nonhomogeneous equation.

(H) (*Method of variation of parameters.*) Let

$$\boldsymbol{\varphi}_1(t), \ldots, \boldsymbol{\varphi}_n(t)$$

be a fundamental system of solutions of the homogeneous equation (3). We shall seek a solution of (18) in the form

$$\mathbf{y} = c^1(t)\boldsymbol{\varphi}_1(t) + \cdots + c^n(t)\boldsymbol{\varphi}_n(t),$$

where the coefficients are indeterminate functions of t. Substituting this value $\mathbf{y}$ into (18), we obtain

$$\begin{aligned}\dot{c}^1(t)\boldsymbol{\varphi}_1(t) + \cdots + \dot{c}^n(t)\boldsymbol{\varphi}_n(t) + c^1(t)\dot{\boldsymbol{\varphi}}_1(t) + \cdots + c^n(t)\dot{\boldsymbol{\varphi}}_n(t)& \\ = A(t)[c^1(t)\boldsymbol{\varphi}_1(t) + \cdots + c^n(t)\boldsymbol{\varphi}_n(t)] + \mathbf{b}(t)&,\end{aligned}$$

whence, remembering that $\boldsymbol{\varphi}_1(t), \ldots, \boldsymbol{\varphi}_n(t)$ are solutions of (3), we obtain

$$\dot{c}^1(t)\boldsymbol{\varphi}_1(t) + \cdots + \dot{c}^n(t)\boldsymbol{\varphi}_n(t) = \mathbf{b}(t). \tag{19}$$

Since $\boldsymbol{\varphi}_1(t), \ldots, \boldsymbol{\varphi}_n(t)$ are linearly independent vectors at every point t, it follows from (19) that the quantities $\dot{c}^1(t), \ldots, \dot{c}^n(t)$ are uniquely determined, and therefore the values $c^1(t), \ldots, c^n(t)$ may be found by integrating. Equation (19) in the quantities $\dot{c}^1(t), \ldots, \dot{c}^n(t)$ may be written in terms of the coordinates and has the form

$$\sum_{j=1}^{n} \varphi_j^i(t)\dot{c}^j(t) = b^i(t), \qquad i = 1, \ldots, n.$$

Matrix form of systems of linear equations. In a number of cases it is preferable to write equation (3) in its *matrix form,* where the fundamental matrix of (3) is the unknown value. We shall give this form here.

(I) If (7) is a fundamental system of solutions of (3), then

$$\dot{\varphi}_j^i(t) = \sum_{\alpha=1}^{n} a_\alpha^i(t)\varphi_j^\alpha(t).$$

This relation takes the matrix form

$$\dot{\Phi}(t) = A(t)\Phi(t), \tag{20}$$

where $\dot{\Phi}(t)$ is the derivative of the fundamental matrix $\Phi(t) = (\varphi_j^i(t))$ with respect to time t, that is, $\dot{\Phi}(t) = (\dot{\varphi}_j^i(t))$. Thus the fundamental matrix $\Phi(t)$ of (3) satisfies the matrix equation (20); in addition, every solution of the matrix equation

$$\dot{X} = A(t)X, \tag{21}$$

where X is an unknown matrix, is the fundamental matrix of (3) only if the determinant of X is not zero. Hereafter, by a *solution* of equation (21) we shall mean only a matrix X which satisfies (21) and whose determinant is different from zero. It is obvious that finding one solution of (21) is equivalent to finding all solutions of (3). We note that if $X = \Phi(t)$ and $X = \hat{\Phi}(t)$ are two solutions of (21), then there exists a constant matrix P such that

$$\hat{\Phi}(t) = \Phi(t)P. \tag{22}$$

Let us prove this last statement. If

$$\Phi(t) = (\varphi_j^i(t)),$$
$$\hat{\Phi}(t) = (\hat{\varphi}_j^i(t)),$$
$$\boldsymbol{\varphi}_j(t) = (\varphi_j^1(t), \ldots, \varphi_j^n(t)),$$
$$\hat{\boldsymbol{\varphi}}_j(t) = (\hat{\varphi}_j^1(t), \ldots, \hat{\varphi}_j^n(t)),$$

then

$$\boldsymbol{\varphi}_1(t), \ldots, \boldsymbol{\varphi}_n(t) \tag{23}$$

is a fundamental system of solutions of (3), and since $\hat{\boldsymbol{\varphi}}_j(t)$ is also a solution of (3), it can be expressed in terms of the fundamental system (23), so that we have

$$\hat{\boldsymbol{\varphi}}_j(t) = \sum_{\alpha=1}^{n} p_j^\alpha \boldsymbol{\varphi}_\alpha(t).$$

If we rewrite these relations in scalar form, we obtain

$$\hat{\varphi}_j^i(t) = \sum_{\alpha=1}^{n} \varphi_\alpha^i(t) p_j^\alpha. \tag{24}$$

The relation (22) represents the matrix form of (24) for $P = (p_j^i)$.

(J) We shall introduce a new unknown vector $\mathbf{y}$ in equation (3) by using the transformation

$$\mathbf{y} = S(t)\mathbf{x}, \tag{25}$$

where $S(t) = (s_j^i(t))$ is a nonsingular matrix which depends on t. The equation for the new unknown vector function $\mathbf{y}$ has the form

$$\dot{\mathbf{y}} = (\dot{S}(t) + S(t)A(t))S^{-1}(t)\mathbf{y}, \tag{26}$$

i.e., we obtain again an equation of type (3). To the transformation (25) of the vector variable corresponds the transformation

$$Y = S(t)X \tag{27}$$

of the matrix variable [see (I)].

We shall first derive equation (26) for the unknown $\mathbf{y}$. We have

$$\dot{\mathbf{y}} = \frac{d}{dt}(S(t)\mathbf{x}) = \dot{S}(t)\mathbf{x} + S(t)\dot{\mathbf{x}}$$
$$= (\dot{S}(t) + S(t)A(t))\mathbf{x} = (\dot{S}(t) + S(t)A(t))S^{-1}(t)\mathbf{y}.$$

In order to show that the transformation (26) of the matrix variable corresponds to the transformation (25) of the vector variable, we shall rewrite (25) in scalar form

$$y^i = \sum_{\alpha=1}^{n} s_\alpha^i(t) x^\alpha.$$

By means of the formula

$$\psi_j^i(t) = \sum_{\alpha=1}^{n} s_\alpha^i(t) \varphi_j^\alpha(t)$$

the transformation (25) sets up a correspondence between the vector $\boldsymbol{\psi}_j(t) = (\psi_j^1(t), \ldots, \psi_j^n(t))$ and the vector $\boldsymbol{\varphi}_j(t) = (\varphi_j^1(t), \ldots, \varphi_j^n(t))$ of the fundamental system (3). Thus the fundamental matrix $\Psi(t)$ of (26), given by the formula

$$\Psi(t) = S(t)\Phi(t),$$

corresponds to the fundamental matrix $\Phi(t)$ of (3), so that the matrix indeterminate transforms according to formula (27).

Example

From proposition (C) it is clear that in order to find all solutions of equation (3) it is sufficient to find its fundamental system of solutions which contains n linearly independent solutions. We shall show that, if we know one nontrivial solution of (2), then we can lower the order of (2) by one, i.e., we can reduce the problem to the solution of a linear system of order $n - 1$. Let

$$\boldsymbol{\varphi}(t) = (\varphi^1(t), \ldots, \varphi^n(t))$$

be a solution of (3) or, what is the same thing, of (2). We seek a solution of (3) in the form

$$\mathbf{x} = u\boldsymbol{\varphi}(t) + \mathbf{y}, \tag{28}$$

where u is an unknown function and $\mathbf{y}$ an unknown vector, whose first component we can assume to be zero:

$$\mathbf{y} = (0, y^2, \ldots, y^n).$$

Substitution of the vector $\mathbf{x}$ from formula (28) into (3) gives

$$\dot{u}\boldsymbol{\varphi}(t) + u\dot{\boldsymbol{\varphi}}(t) + \dot{\mathbf{y}} = A(t)(u\boldsymbol{\varphi}(t) + \mathbf{y}).$$

Since $\boldsymbol{\varphi}(t)$ is a solution of (3), we obtain

$$\dot{u}\boldsymbol{\varphi}(t) + \dot{\mathbf{y}} = A(t)\mathbf{y}.$$

We express this equation in terms of the coordinates, writing separately the first of the equations so obtained:

$$\dot{u}\varphi^1(t) = \sum_{j=2}^{n} a_j^1(t)y^j, \tag{29}$$

$$\dot{y}^i = \sum_{j=2}^{n} a_j^i(t)y^j - \dot{u}\varphi^i(t), \qquad i = 2, \ldots, n. \tag{30}$$

If we determine $\dot{u}$ from (29) and substitute the value obtained into (30), we have

$$\dot{y}^i = \sum_{j=2}^{n} b_j^i(t)y^j, \qquad i = 2, \ldots, n, \tag{31}$$

where

$$b_j^i(t) = a_j^i(t) - \frac{\varphi^i(t)}{\varphi^1(t)}\, a_j^1(t).$$

It must be remembered that substitution of the quantity $\dot{u}$ from (29) into (30) is possible only on the interval where the function $\varphi^1(t)$ does not vanish. Now if $\boldsymbol{\psi}(t) = \{\psi^2(t), \ldots, \psi^n(t)\}$ is any solution of (31), then, by determining u from the relation

$$\dot{u}\varphi^1(t) = \sum_{j=2}^{n} a_j^1(t)\psi^j(t)$$

by means of quadratures, we obtain a solution of the original system (2) in the form

$$\mathbf{x} = u\boldsymbol{\varphi}(t) + \boldsymbol{\psi}(t).$$

18. The linear equation of nth order. Here we shall study the nth-order linear equation

$$y^{(n)} + a_1(t)y^{(n-1)} + \cdots + a_n(t)y = b(t), \tag{1}$$

whose coefficients $a_i(t)$ and free term $b(t)$ we assume to be defined and continuous on the interval $q_1 < t < q_2$. In our study of equation (1) we shall reduce it to a normal system of linear equations by the method indicated in §4.

The fundamental system of solutions. (A) In order to reduce equation (1) to a normal linear system, we introduce the new unknown functions

$$x^1 = y, \qquad x^2 = \dot{y}, \qquad \ldots, \qquad x^n = y^{(n-1)}.$$

These new unknowns $x^1, \ldots, x^n$ satisfy the linear system [see §4, (A)]

$$\begin{aligned}
\dot{x}^1 &= x^2, \\
\dot{x}^2 &= x^3, \\
&\vdots \\
\dot{x}^{n-1} &= x^n, \\
\dot{x}^n &= -a_n(t)x^1 - a_{n-1}(t)x^2 - \cdots - a_1(t)x^n + b(t).
\end{aligned}$$

We write the system obtained in vector form

$$\dot{\mathbf{x}} = A(t)\mathbf{x} + \mathbf{b}(t), \tag{2}$$

where the matrix $A(t)$ has the form

$$\begin{pmatrix}
0 & 1 & 0 & \cdots & 0 \\
0 & 0 & 1 & \cdots & 0 \\
0 & 0 & 0 & \cdots & 0 \\
\vdots & & & & \\
0 & 0 & 0 & \cdots & 1 \\
-a_n(t) & -a_{n-1}(t) & -a_{n-2}(t) & \cdots & -a_1(t)
\end{pmatrix}, \tag{3}$$

and the vector $\mathbf{b}(t)$ is defined by the formula

$$\mathbf{b}(t) = \big(0, 0, \ldots, b(t)\big).$$

Equations (1) and (2) are equivalent; that is, to every solution $y = \psi(t)$ of (1) corresponds the solution

$$\mathbf{x} = \boldsymbol{\varphi}(t) = \big(\psi(t), \dot{\psi}(t), \ldots, \psi^{(n-1)}(t)\big)$$

of (2), and conversely, to each solution

$$\mathbf{x} = \boldsymbol{\varphi}(t) = \big(\varphi^1(t), \varphi^2(t), \ldots, \varphi^n(t)\big)$$

of (2) corresponds the solution

$$y = \varphi^1(t)$$

of (1), this correspondence being one-to-one. If the solutions $\psi(t)$ of (1) and $\boldsymbol{\varphi}(t)$ of (2) correspond in such a way, then we write

$$\psi(t) \rightleftarrows \boldsymbol{\varphi}(t).$$

In particular, it follows from the equivalence of equations (1) and (2) that the maximum interval of existence of every solution of (1) is the entire interval $q_1 < t < q_2$ (see Theorem 3), so that henceforth we may assume that every solution of (1) under consideration is defined on this interval and every value t under discussion belongs to it.

We shall study first the homogeneous equation

$$y^{(n)} + a_1(t)y^{(n-1)} + \cdots + a_n(t)y = 0. \tag{4}$$

Let

$$\dot{\mathbf{x}} = A(t)\mathbf{x} \tag{5}$$

be a system of equations in vector form corresponding to (4), where the matrix $A(t)$ is defined by formula (3).

(B) Let

$$\psi_1(t), \psi_2(t), \ldots, \psi_r(t) \tag{6}$$

be a certain system of solutions of equation (4). It can be verified directly that the function

$$\psi(t) = c^1\psi_1(t) + \cdots + c^r\psi_r(t),$$

where $c^1, \ldots, c^r$ are constants, is a solution of equation (4). The system (6) is called *linearly dependent* if there exist constants $c^1, \ldots, c^r$, not all zero, such that

$$c^1\psi_1(t) + \cdots + c^r\psi_r(t) = 0. \tag{7}$$

Thus if

$$\varphi_1(t), \ldots, \varphi_r(t) \tag{8}$$

are the solutions of (5) corresponding to the solutions (6):

$$\psi_i(t) \rightleftarrows \varphi_i(t), \qquad i = 1, \ldots, r,$$

[see (A)], then the solutions (8) are linearly dependent [see §17, (B)] if and only if the solutions (6) are linearly dependent.

We shall prove this. Let us assume that the solutions (6) are linearly dependent so that equation (7) is valid. Writing out (7) and the relations obtained from it by differentiation, we obtain

$$\begin{aligned} c^1\psi_1(t) + \cdots + c^r\psi_r(t) \quad &= 0, \\ c^1\dot{\psi}_1(t) + \cdots + c^r\dot{\psi}_r(t) \quad &= 0, \\ \vdots \qquad\qquad\quad \vdots \qquad & \\ c^1\psi_1^{(n-1)}(t) + \cdots + c^r\psi_r^{(n-1)}(t) &= 0. \end{aligned} \tag{9}$$

If we remember that

$$\varphi_i(t) = \left(\psi_i(t), \dot{\psi}_i(t), \ldots, \psi_i^{(n-1)}(t)\right),$$

we see that the vector form of (9) is

$$c^1\varphi_1(t) + \cdots + c^r\varphi_r(t) = 0, \tag{10}$$

so that a linear dependence also exists between the solutions (8). Let us assume, conversely, that the solutions (8) are linearly dependent so that the relations (10) are valid. By replacing every vector $\varphi_i(t)$ of (10) by its first component, we obtain (7), so that the solutions (6) are linearly dependent.

(C) The system of solutions

$$\psi_1(t), \ldots, \psi_n(t) \tag{11}$$

of equation (4) is called a *fundamental system* if it is linearly independent. [The notation indicates that the number of solutions of (11) is equal to the order of (4).] Thus we find that fundamental systems of solutions (4) exist and that if (11) is fundamental, then every solution (4) can be written in the form

$$\psi(t) = c^1\psi_1(t) + \cdots + c^n\psi_n(t),$$

where $c^1, \ldots, c^n$ are constants. It is now clear that in order to find all solutions of (4) we need only find its fundamental system of solutions. (The fundamental system of solutions for a linear homogeneous equation with constant coefficients was constructed in §7 and §8.)

We shall first show that the fundamental system of solutions of (4) exists. In order to do this we shall utilize the existence of a fundamental system of solutions of (5) [see §17, (C)]. Let

$$\varphi_1(t), \ldots, \varphi_n(t) \tag{12}$$

be a fundamental system of solutions of (5), and let

$$\psi_1(t), \ldots, \psi_n(t) \tag{13}$$

be the solutions of (4) which correspond to the solutions of (12):

$$\psi_i(t) \rightleftarrows \varphi_i(t), \qquad i = 1, \ldots, n$$

[see (A)]. Since the solutions (12) are linearly independent, it follows from (B) that the solutions (13) are also linearly independent, so that they constitute a fundamental system. Let us now assume that the system (11) is fundamental for (4). Let the solutions (12) correspond to the solutions (11), and let $\psi(t)$ be an arbitrary solution of (4) and $\varphi(t)$ the corresponding solution of (5). Since, by assumption, the system (11) is fundamental, i.e., linearly independent, the corresponding system (12) is also linearly independent, i.e., fundamental. Thus, it follows from (C) of §17 that

$$\varphi(t) = c^1\varphi_1(t) + \ldots + c^n\varphi_n(t).$$

If we replace each vector in this system by its first component, we obtain

$$\psi(t) = c^1\psi_1(t) + \cdots + c^n\psi_n(t),$$

so that (C) is proved.

(D) The determinant

$$W(t) = \begin{vmatrix} \psi_1(t) & \cdots & \psi_n(t) \\ \dot{\psi}_1(t) & \cdots & \dot{\psi}_n(t) \\ \vdots & & \vdots \\ \psi_1^{(n-1)}(t) & \cdots & \psi_n^{(n-1)}(t) \end{vmatrix} \tag{14}$$

is called the *Wronskian determinant* of the system of solutions (11) of equation (4). If the solutions (12) of (5) correspond to the solutions (11) [see (A)], then it is obvious that the Wronskian [see §17, (D)] of the system of solutions (12) of (5) coincides with (14). Thus, what is true for the Wronskian of (12) is also true for (14). Hence, by (D) of §17, we conclude that the determinant (14) either does not vanish at any point or else it vanishes identically; in order that the system of solutions (11) be linearly independent, i.e., fundamental, it is necessary and sufficient that (14) not vanish. From formula (16) of §17 we can obtain the Liouville formula for determinant (14)

$$W(t) = W(t_0) \exp\left(-\int_{t_0}^{t} a_1(\tau)\, d\tau\right), \tag{15}$$

since the trace, i.e., the sum of the diagonal elements of the matrix (3), is equal to $-a_1(t)$. Below, in Example 2, we shall give a simple direct proof of (15).

(E) Let

$$z^{(n)} + a_1(t)z^{(n-1)} + \cdots + a_n(t)z = b(t) \tag{16}$$

be a nonhomogeneous equation and let

$$y^{(n)} + a_1(t)y^{(n-1)} + \cdots + a_n(t)y = 0 \tag{17}$$

be its corresponding homogeneous equation. From the propositions of §6 it follows directly that if $\chi_0(t)$ is a particular solution of (16), then an arbitrary solution of (16) has the form

$$z = \psi(t) + \chi_0(t),$$

where $\psi(t)$ is a solution of (17).

(F) (*Method of variation of parameters.*) Let

$$\psi_1(t), \ldots, \psi_n(t) \tag{18}$$

be any fundamental system of solutions of equation (17). Then a solution of (16) can be obtained in the form

$$z = c^1(t)\psi_1(t) + \cdots + c^n(t)\psi_n(t), \tag{19}$$

where the functions

$$\dot{c}^1(t), \ldots, \dot{c}^n(t) \tag{20}$$

are obtained as solutions of the system of algebraic equations:

$$\begin{aligned} \psi_1(t)\dot{c}^1(t) + \cdots + \psi_n(t)\dot{c}^n(t) &= 0, \\ \dot{\psi}_1(t)\dot{c}^1(t) + \cdots + \dot{\psi}_n(t)\dot{c}^n(t) &= 0, \\ \vdots \qquad\qquad\qquad \vdots \quad & \\ \psi_1^{(n-2)}(t)\dot{c}^1(t) + \cdots + \psi_n^{(n-2)}(t)\dot{c}^n(t) &= 0, \\ \psi_1^{(n-1)}(t)\dot{c}^1(t) + \cdots + \psi_n^{(n-1)}(t)\dot{c}^n(t) &= b(t). \end{aligned} \tag{21}$$

Since the determinant of the system of equations (21) with respect to the indeterminates (20) is the Wronskian of system (18), it follows from (D) that it does not vanish for any value of t; therefore, we can find the quantities (20) from (21) and then integrate these quantities to determine the required functions

$$c^1(t), \dots, c^n(t).$$

The proof of (F) follows from (H) of §17. It can also be derived directly, which we shall proceed to do. Assuming that the quantities (20) are restricted by the conditions (21), we obtain the following relations by differentiating formula (19):

$$\begin{aligned}
z &= c^1(t)\psi_1(t) + \cdots + c^n(t)\psi_n(t),\\
\dot{z} &= c^1(t)\dot{\psi}_1(t) + \cdots + c^n(t)\dot{\psi}_n(t),\\
&\;\vdots\\
z^{(n-1)} &= c^1(t)\psi_1^{(n-1)}(t) + \cdots + c^n(t)\psi_n^{(n-1)}(t),\\
z^{(n)} &= c^1(t)\psi_1^{(n)}(t) + \cdots + c^n(t)\psi_n^{(n)}(t)\\
&\quad + \dot{c}^1(t)\psi_1^{(n-1)}(t) + \cdots + \dot{c}^n(t)\psi_n^{(n-1)}(t)\\
&= c^1(t)\psi_1^{(n)}(t) + \cdots + c^n(t)\psi_n^{(n)}(t) + b(t).
\end{aligned}$$

Substituting these expressions into (16), we obtain an identity. Thus, if the functions $c^1(t), \dots, c^n(t)$ satisfy (21), then the function (19) is a solution of (16).

Examples

1. If a nontrivial (not identically zero) solution $\psi(t)$ of equation (4) is known, then the order of this equation can be decreased by one, i.e., its solution can be reduced to the solution of a linear equation of order $n - 1$. In order to do this we make the substitution

$$y = \psi(t)v, \tag{22}$$

where v is a new unknown function. We shall now show that the result of substituting (22) into the left-hand side of (4) leads to the equation

$$b_0(t)v^{(n)} + b_1(t)v^{(n-1)} + \cdots + b_{n-1}(t)\dot{v} + b_n(t)v = 0 \tag{23}$$

for v, where

$$b_0(t) = \psi(t), \quad b_n(t) \equiv 0. \tag{24}$$

Since our whole investigation is correct for an nth-order equation in which the coefficient of the nth derivative is one, it is necessary to divide (23) by $b_0(t) = \psi(t)$, so that (4) reduces to (23) only on an interval where $\psi(t)$ does not vanish. If we set

$$\frac{b_i(t)}{\psi(t)} = l_i(t)$$

and replace the unknown function v by a new unknown function

$$w = \dot{v},$$

we arrive at an equation

$$w^{(n-1)} + l_1(t)w^{(n-2)} + \cdots + l_{n-1}(t)w = 0$$

of order $n - 1$. If a solution $\chi(t)$ of this equation exists, then we obtain the solution v of (23) by quadrature:

$$v = \int \chi(t)\, dt,$$

and the solution y of (4) may be obtained by substituting this function v into (22).

We shall prove that the substitution (22) reduces (4) to the form (23), while preserving the relations (24). Differentiating (22), we obtain

$$y^{(k)} = \psi(t)v^k + \cdots,$$

where terms containing derivatives of v of orders lower than k are not written. Therefore equation (4) takes the form (23), where $b_0(t) = \psi(t)$. Since $\psi(t)$ is a solution of (4), $v = 1$ is a solution of (23). Substituting $v = 1$ into (23), we obtain $b_n(t) = 0$. Thus the relations (24) are proved.

2. We shall prove Liouville's formula (15) for one nth-order equation without using Liouville's formula for a system [see formula (16), §17]. Here we shall use the differentiation rule for a determinant given in §17 [see §17, (F)]. Applying this differentiation rule to (14), we obtain

$$\dot{W}(t) = W_1(t) + \cdots + W_i(t) + \cdots + W_n(t),$$

where $W_i(t)$ is the Wronskian $W(t)$ in which the ith row is differentiated. If $i < n$, then, as a result of differentiating of the ith row, we obtain a row coinciding with the $(i + 1)$th row of the determinant $W(t)$, and therefore we have

$$W_1(t) = W_2(t) = \cdots = W_{n-1}(t) = 0.$$

By differentiation of the nth row we obtain the row

$$\psi_1^{(n)}(t), \ldots, \psi_n^{(n)}(t),$$

which by (4) is a linear combination of rows of $W(t)$, the nth row being taken with the coefficient $-a_1(t)$. Since the determinant $W_n(t)$ contains rows with the indices $1, \ldots, n - 1$ of the Wronskian, these rows can be rejected in the linear combination, leaving only the nth row with the coefficient $-a_1(t)$. Thus for the Wronskian we obtain the differential equation

$$\dot{W}(t) = -a_1(t)W(t),$$

the solution of which gives us Liouville's formula (15).

19. The normal linear homogeneous system with periodic coefficients. Among linear equations with variable coefficients a particularly important role is played by equations with *periodic* coefficients. The present section is devoted to an account of some of the properties of normal linear homogeneous systems of differential equations with periodic coefficients. Lyapunov's theorem is the most essential of these properties. The proof of Lyapunov's theorem is given here, but it is less elementary than any of the previous developments in this book. It is based on the *matrix calculus*, the necessary details of which are presented in the last chapter of the book.

Let

$$\dot{X} = A(t)X \tag{1}$$

be a normal linear homogeneous system of equations written in matrix form [see §17, (I)]. We shall assume that the coefficients of this system are periodic functions of time t with period τ, i.e., the matrix $A(t)$ satisfies the condition

$$A(t + \tau) = A(t).$$

(A) For any (matrix) solution

$$X = \Phi(t) \tag{2}$$

of equation (1) [see §17, (I)], a constant (nonsingular) matrix C can be found such that

$$\Phi(t + \tau) = \Phi(t)C.$$

We shall call the matrix C the *fundamental matrix* for the solution (2). If $X = \hat{\Phi}(t)$ is any other solution of (1), and $\hat{C}$ is its fundamental matrix, then we have

$$\hat{C} = P^{-1}CP, \tag{3}$$

where P is a certain nonsingular constant matrix.

To prove the existence of the matrix C we observe that, together with the solution (2), the matrix $\Phi(t + \tau)$ is also a solution of equation (1). In fact,

$$\dot{\Phi}(t + \tau) = A(t + \tau)\Phi(t + \tau) = A(t)\Phi(t + \tau).$$

Thus by formula (22) of §17, we have

$$\Phi(t + \tau) = \Phi(t)C,$$

where C is a constant matrix.

To prove formula (3) we shall also utilize formula (22) of §17. Because $\hat{\Phi}(t)$ is a solution of equation (1), it follows from the formula above that

$$\hat{\Phi}(t) = \Phi(t)P.$$

Hence

$$\hat{\Phi}(t + \tau) = \Phi(t + \tau)P = \Phi(t)CP = \hat{\Phi}(t)P^{-1}CP,$$

which also gives (3).

(B) Equation (1) and the equation

$$\dot{Y} = B(t)Y \tag{4}$$

with a periodic matrix $B(t)$ of the same period τ as the matrix $A(t)$ are called *equivalent* if there exists a linear transformation

$$Y = S(t)X$$

[see §17, (J)] with a periodic matrix $S(t)$ of period τ which transforms (1) into (4). Hence formulas (1) and (4) are equivalent if and only if there exist solutions $X = \Phi(t)$ and $Y = \Psi(t)$ of these equations with the same fundamental matrix.

Let us prove this assertion. Let us first assume that (1) and (4) are equivalent. Let $X = \Phi(t)$ be an arbitrary solution of (1) with a fundamental C; then $Y = \Psi(t) = S(t)\Phi(t)$ is a solution of (4), and we have

$$\Psi(t + \tau) = S(t + \tau)\Phi(t + \tau) = S(t)\Phi(t + \tau) = S(t)\Phi(t)C = \Psi(t)C.$$

Thus the fundamental matrix C of solution $\Phi(t)$ is also fundamental for the solution $\Psi(t)$.

We shall now assume that the solutions $X = \Phi(t)$ and $Y = \Psi(t)$ of (1) and (4) exist with the same fundamental matrix C; then we have

$$\Phi(t + \tau) = \Phi(t)C, \qquad \Psi(t + \tau) = \Psi(t)C.$$

Dividing the second of these relations by the first, we obtain

$$\Psi(t + \tau)\Phi^{-1}(t + \tau) = \Psi(t)\Phi^{-1}(t).$$

Thus the matrix

$$S(t) = \Psi(t)\Phi^{-1}(t)$$

is periodic with period τ, and we have

$$\Psi(t) = S(t)\Phi(t). \tag{5}$$

Since each of the solutions $\Phi(t)$ and $\Psi(t)$ is uniquely determined [see §17(E)], it follows from (5) that equation (4) is obtained from (1) by means of a transformation with matrix $S(t)$.

As is evident from (A) and (B), to every equation of the form (1) which is considered unique up to an equivalence [see (B)], there corresponds a matrix C determined uniquely up to a transformation [see (3)]. In addition, the set of all invariants of the matrix C with respect to transformations of the form (3) constitutes a *complete system of invariants* of equation (1), which is uniquely determined up to an equivalence.

It should be noted that everything said in (A) and (B) is true both for the case of real matrices and for the case of complex matrices. In the following important theorem of Lyapunov (Theorem 12), we shall distinguish between real and complex cases.

THEOREM 12. Any equation (1) is equivalent [see(B)] to the equation

$$\dot{Y} = BY,$$

where B is a constant matrix. (The matrix B, generally speaking, is complex.) If in equation (1) the matrix $A(t)$ is real with period τ then this equation, considered as periodic with period 2τ, is equivalent to the equation

$$\dot{Y} = B_1 Y,$$

where the matrix B_1 is constant and real, and the matrix $S(t)$ taking (1) into $\dot{Y} = B_1 Y$ is also real.

We shall preface the proof of Theorem 12 by the following proposition:

(C) Let

$$\dot{Y} = BY \tag{6}$$

be a system of linear homogeneous equations with constant coefficients written in matrix form. Here the matrix B is constant. It turns out that the matrix

$$Y = e^{tB} \tag{7}$$

[see (D), §33] is a solution of (6).

To prove that (7) is a solution of (6), we shall write out the function e^{tB} explicitly in the form

$$e^{tB} = E + tB + \frac{t^2}{2!}B^2 + \frac{t^3}{3!}B^3 + \cdots.$$

Hence for the derivative $(d/dt)e^{tB}$ we obtain

$$\frac{d}{dt}\,e^{tB} = B\left(E + tB + \frac{t^2}{2!}\,B^2 + \cdots\right) = Be^{tB}.$$

Proof of Theorem 12. Let C be the fundamental matrix of some solution $X = \Phi(t)$ of equation (1). By (D) of §33 there exists a matrix B which satisfies the condition

$$e^{\tau B} = C.$$

We shall prove that equations (1) and

$$\dot{Y} = BY \tag{8}$$

are equivalent. In fact, proposition (B) implies that the matrix $Y = e^{tB}$ is a solution of (8). Thus, if (8) is considered an equation with periodic coefficients with period τ, then the fundamental matrix of the solution $Y = e^{tB}$ is C, that is [see formula (20) of §33]

$$e^{(t+\tau)B} = e^{tB}e^{\tau B} = e^{tB}C.$$

Since the fundamental matrices of the solutions of equation (1) and of equation (8) coincide, these equations are equivalent [see (B)], and the first part of Theorem 12 is proved.

We shall now assume that $A(t)$ is a real matrix, $\Phi(t)$ is a certain real solution of equation (1), and C is the fundamental matrix of this solution, so that

$$\Phi(t + \tau) = \Phi(t)C. \tag{9}$$

Since $\Phi(t)$ is a real, the matrix C is also real. From (9) it follows that

$$\Phi(t + 2\tau) = \Phi(t + \tau)C = \Phi(t)C^2. \tag{10}$$

By (D) of §33 there exists a real matrix B_1 which satisfies the condition

$$e^{2\tau B_1} = C^2.$$

We shall prove that equation (1) and the equation

$$\dot{Y} = B_1 Y, \tag{11}$$

considered as equations with period 2τ, are equivalent. Actually, the matrix e^{tB_1} is a solution of (11), so that if (11) is considered as an equation with periodic coefficients of period 2τ, then the fundamental matrix of the solution $Y = e^{tB_1}$ is C^2. Since the fundamental matrices of the solutions of (1) and (11) coincide [see (10)], these equations must be equivalent. Theorem 12 is thus proved.

(D) Let C be an arbitrary nth-order square matrix, of which the absolute value of all the eigenvalues is smaller than a certain positive number ρ. The elements of the matrix C^m, where m is an integer, will be denoted by ${}^m c_j^i$, so that $C^m = ({}^m c_j^i)$. Then there exists a positive number r, which does not depend on i, j, m, such that

$$|{}^m c_j^i| < r\rho^m. \tag{12}$$

Hence, in particular, it follows that for an arbitrary vector $\mathbf{x}$ the inequality

$$|C^m \mathbf{x}| \leq n^2 r \rho^m |\mathbf{x}| \tag{13}$$

is valid. In order to obtain the bound (12) we consider the series

$$f(z) = 1 + \frac{z}{\rho} + \frac{z^2}{\rho^2} + \cdots + \frac{z^m}{\rho^m} + \cdots,$$

whose radius of convergence is clearly ρ. From Theorem 27 (see §33) it follows that the matrix series

$$f(C) = E + \frac{C}{\rho} + \frac{C^2}{\rho^2} + \cdots + \frac{C^m}{\rho^m} + \cdots$$

is convergent and, in particular, the numerical series

$$\delta_j^i + \frac{{}^1 c_j^i}{\rho} + \frac{{}^2 c_j^i}{\rho^2} + \cdots + \frac{{}^m c_j^i}{\rho^m} + \cdots$$

is also convergent. Since this series converges, its terms do not exceed a certain number r which may be chosen uniformly for all pairs of numbers (i, j), so that the bound (12) holds.

(E) Let

$$\dot{\mathbf{x}} = A(t)\mathbf{x} \tag{14}$$

be the vector form of the matrix equation (1), and let C be the fundamental matrix of some solution $\Phi(t)$ of (1). An eigenvalue λ of multiplicity k of the matrix C is called a *characteristic number* of multiplicity k of (1) and (14). Since, up to a linear transformation, the matrix C does not depend on a random selection of the solution $\Phi(t)$ of (1) [see (A)], the characteristic numbers of (14) and their multiplicities are determined here in an invariant way. If λ is a characteristic number of multiplicity k of (14), then the number $(1/\tau) \ln \lambda$ is called the *characteristic exponent* of multiplicity k of (14). Let us assume that all real parts of the characteristic exponents of (14) are smaller than some number γ; then there exists a positive number R such that for any solution $\boldsymbol{\varphi}(t)$ of (14) the bound

$$|\boldsymbol{\varphi}(t)| \leq R|\boldsymbol{\varphi}(0)|e^{\gamma t} \tag{15}$$

holds for $t \geq 0$.

We shall prove (15). Let $\Phi(t)$ be a solution of (1) with initial condition $\Phi(0) = E$; then any solution $\boldsymbol{\varphi}(t)$ of (14) may be written in the form

$$\boldsymbol{\varphi}(t) = \Phi(t)\boldsymbol{\varphi}(0). \tag{16}$$

This may be verified by substituting (16) into (14). Furthermore we have

$$\begin{aligned} \Phi(t+\tau) &= \Phi(t)C, \\ \Phi(t+2\tau) &= \Phi(t)C^2, \\ &\vdots \\ \Phi(t+m\tau) &= \Phi(t)C^m, \quad \ldots . \end{aligned} \tag{17}$$

Since the elements of the matrix $\Phi(t_1)$ are bounded on the interval $0 \leq t_1 \leq \tau$, there exists a positive number σ such that

$$|\Phi(t_1)\mathbf{x}| \leq \sigma|\mathbf{x}| \qquad \text{for} \qquad 0 \leq t_1 \leq \tau. \tag{18}$$

Since, further, all the eigenvalues of the matrix C are smaller than $e^{\tau\gamma}$ in absolute value, it follows from (13) that the bound

$$|C^m\mathbf{x}| \leq n^2 r e^{\tau m \gamma}|\mathbf{x}| \tag{19}$$

holds for an arbitrary vector $\mathbf{x}$. Now let t be any positive number; we can find a nonnegative integer m such that $t = m\tau + t_1$, $0 \leq t_1 < \tau$. By virtue of (16) and (17) we have

$$\boldsymbol{\varphi}(t) = \Phi(m\tau + t_1)\boldsymbol{\varphi}(0) = \Phi(t_1)C^m\boldsymbol{\varphi}(0).$$

Hence in accordance with (18) and (19) we obtain

$$|\boldsymbol{\varphi}(t)| \leq \sigma n^2 r e^{\tau m\gamma}|\boldsymbol{\varphi}(0)|.$$

Since for $0 \leq t_1 \leq \tau$ the number $e^{t_1\gamma}$ is not less than a certain constant $c > 0$, the last inequality can be written in the form

$$|\boldsymbol{\varphi}(t)| \leq \frac{\sigma n^2 r}{c}\, e^{\gamma t}|\boldsymbol{\varphi}(0)|.$$

This yields the bound (15).

CHAPTER 4

EXISTENCE THEOREMS

Here we shall first prove the existence and uniqueness Theorems 1, 2, and 3 which were formulated earlier. Further study is given to the question of the dependence of a solution on the initial values and on the parameters, if the latter actually appear in the equations. We shall investigate the dependence of a solution with fixed initial values on the parameters; then, by a quite simple technique, we shall convert the initial values into parameters. Thus the problem will be reduced to one of describing the dependence of a solution on the parameters. For both the case of initial values and the case of parameters, we shall prove that the dependence of the solution on these variables is continuous and that the solution is differentiable with respect to these variables. In both cases it is necessary to distinguish between *local* theorems and what we shall call *integral* theorems. In the first case, one can assert that certain properties of a solution hold on the time interval $|t - t_0| < r$, where r is a positive number which is generally speaking "small." In the second case, it is assumed that for fixed values of the parameter or for fixed initial values there exists a solution defined on the time interval $r_1 \leq t \leq r_2$ which can be "large." It is then asserted that for initial values, or for values of the parameter which are very nearly fixed, a solution exists on the entire interval $r_1 \leq t \leq r_2$ which is continuous for all values of these variables or differentiable with respect to all the variables.

In addition to this body of material, we include in the present chapter a section on the first integrals of a system of ordinary differential equations and a study of a linear partial differential equation which is closely related to the concept of a first integral. The results of this section will not be used further in this book.

20. Proof of the existence and uniqueness theorem for one equation. In this section we shall give a proof of Theorem 1, the theorem of existence and uniqueness formulated in §1, for the first-order equation

$$\dot{x} = f(t, x), \tag{1}$$

whose right-hand side is defined and continuous, together with the partial derivative $\partial f/\partial x$, in a certain domain Γ of the tx-plane P. A more complex and general form of the proof of Theorem 1 is found in the proof of Theorem 2 in the following section. In presenting the proof first for the case of one equation, it is our aim to bring out the basic *ideas* of this type of proof, which in the more general application is encumbered by details of a secondary character. We shall prove Theorems 1 and 2 by the *method of successive approximations* developed by Picard and applied in analysis for the proofs of many existence theorems. This method can also be used for the approximate calculation of a solution and is therefore of great practical value. In certain cases the method of successive approximations can be interpreted as a *method of contraction mappings*. The proof will be carried out with such an interpretation in order to show the relation between these two methods. The difference between these two methods will be brought out in the proof of Theorem 3.

Basic ideas of the proof. The first step in proving Theorem 1 by the method of successive approximations is the transition from the differential equation to an *integral* equation, which we shall formulate as a separate proposition.

(A) Let $x = \varphi(t)$ be a certain solution of equation (1) defined on the interval $r_1 < t < r_2$, so that the identity

$$\dot{\varphi}(t) = f(t, \varphi(t)), \tag{2}$$

is satisfied, and let

$$\varphi(t_0) = x_0 \tag{3}$$

be a certain initial condition which this solution satisfies. We then find that for the function $\varphi(t)$ the integral identity

$$\varphi(t) = x_0 + \int_{t_0}^{t} f(\tau, \varphi(\tau))\, d\tau \tag{4}$$

is satisfied on the entire interval $r_1 < t < r_2$. Conversely, if a continuous function $\varphi(t)$ satisfies the identity (4) on the interval $r_1 < t < r_2$, then the differentiable function $x = \varphi(t)$ is a solution of the equation (1) and satisfies the initial condition (3). In other words, the integral equation (4) is *equivalent* to the differential equation (2), together with the initial condition (3).

We shall prove this. Let us assume first that the relation (4) is satisfied. If we replace the variable t by its value t_0, we obtain $\varphi(t_0) = x_0$, so that (3) follows from (4). Moreover, the right-hand side of (4) is obviously differentiable with respect to t, whence the left-hand side is also differentiable with respect to t. By differentiating (4) we obtain the identity (2).

Let us now assume that the relations (2) and (3) are satisfied. By integrating (2) from t_0 to t we obtain

$$\varphi(t) - \varphi(t_0) = \int_{t_0}^{t} f(\tau, \varphi(\tau))\, d\tau.$$

We then obtain equation (4) from the last equality by virtue of (3). Thus proposition (A) is proved.

We shall now introduce a certain notation which we shall use below in the proof of Theorem 1.

(B) Let $x = \varphi(t)$ be a continuous function defined on some interval $r_1 \leq t \leq r_2$ such that its graph is contained completely in Γ, and let t_0 be a certain point of the interval $r_1 \leq t \leq r_2$. Then, if we use the right-hand side of (4), we can set up a correspondence between the function $\varphi(t)$ and the function $\varphi^*(t)$, which is also defined on $r_1 \leq t \leq r_2$, by means of the equality

$$\varphi^*(t) = x_0 + \int_{t_0}^{t} f(\tau, \varphi(\tau))\, d\tau. \tag{5}$$

(The graph of the function $\varphi^*(t)$, of course, need not pass through Γ.) Thus the right-hand side of (4) can be regarded as an *operator* establishing the correspondence between the functions φ^* and φ. Denoting this operator by the single letter A, we write (5) in the form

$$\varphi^* = A\varphi. \tag{6}$$

By using the operator A, the integral equation (4) can be written in the form

$$\varphi = A\varphi. \tag{7}$$

(C) Let $\varphi(t)$ be a certain continuous function defined on the interval $r_1 \leq t \leq r_2$. The maximum modulus of this function,

$$\|\varphi\| = \max_{r_1 \leq t \leq r_2} |\varphi(t)|,$$

will be called its *norm* $\|\varphi\|$. If $\psi(t)$ and $\chi(t)$ are two continuous functions defined on the interval $r_1 \leq t \leq r_2$, then the norm $\|\psi - \chi\|$ of their difference $\psi(t) - \chi(t)$ is a nonnegative number which expresses the extent to which these functions differ: if the number $\|\psi - \chi\|$ is small, then the functions ψ and χ are "close" to each other. The equality $\|\psi - \chi\| = 0$ is valid if and only if ψ and χ coincide identically. The notion of uniform convergence, which is familiar from a course in analysis, may be formulated easily in terms of the norm. Let

$$\varphi_0(t), \quad \varphi_1(t), \ldots, \varphi_i(t), \ldots \tag{8}$$

be a sequence of continuous functions defined on the interval $r_1 \leq t \leq r_2$. The sequence (8) converges uniformly to a function φ, defined on the same interval $r_1 \leq t \leq r_2$, if

$$\lim_{i \to \infty} \|\varphi - \varphi_i\| = 0.$$

In order that the sequence (8) converge uniformly, it is sufficient that the inequalities

$$\|\varphi_{i+1} - \varphi_i\| \leq a_i$$

be valid, where the numbers $a_0, a_1, \ldots, a_i, \ldots$ form a convergent series.

Before proving Theorem 1 in detail, we shall outline briefly the method of successive approximations, which is applicable to the solution of equation (7). We form the sequence

$$\varphi_0, \varphi_1, \ldots, \varphi_i, \ldots \tag{9}$$

of continuous functions which are defined on a certain interval $r_1 \leq t \leq r_2$ containing the point t_0. Every function of the sequence (9) is determined by the preceding one by means of the formula

$$\varphi_{i+1} = A\varphi_i, \qquad i = 0, 1, 2, \ldots. \tag{10}$$

If the graph of the function φ_i passes through Γ, then φ_{i+1} is defined by (10), but in order to define the next function φ_{i+2} it is necessary that the graph of φ_{i+1} pass through Γ. This, as we shall show, can be achieved by selecting a sufficiently short interval $r_1 \leq t \leq r_2$. Furthermore, by decreasing the length of the interval $r_1 \leq t \leq r_2$, we may also show that the elements of the sequence (9) satisfy the inequalities

$$\|\varphi_{i+1} - \varphi_i\| \leq k\|\varphi_i - \varphi_{i-1}\|, \qquad i = 1, 2, \ldots, \tag{11}$$

where $0 < k < 1$. From (11) follow the inequalities

$$\|\varphi_{i+1} - \varphi_i\| \leq \|\varphi_1 - \varphi_0\| \cdot k^i, \qquad i = 1, 2, \ldots,$$

so that the sequence (9) is uniformly convergent [see (C)]. Furthermore, it is also easily established that the limit φ of (9) satisfies equation (7).

The same construction can be described in a somewhat different way by the method of contraction mappings. We shall select a certain family Ω of functions defined on the interval $r_1 \leq t \leq r_2$ (where $r_1 < t_0 < r_2$) so that the graphs of these functions pass through the domain Γ. We shall assume, in addition, that with respect to the operator A the family Ω satisfies the following two conditions: (a) whenever A is applied to any function of Ω, we again obtain a function of Ω; (b) there exists a number k, $0 < k < 1$, such that for two arbitrary functions ψ and χ of Ω the inequality

$$\|A\psi - A\chi\| \leq k\|\psi - \chi\|$$

is satisfied. In this sense the "mapping" A is a contraction mapping (it would be more correct to say "contracting").

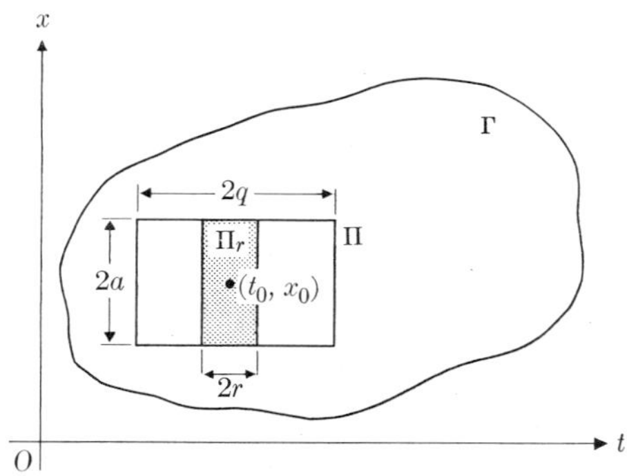

FIGURE 44

It is easy to see that, if the conditions formulated are satisfied for the family Ω, then, by starting from an arbitrary function φ_0 of Ω, we obtain by the induction formula (10) an infinite sequence (9) satisfying (11) and, as was noted above, converging uniformly to a solution φ of (7).

We shall now turn to the proof of Theorem 1 on the basis of these remarks.

Proof of Theorem 1. The initial values t_0 and x_0 of the solution of equation (1) to be found are coordinates of the point (t_0, x_0) in Γ. First of all, we choose any rectangle Π with its center at (t_0, x_0), its sides parallel to the axes, and with its boundary entirely contained in Γ (Fig. 44). We shall denote the length of a horizontal side of Π (parallel to the t axis) by $2q$ and the length of a vertical side by $2a$. Thus the point (t, x) belongs to Π if and only if the equalities

$$|t - t_0| \leq q, \qquad |x - x_0| \leq a \tag{12}$$

are satisfied. Since Π is a closed set which is contained in Γ, the continuous functions $f(t, x)$ and $\partial f(t, x)/\partial x$ are bounded on Γ and therefore two positive numbers M and K exist such that for any t and x satisfying (12), the inequalities

$$|f(t, x)| \leq M, \qquad \left|\frac{\partial f(t, x)}{\partial x}\right| \leq K \tag{13}$$

are valid.

Along with the rectangle Π we shall consider a "narrower" rectangle Π_r, defined by the inequalities

$$|t - t_0| \leq r, \qquad |x - x_0| \leq a,$$

where

$$r \leq q \tag{14}$$

(see Fig. 44). We shall determine the number r more precisely. We denote by Ω_r the family of all continuous functions defined on the interval $|t - t_0| \leq r$ whose graphs pass through Π_r. Thus the function φ, which is defined on $|t - t_0| \leq r$, belongs to Ω_r if and only if the inequality

$$|\varphi(t) - x_0| \leq a \tag{15}$$

is satisfied for any t belonging to this interval.

We shall now endeavor to select the number r in such a way that the following two conditions will be satisfied:

(a) If the function φ belongs to Ω_r, then the function $\varphi^* = A\varphi$ [see (5) and (6)] also belongs to Ω_r.

(b) There exists a number k, $0 < k < 1$, such that the inequality

$$\|A\psi - A\chi\| \leq k\|\psi - \chi\| \tag{16}$$

is valid for any two functions ψ and χ of Ω_r.

Let us consider the condition (a). In order that $\varphi^* = A\varphi$ belong to Ω_r, it is necessary and sufficient that the inequality

$$|\varphi^*(t) - x_0| \leq a$$

be satisfied for $|t - t_0| \leq r$. By virtue of (5) and (13) we have

$$|\varphi^*(t) - x_0| = \left|\int_{t_0}^{t} f(\tau, \varphi(\tau))\, d\tau\right| \leq Mr,$$

from which it is evident that condition (a) is satisfied for

$$r \leq \frac{a}{M}\cdot \tag{17}$$

Let us now consider condition (b). We have

$$\psi^*(t) = x_0 + \int_{t_0}^{t} f(\tau, \psi(\tau))\, d\tau,$$

$$\chi^*(t) = x_0 + \int_{t_0}^{t} f(\tau, \chi(\tau))\, d\tau.$$

Subtracting the second equality from the first, we obtain

$$\begin{aligned}|\psi^*(t) - \chi^*(t)| &= \left|\int_{t_0}^{t} [f(\tau, \psi(\tau)) - f(\tau, \chi(\tau))]\, d\tau\right| \\ &\leq \left|\int_{t_0}^{t} |f(\tau, \psi(\tau)) - f(\tau, \chi(\tau))\, d\tau\right|\cdot\end{aligned} \tag{18}$$

Let us now estimate the size of the last integrand by means of Lagrange's formula and the second of the inequalities (13):

$$|f(\tau, \psi(\tau)) - f(\tau, \chi(\tau))| = \left|\frac{\partial f(\tau, \theta)}{\partial x} (\psi(\tau) - \chi(\tau))\right| \leq K \cdot |\psi(\tau) - \chi(\tau)|; \tag{19}$$

here θ is a number between $\psi(\tau)$ and $\chi(\tau)$ and, consequently, it satisfies the inequality $|\theta - x_0| \leq a$. From (18) and (19) we have

$$\|A\psi - A\chi\| = \|\psi^* - \chi^*\| \leq Kr\|\psi - \chi\|.$$

Thus (b) is satisfied if the number $k = Kr$ is less than one, i.e., if

$$r < \frac{1}{K}\cdot \tag{20}$$

Thus, if r satisfies (14), (17), and (20), then the conditions (a) and (b) are satisfied for the family Ω_r. In what follows we shall assume that r has been chosen in such a way that (14,) (17), and (20) are satisfied.

We shall now construct a sequence

$$\varphi_0, \varphi_1, \ldots, \varphi_i, \ldots \tag{21}$$

of functions defined on the interval $|t - t_0| \leq r$ by setting

$$\varphi_0(t) \equiv x_0, \tag{22}$$

$$\varphi_{i+1} = A\varphi_i, \qquad i = 0, 1, 2, \ldots. \tag{23}$$

Since the function (22) belongs to the family Ω_r, all functions of (21) also belong to the same family [see condition (a)]. In addition, we have

$$\|\varphi_1 - \varphi_0\| = \max_{|t-t_0|\leq r} |\varphi_1 - x_0| \leq a$$

[see (15)]. By virtue of (16) we obtain

$$\|\varphi_{i+1} - \varphi_i\| = \|A\varphi_i - A\varphi_{i-1}\| \leq k\,\|\varphi_i - \varphi_{i-1}\|,$$

whence

$$\|\varphi_{i+1} - \varphi_i\| \leq ak^i, \qquad i = 0, 1, 2, \ldots.$$

Thus by (C) the sequence (21) converges uniformly on the interval $|t - t_0| \leq r$ to a continuous function φ. Since all functions of (21) belong to Ω_r, the function φ also belongs to Ω_r [see (15)]. We now show that the function φ satisfies (7). To do this, we note that the sequence

$$A\varphi_0, A\varphi_1, \ldots, A\varphi_i, \ldots$$

converges uniformly to the function $A\varphi$; actually, we have

$$\|A\varphi - A\varphi_i\| \leq k\|\varphi - \varphi_i\|.$$

Taking the limit as $i \to \infty$ in relation (23) we obtain

$$\varphi = A\varphi.$$

Thus, *we have proved that a solution* $x = \varphi(t)$ *of equation* (1) *exists which satisfies the initial condition* (3); *moreover, we have established that the solution* $x = \varphi(t)$ *is defined on the interval* $|t - t_0| < r$, *where* r *is an arbitrary number which satisfies* (14), (17), *and* (20).

We now proceed to the proof of uniqueness. Let $x = \psi(t)$ and $x = \chi(t)$ be two solutions of equation (1) with common initial values t_0, x_0, and let $r_1 < t < r_2$ be the interval which is the intersection of the intervals of existence of the solutions ψ and χ; it is clear that $r_1 < t_0 < r_2$. Let us denote by N the set of all such points of the interval $r_1 < t < r_2$ at which the solutions ψ and χ coincide. The set N is nonempty since it contains the point t_0. We shall show that the set N is *open* and that it is *closed* on the interval $r_1 < t < r_2$.

We shall prove first that the set N is open. Let t_1 be an arbitrary point of N; since the solutions $\psi(t)$ and $\chi(t)$ coincide at this point, that is, since $\psi(t_1) = \chi(t_1) = x_1$, the point (t_1, x_1) can be regarded as the common initial value for both solutions. In this sense (t_1, x_1) is no different from (t_0, x_0), and therefore we shall preserve for (t_1, x_1) the notation (t_0, x_0); this will allow us to keep the previous notation. Going over from the differential equation (1) to the integral equation (4), we obtain for both $\psi(t)$ and $\chi(t)$ integral equalities which in terms of operators can be written in the form

$$\psi = A\psi, \qquad \chi = A\chi. \tag{24}$$

As before, we now choose first a rectangle Π in Γ with its center at the point (t_0, x_0), and then a rectangle Π_r in such a way that the number r will satisfy still another condition in addition to the inequalities (14), (17), and (20): namely that the functions ψ and χ are defined for $|t - t_0| \leq r$ and satisfy the inequalities

$$|\psi(t) - x_0| \leq a, \qquad |\chi(t) - x_0| \leq a;$$

this is possible since $\psi(t)$ and $\chi(t)$ are continuous. Then on the interval $|t - t_0| \leq r$ the functions $\psi(t)$ and $\chi(t)$ belong to the family Ω_r, and consequently, by (16) and (24), we obtain

$$\|\psi - \chi\| = \|A\psi - A\chi\| \leq k\|\psi - \chi\|,$$

which is possible only when $\|\psi - \chi\| = 0$, i.e., when ψ and χ coincide on the interval $|t - t_0| \leq r$. Consequently, we have established that for every point t_0 of N, the set N also contains some interval $|t - t_0| < r$ with its center at t_0, so that the set N is open.

We shall now prove that the set N is closed on the interval $r_1 < t < r_2$. This means that we must show that if the sequence $t_1, t_2, \ldots, t_i, \ldots$ of points of N converges to some point $\bar{t}$ of the interval $r_1 < t < r_2$, then $\bar{t}$ also belongs to N. We have

$$\psi(t_i) = \chi(t_i), \qquad i = 1, 2, \ldots.$$

Since $\psi(t)$ and $\chi(t)$ are defined and continuous at $\bar{t}$, we may pass to the limit as $i \to \infty$ to obtain $\psi(\bar{t}) = \chi(\bar{t})$, so that the point $\bar{t}$ also belongs to the set N.

In proposition (D) below it will be shown that since the nonempty set N is both closed and open on a certain interval, it must coincide with this interval. Thus we shall have established that the solutions $\psi(t)$ and $\chi(t)$ coincide on the entire interval $r_1 < t < r_2$. Theorem 1 will thus be proved.

(D) Let $r_1 < t < r_2$ be any interval of real numbers and N a nonempty subset which is both open and closed on the interval $r_1 < t < r_2$; then the set N coincides with the interval $r_1 < t < r_2$.

Since the set N is nonempty, it contains at least one point t_0. Let us assume that there exists on the interval $r_1 < t < r_2$ a point t_1 which does not belong to N. To be definite, we assume that $t_1 > t_0$. We denote by N_1 the set of all points t of the set N which satisfy the inequality $t \leq t_1$, and let $\bar{t}$ be the least upper bound of the set N_1. Since N_1 is obviously closed on the interval $r_1 < t < r_2$, the point $\bar{t}$ belongs to N_1 and therefore cannot coincide with t_1, so that $\bar{t} < t_1$. But since N is open, it must contain an entire interval with the point $\bar{t}$ and therefore $\bar{t}$ cannot be the least upper bound of the set N_1. This contradiction proves proposition (D).

Example

Let us find a solution by the method of successive approximations for the simple equation

$$\dot{x} = x.$$

We shall seek the solution with initial values

$$t_0 = 0, \qquad x_0 = 1.$$

The corresponding integral equation can be written in the form

$$\varphi(t) = 1 + \int_0^t \varphi(\tau)\, d\tau.$$

If we form the sequence

$$\varphi_0, \varphi_1, \ldots, \varphi_i, \ldots,$$

we have

$$\varphi_0(t) \equiv 1,$$

$$\varphi_1(t) = 1 + \int_0^t d\tau = 1 + t,$$

$$\varphi_2(t) = 1 + \int_0^t (1 + \tau)\, d\tau = 1 + t + \frac{1}{2!} t^2,$$

$$\varphi_3(t) = 1 + \int_0^t \left(1 + \tau + \frac{1}{2!}\tau^2\right) d\tau = 1 + t + \frac{1}{2!} t^2 + \frac{1}{3!} t^3,$$

$$\vdots$$

$$\varphi_n(t) = 1 + t + \frac{1}{2!} t^3 + \frac{1}{3!} t^3 + \cdots + \frac{1}{n!} t^n.$$

The limit of this sequence (which is uniformly convergent on any interval on the number axis) is the function $\varphi(t) = e^t$.

21. Proof of the existence and uniqueness theorem for a normal system of equations. Here we shall prove the theorem of existence and uniqueness, Theorem 2 which was formulated in §3 for a normal system of equations

$$\dot{x}^i = f^i(t, x^1, x^2, \ldots, x^n), \qquad i = 1, \ldots, n; \tag{1}$$

the right-hand sides $f^i(t, x^1, x^2, \ldots, x^n)$ of (1), together with the partial derivatives $\partial f^i(t, x^1, \ldots, x^n)/\partial x^j$, $i, j = 1, \ldots, n$, are defined and continuous in a domain Γ of the space of the variables $t, x^1, \ldots, x^n$. If we set

$$\mathbf{x} = (x^1, \ldots, x^n), \tag{2}$$

$$\mathbf{f}(t, \mathbf{x}) = \big(f^1(t, \mathbf{x}), f^2(t, \mathbf{x}), \ldots, f^n(t, \mathbf{x})\big),$$

we may rewrite (1) in the vector form [compare §14, (A)]

$$\dot{\mathbf{x}} = \mathbf{f}(t, \mathbf{x}). \tag{3}$$

We shall carry out the proof in vector form by the method of successive approximations; this will be an almost verbatim repetition of the proof of

Theorem 1 given in the preceding section. In addition to the proof of Theorem 2, we shall also present here the proof of Theorem 3 by a somewhat modified form of the method of successive approximations.

Auxiliary propositions. In order that we may use vector notation freely, we shall establish first of all certain natural definitions and simple inequalities for vectors and vector functions.

The *length* or *modulus* $|\mathbf{x}|$ of the vector (2), as we know, is defined by the formula

$$|\mathbf{x}| = +\sqrt{(x^1)^2 + \cdots + (x^n)^2}.$$

It is an easily proved fact that if $\mathbf{x}$ and $\mathbf{y}$ are two vectors, then the inequality

$$|\mathbf{x} + \mathbf{y}| \leq |\mathbf{x}| + |\mathbf{y}|$$

is valid. From this inequality follows a similar inequality for an arbitrary number of vectors $\mathbf{x}_1, \ldots, \mathbf{x}_l$, namely,

$$|\mathbf{x}_1 + \cdots + \mathbf{x}_l| \leq |\mathbf{x}_1| + \cdots + |\mathbf{x}_l|. \tag{4}$$

Let $\boldsymbol{\varphi}(t) = (\varphi^1(t), \ldots, \varphi^n(t))$ be a continuous vector function of a real variable t, i.e., a vector whose coordinates are continuous functions of t. If the function $\boldsymbol{\varphi}(t)$ is defined on the interval $r_1 < t < r_2$, then for $r_1 < t_0 < r_2$ on the same interval we can define the vector function

$$\boldsymbol{\psi}(t) = \int_{t_0}^{t} \boldsymbol{\varphi}(\tau)\, d\tau,$$

whose components $\psi^1(t), \ldots, \psi^n(t)$ are defined by the formulas

$$\psi^i(t) = \int_{t_0}^{t} \varphi^i(\tau)\, d\tau;$$

here the inequality

$$\left| \int_{t_0}^{t} \boldsymbol{\varphi}(\tau)\, d\tau \right| \leq \left| \int_{t_0}^{t} |\boldsymbol{\varphi}(\tau)|\, d\tau \right| \tag{5}$$

is valid. To prove this inequality we shall divide the interval of integration into m equal parts by setting

$$\Delta = \frac{t - t_0}{m}; \qquad t_k = t_0 + k\Delta, \qquad k = 1, \ldots, m,$$

where the number Δ will be positive for $t > t_0$ and negative for $t < t_0$. Then, by the definition of the integral of a vector function and by virtue of (4), we have

$$\left|\int_{t_0}^{t} \varphi(\tau)\, d\tau\right| = \left|\lim_{m\to\infty} \sum_{k=1}^{m} \varphi(t_k)\Delta\right| \le \lim_{m\to\infty} \sum_{k=1}^{m} |\varphi(t_k)| \cdot |\Delta|$$
$$= \left|\int_{t_0}^{t} |\varphi(\tau)|\, d\tau\right| .$$

We shall establish an inequality for the *vector function*

$$\mathbf{g}(\mathbf{x}) = (g^1(x^1, \ldots, x^n), \ldots, g^n(x^1, \ldots, x^n))$$

of the *vector variable* $\mathbf{x}$ defined on a convex set Δ of the space of variables $x^1, \ldots, x^n$. We shall assume that the inequalities

$$\left|\frac{\partial g^i(x^1, \ldots, x^n)}{\partial x^j}\right| \le K$$

are valid, where K is a positive number. Then for any two points $\mathbf{x}$ and $\mathbf{y}$ of the set Δ the inequalities

$$|\mathbf{g}(\mathbf{x}) - \mathbf{g}(\mathbf{y})| \le n^2 K |\mathbf{x} - \mathbf{y}| \tag{6}$$

are valid.

To prove inequality (6) we introduce the notion of the *segment* joining $\mathbf{x}$ and $\mathbf{y}$, namely, we set

$$\mathbf{z}(s) = \mathbf{y} + s(\mathbf{x} - \mathbf{y}).$$

Whenever s runs over all values of $0 \le s \le 1$, then $\mathbf{z}(s)$ takes all values of the segment joining the points $\mathbf{x}$ and $\mathbf{y}$, and always remains in this segment because of the convexity of the set Δ. We obtain (by applying the Lagrange formula)

$$g^i(\mathbf{x}) - g^i(\mathbf{y}) = g^i(\mathbf{z}(1)) - g^i(\mathbf{z}(0)) = \left.\frac{dg^i(\mathbf{z}(s))}{ds}\right|_{s=\theta} .$$

By calculating $dg^i(\mathbf{z}(s))/ds$ by the formula for the derivative of a composite function, we obtain

$$\frac{dg^i(\mathbf{z}(s))}{ds} = \frac{dg^i(z^1(s), \ldots, z^n(s))}{ds} = \sum_{k=1}^{n} \frac{\partial g^i(z^1(s), \ldots, z^n(s))}{\partial x^k} \cdot \frac{dz^k(s)}{ds}$$
$$= \sum_{k=1}^{n} \frac{\partial g^i(z^1(s), \ldots, z^n(s))}{\partial x^k} (x^k - y^k);$$

and therefore

$$|g^i(\mathbf{x}) - g^i(\mathbf{y})| \le \sum_{k=1}^{n} K|x^k - y^k| \le \sum_{k=1}^{n} K|\mathbf{x} - \mathbf{y}| \le nK|\mathbf{x} - \mathbf{y}|.$$

By squaring the last inequality, summing over i, and extracting the root, we obtain

$$|\mathbf{g}(\mathbf{x}) - \mathbf{g}(\mathbf{y})| \leq n^{3/2}K|\mathbf{x} - \mathbf{y}| \leq n^2K|\mathbf{x} - \mathbf{y}|.$$

We proceed from the differential equation (1) to an integral equation in the same way as in the proof of Theorem 1.

(A) Let $\mathbf{x} = \boldsymbol{\varphi}(t)$ be a certain solution of the differential equation (3), so that the identity

$$\dot{\boldsymbol{\varphi}}(t) = \mathbf{f}(t, \boldsymbol{\varphi}(t)) \tag{7}$$

is valid, and let

$$\boldsymbol{\varphi}(t_0) = \mathbf{x}_0 \tag{8}$$

be the initial condition which this solution satisfies.

We shall now prove that the pair of relations (7) and (8) is equivalent to one relation,

$$\boldsymbol{\varphi}(t) = \mathbf{x}_0 + \int_{t_0}^{t} \mathbf{f}(\tau, \boldsymbol{\varphi}(\tau))\, d\tau. \tag{9}$$

Assuming that the integral identity (9) is satisfied and substituting $t = t_0$ into it, we obtain (8), which, when differentiated with respect to t, yields (7). We shall now assume that (7) and (8) are satisfied. Integrating (7) from t_0 to t and taking into account (8), we obtain (9).

(B) By using the right-hand side of (9), we can make every vector function $\boldsymbol{\varphi}(t)$ whose graph passes through domain Γ correspond with the function $\boldsymbol{\varphi}^*(t)$ by setting

$$\boldsymbol{\varphi}^*(t) = \mathbf{x}_0 + \int_{t_0}^{t} \mathbf{f}(\tau, \boldsymbol{\varphi}(\tau))\, d\tau. \tag{10}$$

We may write the same relation briefly in the operator form

$$\boldsymbol{\varphi}^* = A\boldsymbol{\varphi}. \tag{11}$$

Equation (9) may now be written

$$\boldsymbol{\varphi} = A\boldsymbol{\varphi}. \tag{12}$$

(C) Let $\boldsymbol{\varphi}(t)$ be a continuous vector function defined on the interval $r_1 \leq t \leq r_2$. We define the *norm* $\|\boldsymbol{\varphi}\|$ of this function by setting

$$\|\boldsymbol{\varphi}\| = \max_{r_1 \leq t \leq r_2} |\boldsymbol{\varphi}(t)|.$$

By using the concept of the norm, we may formulate the condition of uniform convergence of the sequence

$$\boldsymbol{\varphi}_0, \boldsymbol{\varphi}_1, \ldots, \boldsymbol{\varphi}_i, \ldots \tag{13}$$

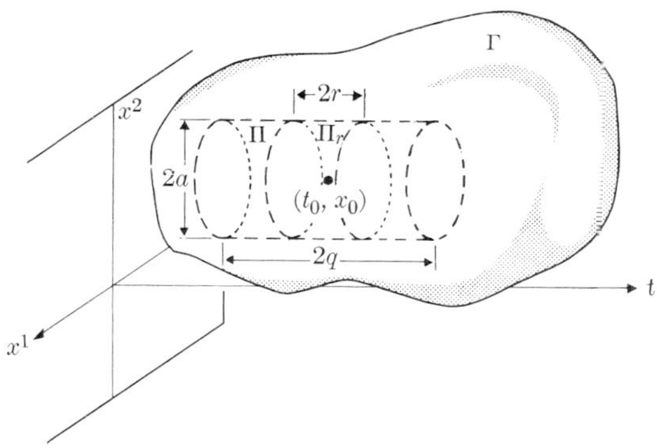

FIGURE 45

of vector functions defined on the interval $r_1 \leq t \leq r_2$. The sequence (13) of vector functions converges uniformly to a vector function $\boldsymbol{\varphi}$ defined on the same interval $r_1 \leq t \leq r_2$, if

$$\lim_{i \to \infty} \|\boldsymbol{\varphi} - \boldsymbol{\varphi}_i\| = 0.$$

In order that the sequence (13) be uniformly convergent, it is sufficient that the inequalities

$$\|\boldsymbol{\varphi}_{i+1} - \boldsymbol{\varphi}_i\| \leq a_i$$

be satisfied, where the numbers $a_0, a_1, \ldots, a_i, \ldots$ form a convergent series.

We now proceed to the proof of Theorem 2.

Proof of Theorem 2. Since the point $(t_0, \mathbf{x}_0) = (t_0, x_0^1, x_0^2, \ldots, x_0^n)$ belongs to Γ, there exist positive numbers q and a such that all points $(t, \mathbf{x})$, which satisfy the conditions

$$|t - t_0| \leq q, \qquad |\mathbf{x} - \mathbf{x}_0| \leq a, \tag{14}$$

are in Γ. Since the set Π consisting of all points $(t, \mathbf{x})$ satisfying (14) is closed and bounded (Fig. 45), the continuous functions

$$|\mathbf{f}(t, \mathbf{x})| \qquad \text{and} \qquad \left|\frac{\partial f^i(t, \mathbf{x})}{\partial x^j}\right|, \qquad i, j = 1, \ldots, n,$$

are bounded on Π, i.e., there exist positive numbers M and K such that

$$|\mathbf{f}(t, \mathbf{x})| \leq M, \qquad \left|\frac{\partial f^i(t, \mathbf{x})}{\partial x^j}\right| \leq K, \qquad i, j = 1, \ldots, n, \tag{15}$$

on the set Π.

Along with Π we consider its subset Π_r which is defined by the inequalities

$$|t - t_0| \leq r, \qquad |\mathbf{x} - \mathbf{x}_0| \leq a,$$

where

$$r \leq q \tag{16}$$

(Fig. 45). We shall denote by Ω_r the family of all continuous vector functions defined on the interval $|t - t_0| \leq r$ whose graphs pass through Π_r. Thus the function $\boldsymbol{\varphi}$, defined on the interval $|t - t_0| \leq r$, belongs to the family Ω_r if and only if the inequality

$$|\boldsymbol{\varphi}(t) - \mathbf{x}_0| \leq a \tag{17}$$

is satisfied for any t on this interval.

We now choose r in such a way that the following two conditions will be satisfied:

(a) If the function $\boldsymbol{\varphi}$ belongs to the family Ω_r, then the function $\boldsymbol{\varphi}^* = A\boldsymbol{\varphi}$ [see (10) and (11)] also belongs to Ω_r.

(b) There exists a number k on the interval $0 < k < 1$ such that for any two function $\boldsymbol{\psi}$ and $\boldsymbol{\chi}$ of the family Ω_r the inequality

$$\|A\boldsymbol{\psi} - A\boldsymbol{\chi}\| \leq k\|\boldsymbol{\psi} - \boldsymbol{\chi}\| \tag{18}$$

is valid.

Let us consider condition (a). In order that the function $\boldsymbol{\varphi}^* = A\boldsymbol{\varphi}$ belong to Ω_r, it is necessary and sufficient that for $|t - t_0| \leq r$ the inequality

$$|\boldsymbol{\varphi}^*(t) - \mathbf{x}_0| \leq a$$

be satisfied. By virtue of (10), (5), and (15), we have

$$|\boldsymbol{\varphi}^*(t) - \mathbf{x}_0| = \left|\int_{t_0}^{t} \mathbf{f}(\tau, \boldsymbol{\varphi}(\tau))\, d\tau\right| \leq \left|\int_{t_0}^{t} |\mathbf{f}(\tau, \boldsymbol{\varphi}(\tau))|\, d\tau\right| \leq Mr.$$

From this it is evident that condition (a) is satisfied for

$$r \leq \frac{a}{M}\cdot \tag{19}$$

Let us now consider condition (b). We have

$$|\boldsymbol{\psi}^*(t) - \boldsymbol{\chi}^*(t)| = \left|\int_{t_0}^{t} (\mathbf{f}(\tau, \boldsymbol{\psi}(\tau)) - \mathbf{f}(\tau, \boldsymbol{\chi}(\tau)))\, d\tau\right|$$
$$\leq \left|\int_{t_0}^{t} |\mathbf{f}(\tau, \boldsymbol{\psi}(\tau)) - \mathbf{f}(\tau, \boldsymbol{\chi}(\tau))|\, d\tau\right| \cdot \tag{20}$$

We shall now find an estimate for the last integrand by using inequalities (6) and (15):

$$|\mathbf{f}(\tau, \boldsymbol{\psi}(\tau)) - \mathbf{f}(\tau, \boldsymbol{\chi}(\tau))| \leq n^2K|\boldsymbol{\psi}(\tau) - \boldsymbol{\chi}(\tau)|. \tag{21}$$

From (20) and (21) it follows that

$$\|A\boldsymbol{\psi} - A\boldsymbol{\chi}\| = \|\boldsymbol{\psi}^* - \boldsymbol{\chi}^*\| \leq n^2Kr\|\boldsymbol{\psi} - \boldsymbol{\chi}\|.$$

Thus condition (b) is satisfied if

$$r \leq \frac{k}{n^2K}, \tag{22}$$

where $k < 1$.

Thus if the number r satisfies the inequalities (16), (19), and (22) (which we shall henceforth assume to be satisfied), then the conditions (a) and (b) are satisfied for the family Ω_r.

We now form the sequence of vector functions

$$\boldsymbol{\varphi}_0(t) \equiv \mathbf{x}_0, \qquad \boldsymbol{\varphi}_1(t), \ldots, \boldsymbol{\varphi}_i(t), \ldots; \tag{23}$$

defined on the interval $|t - t_0| \leq r$ by setting

$$\boldsymbol{\varphi}_{i+1} = A\boldsymbol{\varphi}_i, \qquad i = 0, 1, \ldots. \tag{24}$$

Since the function $\boldsymbol{\varphi}_0$ belongs to the family Ω_r, all functions of (23) belong to the same family [see condition (a)]. Further, we have [see (17)]

$$\|\boldsymbol{\varphi}_1 - \boldsymbol{\varphi}_0\| = \max_{|t-t_0| \leq r} |\boldsymbol{\varphi}_1(t) - \mathbf{x}_0| \leq a.$$

From (18) we obtain

$$\|\boldsymbol{\varphi}_{i+1} - \boldsymbol{\varphi}_i\| = \|A\boldsymbol{\varphi}_i - A\boldsymbol{\varphi}_{i-1}\| \leq k\|\boldsymbol{\varphi}_i - \boldsymbol{\varphi}_{i-1}\|,$$

whence

$$\|\boldsymbol{\varphi}_{i+1} - \boldsymbol{\varphi}_i\| \leq ak^i. \tag{25}$$

Thus by virtue of (C), the sequence (23) converges uniformly to a continuous function $\boldsymbol{\varphi}$ belonging to Ω_r. We shall show that the function $\boldsymbol{\varphi}$ satisfies equation (12). For this we observe that the sequence

$$A\boldsymbol{\varphi}_0, A\boldsymbol{\varphi}_1, \ldots, A\boldsymbol{\varphi}_i, \ldots$$

converges uniformly to the function $A\boldsymbol{\varphi}$; actually, we have [see (18)]

$$\|A\boldsymbol{\varphi} - A\boldsymbol{\varphi}_i\| \leq k\|\boldsymbol{\varphi} - \boldsymbol{\varphi}_i\|.$$

If we allow $i \to \infty$ in relation (24), we obtain

$$\boldsymbol{\varphi} = A\boldsymbol{\varphi}.$$

Hence we have proved *the existence of a solution* $\mathbf{x} = \boldsymbol{\varphi}(t)$ *of equation* (3) *satisfying the initial condition* (8); *moreover, the solution* $\mathbf{x} = \boldsymbol{\varphi}(t)$ *is defined on the interval* $|t - t_0| < r$, *where* r *is an arbitrary number satisfying* (16), (19), *and* (22).

We now proceed to the proof of uniqueness. Let $\mathbf{x} = \boldsymbol{\psi}(t)$ and $\mathbf{x} = \boldsymbol{\chi}(t)$ be two solutions of equation (3) with the common initial values t_0, $\mathbf{x}_0$, and let $r_1 < t < r_2$ be the intersection of the intervals of existence of the solutions $\boldsymbol{\psi}$ and $\boldsymbol{\chi}$; it is obvious that $r_1 < t_0 < r_2$. We shall denote by N the set of all points of the interval $r_1 < t < r_2$ at which the solutions $\boldsymbol{\psi}$ and $\boldsymbol{\chi}$ coincide. The set N is nonempty since it contains the point t_0. We shall show that N is *open* and that it is *closed* on the interval $r_1 < t < r_2$. From this it will follow, by proposition (D) of §20, that N coincides with the interval $r_1 < t < r_2$, that is, the solutions $\boldsymbol{\psi}$ and $\boldsymbol{\chi}$ coincide identically on this entire interval.

First we shall prove that the set N is open. Let t_1 be an arbitrary point of N; since the solutions $\boldsymbol{\psi}(t)$ and $\boldsymbol{\chi}(t)$ coincide at this point, so that $\boldsymbol{\psi}(t_1) = \boldsymbol{\chi}(t_1) = \mathbf{x}_1$, then $(t_1, \mathbf{x}_1)$ can be taken as common initial values for both solutions. In this sense, the point $(t_1, \mathbf{x}_1)$ does not differ from the point $(t_0, \mathbf{x}_0)$, and therefore we shall use for $(t_1, \mathbf{x}_1)$ the designation $(t_0, \mathbf{x}_0)$, since this will permit us to retain the previous notation. Going over from the differential equation (3) to the integral equation (9), we obtain for both $\boldsymbol{\psi}(t)$ and $\boldsymbol{\chi}(t)$ integral equalities, which, in operator form, can be written

$$\boldsymbol{\psi} = A\boldsymbol{\psi}, \qquad \boldsymbol{\chi} = A\boldsymbol{\chi}. \tag{26}$$

We choose again the set Π which is contained in the domain Γ with its center at the point $(t_0, \mathbf{x}_0)$ [see (14)]; we then choose the set Π_r in such a way that the number r, in addition to satisfying inequalities (16), (19), and (22), also satisfies the condition that the functions $\boldsymbol{\psi}$ and $\boldsymbol{\chi}$ are defined for $|t - t_0| \leq r$ and satisfy the inequalities

$$|\boldsymbol{\psi}(t) - \mathbf{x}_0| \leq a,$$

$$|\boldsymbol{\chi}(t) - \mathbf{x}_0| \leq a.$$

This is possible, since $\boldsymbol{\psi}(t)$ and $\boldsymbol{\chi}(t)$ are continuous. Then $\boldsymbol{\psi}(t)$ and $\boldsymbol{\chi}(t)$, $|t - t_0| \leq r$, are contained in the family Ω_r so that by (18) and (26) we obtain

$$\|\boldsymbol{\psi} - \boldsymbol{\chi}\| = \|A\boldsymbol{\psi} - A\boldsymbol{\chi}\| \leq k\|\boldsymbol{\psi} - \boldsymbol{\chi}\|,$$

which is possible only if $\|\psi - \chi\| = 0$, that is, if ψ and χ coincide on the interval $|t - t_0| \leq r$. Hence, whenever t_0 is contained in N, the set N will also contain some interval $|t - t_0| < r$ with its center at t_0, so that N is open.

We shall now prove that the set N is closed on the interval $r_1 < t < r_2$. To do this we must show that if the sequence $t_1, t_2, \ldots, t_i, \ldots$ of points of N converges to a point $\bar{t}$ of the interval $r_1 < t < r_2$, then $\bar{t}$ must also belong to N. We have

$$\psi(t_i) = \chi(t_i), \qquad i = 1, 2, \ldots.$$

Since the functions $\psi(t)$ and $\chi(t)$ are defined and continuous at $\bar{t}$, then by passing to the limit as $i \to \infty$, we obtain $\psi(\bar{t}) = \chi(\bar{t})$, so that $\bar{t}$ also belongs to N. Thus Theorem 2 is proved.

We shall now state as a separate proposition certain facts established in the proof of Theorem 2 which will be needed later.

(D) Let us assume that the right-hand sides of the system (1) [or, in vector form, of equation (3)] are defined and continuous, together with their partial derivatives $\partial f^i/\partial x^j$, in Γ. Let $(t_0, \mathbf{x}_0)$ be a point of Γ, and let q and a be positive numbers such that the set Π consisting of all points satisfying (14) is contained in Γ. Furthermore, let M and K be positive numbers such that the inequalities (15) are satisfied for all points $(t, \mathbf{x})$ satisfying (14). Finally, let r be any positive number which satisfies (16), (19), and (22). Then the solution of equation (3) with the initial values $(t_0, \mathbf{x}_0)$ is defined on the interval $|t - t_0| < r$. In addition, the solution may be obtained on the interval $|t - t_0| \leq r$ as the limit of the sequence (23) which is defined inductively by (24), since inequality (25) is satisfied for these functions.

Proof of Theorem 3. We shall proceed to the proof of Theorem 3, which asserts that for the normal linear system

$$\dot{x}^i = \sum_{j=1}^{n} a_j^i(t)x^j + b^i(t) = f^i(t, x^1, \ldots, x^n), \qquad i = 1, \ldots, n, \tag{27}$$

whose coefficients $a_j(t)$ and free terms $b^i(t)$ are defined and continuous on the interval $q_1 < t < q_2$, there exists a solution with the arbitrary initial values

$$t_0, x_0^1, \ldots, x_0^n, \qquad q_1 < t_0 < q_2, \tag{28}$$

defined on the entire interval $q_1 < t < q_2$. We shall show that the same operator A [see (10) and (11)], which was applied in the proof of Theorem 2 but is constructed here by using the right-hand sides of system (27), gives rise to a sequence of vector functions

$$\boldsymbol{\varphi}_0(t) \equiv \mathbf{x}_0, \quad \boldsymbol{\varphi}_1(t), \ldots, \boldsymbol{\varphi}_i(t), \ldots, \tag{29}$$

which not only converges on the entire interval $q_1 < t < q_2$, but also converges uniformly on any segment contained in that interval. To carry out the method of successive approximations we shall need to find a more precise estimate for the numbers $\|\boldsymbol{\varphi}_{i+1} - \boldsymbol{\varphi}_i\|$, where $i = 0, 1, 2, \ldots$. In this case it can be seen that the method of successive approximations does not fit into the frame of the method of contraction mappings.

Let A be the operator defined by relations (10) and (11), stemming from the system (27) with the initial values (28). Obviously, we shall apply the operator A to an arbitrary continuous function $\boldsymbol{\varphi}(t)$ defined on the interval $q_1 < t < q_2$. By proposition (A) the system (27) with the initial conditions (28) is equivalent to the operator equation

$$\boldsymbol{\varphi} = A\boldsymbol{\varphi}, \tag{30}$$

for which we shall find a solution defined on the entire interval $q_1 < t < q_2$. The functions of the sequence (29), which is defined inductively by the relation

$$\boldsymbol{\varphi}_{i+1} = A\boldsymbol{\varphi}_i, \qquad i = 0, 1, 2, \ldots, \tag{31}$$

are defined on the interval $q_1 < t < q_2$.

Let $r_1 \leq t \leq r_2$ be an arbitrary interval which contains the point t_0 and which is still contained in the interval $q_1 < t < q_2$ so that

$$q_1 < r_1 < t_0 < r_2 < q_2.$$

We shall show that the sequence (29) converges uniformly on the interval $r_1 \leq t \leq r_2$ to the solution of (30). For the right-hand sides of equations (27) we have

$$\frac{\partial f^i}{\partial x^j} = a^i_j(t),$$

and therefore for $r_1 \leq t \leq r_2$ the inequalities

$$\left|\frac{\partial f^i}{\partial x^j}\right| \leq K, \qquad i, j = 1, \ldots, n,$$

are valid, where K is some positive number. Since the function $\boldsymbol{\varphi}_1(t)$ is bounded on the interval $r_1 \leq t \leq r_2$, the inequality

$$|\boldsymbol{\varphi}_1(t) - \boldsymbol{\varphi}_0(t)| \leq C$$

holds over this interval, where C is some constant. Furthermore, on this interval we obtain from (5) and (6) the following relations:

$$|\varphi_2(t) - \varphi_1(t)| = \left| \int_{t_0}^{t} [\mathbf{f}(\tau, \varphi_1(\tau)) - \mathbf{f}(\tau, \varphi_0(\tau))] \, d\tau \right|$$

$$\leq \left| \int_{t_0}^{t} |\mathbf{f}(\tau, \varphi_1(\tau)) - \mathbf{f}(\tau, \varphi_0(\tau))| \, d\tau \right| \leq n^2KC|t - t_0|;$$

$$|\varphi_3(t) - \varphi_2(t)| = \left| \int_{t_0}^{t} [\mathbf{f}(\tau, \varphi_2(\tau)) - \mathbf{f}(\tau, \varphi_1(\tau))] \, d\tau \right|$$

$$\leq \left| \int_{t_0}^{t} |\mathbf{f}(\tau, \varphi_2(\tau)) - \mathbf{f}(\tau, \varphi_1(\tau))| \, d\tau \right| \leq \frac{(n^2K)^2C}{2!} |t - t_0|^2;$$

$$\vdots$$

$$|\varphi_{i+1}(t) - \varphi_i(t)| = \left| \int_{t_0}^{t} [\mathbf{f}(\tau, \varphi_i(\tau)) - \mathbf{f}(\tau, \varphi_{i-1}(\tau))] \, dt \right|$$

$$\leq \left| \int_{t_0}^{t} |\mathbf{f}(\tau, \varphi_i(\tau)) - \mathbf{f}(\tau, \varphi_{i-1}(\tau))| \, d\tau \right| \leq \frac{(n^2K)^iC}{i!} |t - t_0|^i.$$

Hence we have

$$\|\varphi_{i+1} - \varphi_i\| \leq C \frac{(n^2K(r_2 - r_1))^i}{i!}.$$

Since the numbers $C(n^2K(r_2 - r_1))^i/i!$ form a convergent series, the sequence (29) converges uniformly on the interval $r_1 \leq t \leq r_2$ to a certain continuous function $\varphi(t)$. For this function we have

$$\|A\varphi_i - A\varphi\| \leq \max_{r_1 \leq t \leq r_2} \left| \int_{t_0}^{t} |\mathbf{f}(\tau, \varphi_i(\tau)) - \mathbf{f}(\tau, \varphi(\tau))| \, d\tau \right|$$

$$\leq n^2K(r_2 - r_1)\|\varphi_i - \varphi\|,$$

so that the sequence of functions

$$A\varphi_0, A\varphi_1, \ldots, A\varphi_i, \ldots$$

converges uniformly to the function $A\varphi$ on the interval $r_1 \leq t \leq r_2$. Passing to the limit in (31) we obtain

$$\varphi = A\varphi.$$

Since $r_1 \leq t \leq r_2$ is an arbitrary interval containing the point t_0 and contained in the interval $q_1 < t < q_2$, the sequence (29) converges at every point of the interval $q_1 < t < q_2$, so that the function $\varphi(t)$ is defined on the entire interval $q_1 < t < q_2$ and is a solution of equation (30) on the same interval. Thus Theorem 3 is proved.

22. Local theorems of continuity and differentiability of solutions. We shall consider here a normal system of differential equations whose right-hand sides depend on certain parameters, and we shall prove that under certain assumptions, a solution of this system with fixed initial values is a continuous and differentiable function of these parameters, while in the case of variable initial values it is a continuous and differentiable function of the initial values. The normal system under consideration will be written in the form

$$\dot{x}^i = f^i(t, x^1, \ldots, x^n, \mu^1, \ldots, \mu^l), \qquad i = 1, \ldots, n, \tag{1}$$

where $\mu^1, \ldots, \mu^l$ are certain numerical parameters. In what follows we shall assume that the right-hand sides

$$f^i(t, x^1, \ldots, x^n, \mu^1, \ldots, \mu^l), \qquad i = 1, \ldots, n, \tag{2}$$

of these equations, as well as their partial derivatives

$$f^i_j(t, x^1, \ldots, x^n, \mu^1, \ldots, \mu^l) = \frac{\partial f^i(t, x^1, \ldots, x^n, \mu^1, \ldots, \mu^l)}{\partial x^j}, \qquad i, j = 1, \ldots, n, \tag{3}$$

are defined and continuous in some domain $\widetilde{\Gamma}$ of the space of the variables $t, x^1, \ldots, x^n, \mu^1, \ldots, \mu^l$. The system (1) may be rewritten in the vector form

$$\dot{\mathbf{x}} = \mathbf{f}(t, \mathbf{x}, \boldsymbol{\mu}); \tag{4}$$

here

$$\mathbf{x} = (x^1, \ldots, x^n), \qquad \boldsymbol{\mu} = (\mu^1, \ldots, \mu^l), \qquad \mathbf{f} = (f^1, \ldots, f^n)$$

are vectors belonging to vector spaces of dimensions n, l, n, respectively.

Continuous dependence of solutions on parameters. We shall prove Theorem 13 below on the continuous dependence of a solution on parameters by a straightforward repetition of the proof of Theorem 2, in which we shall use only those steps of that proof which are mentioned in proposition (D) of §21.

THEOREM 13. If $(t_0, \mathbf{x}_0, \boldsymbol{\mu}_0)$ is an arbitrary point of the domain $\widetilde{\Gamma}$, there exist positive numbers r and ρ such that for

$$|\boldsymbol{\mu} - \boldsymbol{\mu}_0| < \rho$$

the solution

$$\mathbf{x} = \boldsymbol{\varphi}(t, \boldsymbol{\mu})$$

of equation (4), which satisfies the initial condition

$$\boldsymbol{\varphi}(t_0, \boldsymbol{\mu}) = \mathbf{x}_0, \tag{5}$$

is defined on the interval $|t - t_0| < r$ and is a continuous function of all the variables $t, \mu^1, \ldots, \mu^l$ on which it depends.

Proof. First we shall select positive numbers q, a, ρ such that the set $\tilde{\Pi}$ of all points $(t, \mathbf{x}, \boldsymbol{\mu})$ which satisfy the inequalities

$$|t - t_0| \leq q, \qquad |\mathbf{x} - \mathbf{x}_0| \leq a, \qquad |\boldsymbol{\mu} - \boldsymbol{\mu}_0| \leq \rho \tag{6}$$

belongs to the domain $\tilde{\Gamma}$. Since the functions (2) and (3) are continuous in $\tilde{\Gamma}$ and the set $\tilde{\Pi}$ is closed, there exist positive numbers M and K such that for every point $(t, \mathbf{x}, \boldsymbol{\mu})$ of $\tilde{\Pi}$ the inequalities

$$|\mathbf{f}(t, \mathbf{x}, \boldsymbol{\mu})| \leq M, \qquad |f_j^i(t, \mathbf{x}, \boldsymbol{\mu})| \leq K$$

are valid. We now choose a positive number r which satisfies the inequalities

$$r \leq q, \qquad r \leq \frac{a}{M}, \qquad r \leq \frac{k}{n^2K} \quad (k < 1)$$

[see (16), (19), and (22) of §21 and (D) of §21].

For every fixed value $\boldsymbol{\mu}$ which satisfies the last inequality of (6), we replace the differential equation (4), together with initial condition (5), by the integral equation

$$\boldsymbol{\varphi}(t, \boldsymbol{\mu}) = \mathbf{x}_0 + \int_{t_0}^{t} \mathbf{f}\big(\tau, \boldsymbol{\varphi}(\tau, \boldsymbol{\mu}), \boldsymbol{\mu}\big)\, d\tau \tag{7}$$

[see §21, (A)]. Using the right-hand side of this integral equation, we shall define an operator setting up a correspondence between the function $\boldsymbol{\varphi}(t, \boldsymbol{\mu})$ and the function

$$\boldsymbol{\varphi}^*(t, \boldsymbol{\mu}) = \mathbf{x}_0 + \int_{t_0}^{t} \mathbf{f}\big(\tau, \boldsymbol{\varphi}(\tau, \boldsymbol{\mu}), \boldsymbol{\mu}\big)\, d\tau,$$

or, briefly,

$$\boldsymbol{\varphi}^* = A_{\mu}\boldsymbol{\varphi}.$$

We shall also define a family Ω_r of vector functions $\boldsymbol{\varphi}(t, \boldsymbol{\mu})$ of the variables t, $\boldsymbol{\mu}$ defined on the set $|t - t_0| \leq r$, $|\boldsymbol{\mu} - \boldsymbol{\mu}_0| \leq \rho$, assuming that $\boldsymbol{\varphi}(t, \boldsymbol{\mu})$ belongs to Ω_r if the inequality

$$|\boldsymbol{\varphi}(t, \boldsymbol{\mu}) - \mathbf{x}_0| \leq a$$

is satisfied for $|t - t_0| \leq r$, $|\boldsymbol{\mu} - \boldsymbol{\mu}_0| \leq \rho$. In proposition (D) of §21 it was noted that the sequence of functions

$$\boldsymbol{\varphi}_0(t, \boldsymbol{\mu}) \equiv \mathbf{x}_0, \qquad \boldsymbol{\varphi}_1(t, \boldsymbol{\mu}), \ldots, \boldsymbol{\varphi}_i(t, \boldsymbol{\mu}), \ldots, \tag{8}$$

defined by the inductive relation

$$\boldsymbol{\varphi}_{i+1} = A_{\mu}\boldsymbol{\varphi}_i$$

for every fixed $\boldsymbol{\mu}$, where $|\boldsymbol{\mu} - \boldsymbol{\mu}_0| \leq \rho$, is defined on the interval $|t - t_0| \leq r$ and converges uniformly to the solution $\boldsymbol{\varphi}(t, \boldsymbol{\mu})$ of equation (7), where the inequalities

$$\|\boldsymbol{\varphi}_{i+1} - \boldsymbol{\varphi}_i\| < ak^i$$

are satisfied. These inequalities are satisfied for any values of $t, \mu^1, \ldots, \mu^l$ which satisfy the inequalities $|t - t_0| \leq r$, $|\boldsymbol{\mu} - \boldsymbol{\mu}_0| \leq \rho$. From these inequalities it follows that the sequence (8) of continuous functions depending on $t, \mu^1, \ldots, \mu^l$ converges uniformly in these variables, so that its limit $\boldsymbol{\varphi}(t, \boldsymbol{\mu})$ is a continuous function of the variables $t, \mu^1, \ldots, \mu^l$. Thus Theorem 13 is proved.

Differentiability of the solutions with respect to the parameters. We shall now proceed to examine the differentiability of the solution $\boldsymbol{\varphi}(t, \boldsymbol{\mu})$ of equation (4) with respect to the parameters $\mu^1, \ldots, \mu^l$. As a preliminary, we shall prove an auxiliary proposition (A), usually called *Hadamard's lemma.*

(A) Let $g(t^1, \ldots, t^p, u^1, \ldots, u^q)$ be a function of $p + q$ variables defined in the domain Δ of the space of these variables, which is convex with respect to the variables $u^1, \ldots, u^q$. If we set

$$\mathbf{t} = (t^1, \ldots, t^p), \qquad \mathbf{u} = (u^1, \ldots, u^q),$$

we may write it as a function $g(\mathbf{t}, \mathbf{u})$ of two vectors. We shall assume that in the entire domain of definition $g(\mathbf{t}, \mathbf{u})$ and its partial derivatives $\partial g(\mathbf{t}, \mathbf{u})/\partial u^j$, $j = 1, \ldots, q$, are continuous. Then for any pair of points $(\mathbf{t}, \mathbf{u}_1)$, $(\mathbf{t}, \mathbf{u}_2)$ in Δ with the same coordinate $\mathbf{t}$, the relation

$$g(\mathbf{t}, \mathbf{u}_2) - g(\mathbf{t}, \mathbf{u}_1) = \sum_{j=1}^{q} h_j(\mathbf{t}, \mathbf{u}_1, \mathbf{u}_2)(u_2^j - u_1^j) \tag{9}$$

is valid, where the functions $h_j(\mathbf{t}, \mathbf{u}_1, \mathbf{u}_2)$, $j = 1, \ldots, q$, are defined and continuous for all values of the arguments $\mathbf{t}, \mathbf{u}_1, \mathbf{u}_2$ (and, in particular, for $\mathbf{u}_1 = \mathbf{u}_2$).

To prove (A) we set

$$\mathbf{w}(s) = \mathbf{u}_1 + s(\mathbf{u}_2 - \mathbf{u}_1), \quad 0 \leq s \leq 1, \tag{10}$$

so that we have

$$g(\mathbf{t}, \mathbf{u}_2) - g(\mathbf{t}, \mathbf{u}_1) = g(\mathbf{t}, \mathbf{w}(1)) - g(\mathbf{t}, \mathbf{w}(0)) = \int_0^1 \frac{\partial}{\partial s} g(\mathbf{t}, \mathbf{w}(s))\, ds.$$

Let us now find the derivative $(\partial/\partial s)g(\mathbf{t}, \mathbf{w}(s))$; we have

$$\frac{\partial}{\partial s} g(\mathbf{t}, \mathbf{w}(s)) = \frac{\partial}{\partial s} g(\mathbf{t}, w^1(s), \ldots, w^q(s)) = \sum_{j=1}^{q} \frac{\partial g(\mathbf{t}, \mathbf{w}(s))}{\partial w^j} \cdot \frac{\partial w^j(s)}{\partial s}.$$

Since it is clear that (10) implies

$$\frac{\partial w^j(s)}{\partial s} = u_2^j - u_1^j, \qquad j = 1, \ldots, q,$$

then, by setting

$$h_j(\mathbf{t}, \mathbf{u}_1, \mathbf{u}_2) = \int_0^1 \frac{\partial g(\mathbf{t}, \mathbf{w}(s))}{\partial w^j}\, ds,$$

we obtain formula (9). Since by hypothesis the functions $\partial g(\mathbf{t}, \mathbf{u})/\partial u^j$ are continuous, the functions $h_j(\mathbf{t}, \mathbf{u}_1, \mathbf{u}_2)$ are also continuous. Thus proposition (A) is proved.

THEOREM 14. Let the partial derivatives

$$e_k^i(t, \mathbf{x}, \boldsymbol{\mu}) = \frac{\partial}{\partial \mu^k} f^i(t, \mathbf{x}, \boldsymbol{\mu}), \qquad i = 1, \ldots, n, \qquad k = 1, \ldots, l \tag{11}$$

of the right-hand sides of system (1) exist and be continuous in the domain $\widetilde{\Gamma}$. Let $(t_0, \mathbf{x}_0, \boldsymbol{\mu}_0)$ be some point of $\widetilde{\Gamma}$. Then there exist positive numbers r' and ρ' such that for $|t - t_0| < r'$, $|\boldsymbol{\mu} - \boldsymbol{\mu}_0| < \rho'$ the solution $\boldsymbol{\varphi}(t, \boldsymbol{\mu})$ of equation (4), which satisfies the initial condition

$$\boldsymbol{\varphi}(t_0, \boldsymbol{\mu}) = \mathbf{x}_0, \tag{12}$$

has continuous partial derivatives

$$\frac{\partial \boldsymbol{\varphi}(t, \boldsymbol{\mu})}{\partial \mu^k} \tag{13}$$

and mixed partial derivatives

$$\frac{\partial^2 \boldsymbol{\varphi}(t, \boldsymbol{\mu})}{\partial t\, \partial \mu^k}, \tag{14}$$

which are also continuous and do not depend on the order of differentiation. Furthermore, as functions of time t, the partial derivatives (13) satisfy the system of equations

$$\frac{\partial}{\partial t}\left(\frac{\partial \varphi^i(t, \boldsymbol{\mu})}{\partial \mu^k}\right) = \sum_{j=1}^{n} f_j^i(t, \boldsymbol{\varphi}(t, \boldsymbol{\mu}), \boldsymbol{\mu}) \frac{\partial \varphi^j(t, \boldsymbol{\mu})}{\partial \mu^k} + e_k^i(t, \boldsymbol{\varphi}(t, \boldsymbol{\mu}), \boldsymbol{\mu}), \tag{15}$$

with the initial condition

$$\frac{\partial \varphi^i(t_0, \boldsymbol{\mu})}{\partial \mu^k} = 0. \tag{16}$$

The system (15) is called the *system of variational equations* for the system (1).

Proof. To simplify the unwieldy notation, we shall assume in the proof that $k = l$, and we shall decompose the vector $\boldsymbol{\mu} = (\mu^1, \ldots, \mu^l)$ into the vector $\boldsymbol{\lambda} = (\mu^1, \ldots, \mu^{l-1})$ and the scalar $\nu = \mu^l$, so that we have

$$\boldsymbol{\mu} = (\boldsymbol{\lambda}, \nu).$$

By Theorem (13) the solution $\boldsymbol{\varphi}(t, \boldsymbol{\mu}) = \boldsymbol{\varphi}(t, \boldsymbol{\lambda}, \nu)$ is defined and continuous for

$$|t - t_0| < r, \qquad |\boldsymbol{\mu} - \boldsymbol{\mu}_0| < \rho;$$

in the sequel we shall choose only those values of the parameters $\boldsymbol{\lambda}$, ν which satisfy the second of these conditions.

To form the preliminary difference quotient for calculating the derivative $\partial\varphi^i(t, \boldsymbol{\lambda}, \nu)/\partial\nu$, we set

$$\psi^i(t, \boldsymbol{\lambda}, \nu_1, \nu_2) = \frac{1}{\nu_2 - \nu_1} \left(\varphi^i(t, \boldsymbol{\lambda}, \nu_2) - \varphi^i(t, \boldsymbol{\lambda}, \nu_1)\right);$$

this quotient is defined for $\nu_1 \neq \nu_2$. Since

$$\varphi^i(t, \boldsymbol{\lambda}, \nu), \qquad i = 1, \ldots, n,$$

is a solution of the system (1), we have

$$\begin{aligned} \frac{\partial}{\partial t} \psi^i(t, \boldsymbol{\lambda}, \nu_1, \nu_2) &= \frac{1}{\nu_2 - \nu_1} \left(\frac{\partial}{\partial t} \varphi^i(t, \boldsymbol{\lambda}, \nu_2) - \frac{\partial}{\partial t} \varphi^i(t, \boldsymbol{\lambda}, \nu_1)\right) \\ &= \frac{1}{\nu_2 - \nu_1} \left(f^i(t, \boldsymbol{\varphi}(t, \boldsymbol{\lambda}, \nu_2), \boldsymbol{\lambda}, \nu_2) - f^i(t, \boldsymbol{\varphi}(t, \boldsymbol{\lambda}, \nu_1), \boldsymbol{\lambda}, \nu_1)\right). \end{aligned} \tag{17}$$

We apply Hadamard's lemma [see (A)] to the right-hand side of (17), setting

$$\mathbf{t} = (t, \boldsymbol{\lambda}), \quad \mathbf{u}_1 = (\boldsymbol{\varphi}(t, \boldsymbol{\lambda}, \nu_1), \nu_1), \quad \mathbf{u}_2 = (\boldsymbol{\varphi}(t, \boldsymbol{\lambda}, \nu_2), \nu_2),$$
$$g(\mathbf{t}, \mathbf{u}) = f^i(t, \boldsymbol{\varphi}(t, \boldsymbol{\lambda}, \nu), \boldsymbol{\lambda}, \nu).$$

By virtue of (A) we obtain

$$\begin{aligned} &f^i(t, \boldsymbol{\varphi}(t, \boldsymbol{\lambda}, \nu_2), \boldsymbol{\lambda}, \nu_2) - f^i(t, \boldsymbol{\varphi}(t, \boldsymbol{\lambda}, \nu_1), \boldsymbol{\lambda}, \nu_1) \\ &= \sum_{j=1}^{n} h^i_j(\mathbf{t}, \mathbf{u}_1, \mathbf{u}_2)\left(\varphi^j(t, \boldsymbol{\lambda}, \nu_2) - \varphi^j(t, \boldsymbol{\lambda}, \nu_1)\right) + h^i_{n+1}(\mathbf{t}, \mathbf{u}_1, \mathbf{u}_2)(\nu_2 - \nu_1), \end{aligned}$$

where $h^i_j(\mathbf{t}, \mathbf{u}_1, \mathbf{u}_2)$, $j = 1, \ldots, n+1$, are continuous functions in all arguments, i.e., in the arguments t, $\boldsymbol{\lambda}$, ν_1, ν_2:

$$h^i_j(\mathbf{t}, \mathbf{u}_1, \mathbf{u}_2) = \tilde{h}^i_j(t, \boldsymbol{\lambda}, \nu_1, \nu_2).$$

Thus (17) can be rewritten in the form

$$\frac{\partial}{\partial t}\psi^i(t, \boldsymbol{\lambda}, \nu_1, \nu_2) = \sum_{j=1}^{n} \tilde{h}^i_j(t, \boldsymbol{\lambda}, \nu_1, \nu_2)\psi^j(t, \boldsymbol{\lambda}, \nu_1, \nu_2) + \tilde{h}^i_{n+1}(t, \boldsymbol{\lambda}, \nu_1, \nu_2). \tag{18}$$

The last relation represents a system of linear differential equations with respect to the functions $\psi^i(t, \boldsymbol{\lambda}, \nu_1, \nu_2)$, $i = 1, \ldots, n$, where the coefficients of this system and the free terms depend (continuously) not only on t, but also on $\boldsymbol{\lambda}$, ν_1, ν_2.

We shall determine the nature of the initial conditions satisfied by the functions $\psi^i(t, \boldsymbol{\lambda}, \nu_1, \nu_2)$. By virtue of (12) we have

$$\begin{aligned}\psi^i(t_0, \boldsymbol{\lambda}, \nu_1, \nu_2) &= \frac{1}{\nu_2 - \nu_1}(\varphi^i(t_0, \boldsymbol{\lambda}, \nu_2) - \varphi^i(t_0, \boldsymbol{\lambda}, \nu_1)) \\ &= \frac{1}{\nu_2 - \nu_1}(x^i_0 - x^i_0) = 0.\end{aligned}$$

Thus the initial conditions for the functions $\psi^i(t, \boldsymbol{\lambda}, \nu_1, \nu_2)$ are of the form

$$\psi^i(t_0, \boldsymbol{\lambda}, \nu_1, \nu_2) = 0.$$

Our purpose is to prove that the functions $\psi^i(t, \boldsymbol{\lambda}, \nu_1, \nu_2)$ tend to a definite limit as $\nu_2 \to \nu_1$. In order to establish this fact we consider the system of equations

$$\frac{\partial}{\partial t}\psi^i_*(t, \boldsymbol{\lambda}, \nu_1, \nu_2) = \sum_{j=1}^{n} \tilde{h}^i_j(t, \boldsymbol{\lambda}, \nu_1, \nu_2)\psi^j_*(t, \boldsymbol{\lambda}, \nu_1, \nu_2) + \tilde{h}^i_{n+1}(t, \boldsymbol{\lambda}, \nu_1, \nu_2) \tag{19}$$

in the functions

$$\psi^i_*(t, \boldsymbol{\lambda}, \nu_1, \nu_2), \qquad i = 1, \ldots, n, \tag{20}$$

which satisfy the initial conditions

$$\psi^i_*(t_0, \boldsymbol{\lambda}, \nu_1, \nu_2) = 0.$$

By Theorem 13, the functions (20) are defined and continuous for values of t, $\boldsymbol{\lambda}$, ν_1, ν_2 which are sufficiently close to the constants t_0, $\boldsymbol{\lambda}_0 = (\mu^1_0, \ldots, \mu^{l-1}_0)$, μ^l_0, μ^l_0.

The functions $\psi^i(t, \boldsymbol{\lambda}, \nu_1, \nu_2)$ and $\psi^i_*(t, \boldsymbol{\lambda}, \nu_1, \nu_2)$ satisfy the same system of equations (18), (19) and have the same initial values; this fact, when taken together with Theorem 2, shows that they coincide:

$$\psi^i(t, \boldsymbol{\lambda}, \nu_1, \nu_2) = \psi^i_*(t, \boldsymbol{\lambda}, \nu_1, \nu_2).$$

Here the functions on the right-hand side are also defined at $\nu_1 = \nu_2$ and those on the left-hand side are defined only for $\nu_1 \neq \nu_2$. Passing to the limit as $\nu_2 \to \nu_1$ we see that

$$\frac{\partial}{\partial \nu} \varphi^i(t, \boldsymbol{\lambda}, \nu) = \psi^i_*(t, \boldsymbol{\lambda}, \nu, \nu),$$

so that the derivative $(\partial/\partial\nu)\varphi^i(t, \boldsymbol{\lambda}, \nu)$ exists, is a continuous function of all its arguments, and satisfies the system of equations

$$\frac{\partial}{\partial t}\left(\frac{\partial \varphi^i(t, \boldsymbol{\lambda}, \nu)}{\partial \nu}\right) = \sum_{j=1}^{n} \tilde{h}^i_j(t, \boldsymbol{\lambda}, \nu, \nu) \frac{\partial \varphi^j(t, \boldsymbol{\lambda}, \nu)}{\partial \nu} + \tilde{h}^i_{n+1}(t, \boldsymbol{\lambda}, \nu, \nu). \tag{21}$$

From this it is seen that, not only the derivatives

$$\frac{\partial \varphi^i(t, \boldsymbol{\lambda}, \nu)}{\partial \nu}$$

themselves exist and are continuous, but that their derivatives with respect to t, i.e.,

$$\frac{\partial}{\partial t}\left(\frac{\partial \varphi^i(t, \boldsymbol{\lambda}, \nu)}{\partial \nu}\right), \tag{22}$$

also exist and are continuous.

Now that it has been established that the derivatives (14) exist and are continuous, the proof of the first part of the theorem is complete. The second part, however, is more difficult to prove. The system of equations (21), satisfied by the derivatives (13), contains the functions $\tilde{h}^i_j(t, \boldsymbol{\lambda}, \nu, \nu)$ which are determined in a comparatively complicated manner in terms of the functions f^i with the aid of Hadamard's lemma.

To derive the variational equations (15) we shall first write out the fact that the functions $\varphi^i(t, \boldsymbol{\lambda}, \nu)$ satisfy system (1):

$$\frac{\partial}{\partial t} \varphi^i(t, \boldsymbol{\lambda}, \nu) = f^i(t, \varphi^1(t, \boldsymbol{\lambda}, \nu), \ldots, \varphi^n(t, \boldsymbol{\lambda}, \nu), \boldsymbol{\lambda}, \nu). \tag{23}$$

From what has already been proved, the functions $\varphi^i(t, \boldsymbol{\lambda}, \nu)$ are differentiable with respect to ν, so that the right-hand sides of (23) can be differentiated with respect to ν; for this reason the left-hand sides of these relations can also be differentiated with respect to ν. Thus we obtain [see (3) and (11)]

$$\frac{\partial}{\partial \nu}\left(\frac{\partial}{\partial t}\varphi^i(t, \boldsymbol{\lambda}, \nu)\right)$$
$$= \sum_{j=1}^{n} f_j^i(t, \boldsymbol{\varphi}(t, \boldsymbol{\lambda}, \nu), \boldsymbol{\lambda}, \nu)\frac{\partial}{\partial \nu}\varphi^j(t, \boldsymbol{\lambda}, \nu) + e_l^i(t, \boldsymbol{\varphi}(t, \boldsymbol{\lambda}, \nu), \boldsymbol{\lambda}, \nu). \quad (24)$$

From this relation it is evident that the partial derivatives

$$\frac{\partial}{\partial \nu}\left(\frac{\partial}{\partial t}\varphi^i(t, \boldsymbol{\lambda}, \nu)\right) \quad (25)$$

exist and are continuous. On the other hand, since the partial derivatives (22) are also continuous, it follows from a known theorem of analysis that the mixed derivatives (22) and (25) coincide. Changing the order of differentiation with respect to t and ν in equations (24), we obtain the variational equations (15) for $k = l$.

By differentiating the initial conditions (12) with respect to μ^k, we obtain the initial conditions (16) for the functions (13). Thus Theorem 14 is completely proved. Let us now deduce a simple corollary from Theorem 14.

(B) If all partial derivatives of the right-hand sides of system (1) with respect to the variables $x^1, \ldots, x^n, \mu^1, \ldots, \mu^l$ up to the mth order inclusive exist and are continuous, then the functions $\varphi^i(t, \boldsymbol{\mu})$, $i = 1, \ldots, n$, forming the solution of (1) and satisfying the initial conditions (12), also have continuous partial derivatives with respect to the parameters $\mu^1, \ldots, \mu^l$ up to the mth order inclusive.

We shall prove proposition (B) by induction on the number m by means of the variational equations (15). The case $m = 1$ has already been established in Theorem 14. Let us now assume that it is valid for a given number m; we shall prove its validity for derivatives of order $m + 1$. We shall assume that all the partial derivatives of the functions $f^i(t, \mathbf{x}, \boldsymbol{\mu})$ with respect to the variables $x^1, \ldots, x^n, \mu^1, \ldots, \mu^l$ up to order $m + 1$ inclusive are continuous. By the induction hypothesis, the functions $\varphi^i(t, \boldsymbol{\mu})$ have continuous partial derivatives with respect to the variables $\mu^1, \ldots, \mu^l$ up to order m inclusive. Thus the functions $\partial\varphi^i(t, \boldsymbol{\mu})/\partial\mu^k$, which satisfy the system (15), also have continuous partial derivatives up to the mth order inclusive by the induction hypothesis, since the right-hand sides of (15) have partial derivatives up to order m inclusive, both with respect to the functions $\partial\varphi^j(t, \boldsymbol{\mu})/\partial\mu^k$, in which they are linear, and with respect to the parameters $\mu^1, \ldots, \mu^l$. Thus proposition (B) is proved.

Continuous dependence and differentiability of the solutions with respect to the initial values. We shall now investigate the dependence of the solution of equation (4) on the initial values t_0, $\mathbf{x}_0$; in order to emphasize the variability of t_0, $\mathbf{x}_0$, we shall now denote them by τ and ξ, respectively. The solution of (4) with the initial values τ, ξ now depends not only on t and μ but also on τ and ξ, so that it must be written in the form

$$\mathbf{x} = \varphi(t, \mu; \tau, \xi) \tag{26}$$

or, in terms of the coordinates,

$$x^i = \varphi^i(t, \mu; \tau, \xi), \qquad i = 1, \ldots, n.$$

To solve the problem of the range of the variables t, μ, τ, ξ where the function (26) is known to be defined (whether it is continuous and under what conditions it is differentiable), we make a change of variables which leads us to the consideration of a solution with *constant* initial values and which transforms these initial values into parameters. Thus by a change of variables, the entire complex of problems concerning the dependence of a solution on the initial conditions is reduced to a corresponding complex of problems on the dependence of the solution on the parameters. This change of variables is given in the following proposition.

(C) Let τ, ξ, μ be an arbitrary point of the domain $\tilde{\Gamma}$. In place of the independent variable t, which is contained in equation (4), we shall introduce a new independent variable s by the formula

$$t = \tau + s. \tag{27}$$

In place of the unknown vector function $\mathbf{x}$ in (4) we shall introduce a new unknown function $\mathbf{y}$ by the formula

$$\mathbf{x} = \xi + \mathbf{y}. \tag{28}$$

In terms of the new variables, equation (4) can be written

$$\frac{d\mathbf{y}}{ds} = \mathbf{f}(\tau + s, \xi + \mathbf{y}, \mu). \tag{29}$$

Since the function $f(t, \mathbf{x}, \mu)$ of the variables $t, \mathbf{x}, \mu$ is defined in $\tilde{\Gamma}$, the function

$$\mathbf{g}(s, \mathbf{y}, \mu;\ \tau, \xi) = \mathbf{f}(\tau + s; \xi + \mathbf{y}, \mu) \tag{30}$$

of the variables s, $\mathbf{y}$, μ, τ, ξ is defined under the condition that the point $(\tau + s, \xi + \mathbf{y}, \mu)$ belongs to $\tilde{\Gamma}$. This condition, as is easy to see, distinguishes a certain domain Γ^* in the space of variables s, $\mathbf{y}$, μ, τ, ξ, and in this domain the vector function (30) is continuous, and its components have continuous partial derivatives with respect to the variables $y^1, \ldots, y^n$, $\xi^1, \ldots, \xi^n$. Let

$$\mathbf{y} = \boldsymbol{\psi}(s, \boldsymbol{\mu}; \tau, \boldsymbol{\xi}) \tag{31}$$

be a solution of equation (29) which satisfies the initial conditions

$$\boldsymbol{\psi}(0, \boldsymbol{\mu}; \tau, \boldsymbol{\xi}) = 0.$$

If we go back to the old variables by formulas (27) and (28), we obtain the solution

$$\mathbf{x} = \boldsymbol{\varphi}(t, \boldsymbol{\mu}; \tau, \boldsymbol{\xi}) = \boldsymbol{\xi} + \boldsymbol{\psi}(t - \tau, \boldsymbol{\mu}; \tau, \boldsymbol{\xi}) \tag{32}$$

of (4) which obviously satisfies the initial condition

$$\boldsymbol{\varphi}(\tau, \boldsymbol{\mu}; \tau, \boldsymbol{\xi}) = \boldsymbol{\xi}.$$

From the construction given in proposition (C) by applying Theorems 13 and 14 to equation (29), it is easy to deduce that the solution (32) is continuous with respect to all the variables t, $\boldsymbol{\mu}$, τ, $\boldsymbol{\xi}$ and differentiable with respect to $\xi^1, \ldots, \xi^n$. In the case that the functions $f^i(t, \mathbf{x}, \boldsymbol{\mu})$, $i = 1, \ldots, n$, have continuous derivatives with respect to t, $\mu^1, \ldots, \mu^l$, the solution (32) is also differentiable with respect to τ, $\mu^1, \ldots, \mu^l$. Further conclusions may be drawn by applying proposition (B) to equation (29). Here, however, we shall formulate and prove separately only the most essential of the results noted.

THEOREM 15. Let

$$\dot{x}^i = f^i(t, x^1, \ldots, x^n), \qquad i = 1, \ldots, n, \tag{33}$$

be a normal system of differential equations whose right-hand sides are defined and continuous, together with their partial derivatives

$$f^i_j(t, \mathbf{x}) = \frac{\partial}{\partial x^j} f^i(t, \mathbf{x}),$$

in a certain domain Γ of the space of the variables $t, x^1, \ldots, x^n$. Furthermore, let

$$\dot{\mathbf{x}} = f(t, \mathbf{x}) \tag{34}$$

be the vector form of this system. If $(t_0, \mathbf{x}_0)$ is an arbitrary point of Γ, then there exist positive numbers r' and σ' such that the solution

$$\mathbf{x} = \boldsymbol{\varphi}(t; \tau, \boldsymbol{\xi}) \tag{35}$$

of equation (34) which satisfies the initial condition

$$\boldsymbol{\varphi}(\tau; \tau, \boldsymbol{\xi}) = \boldsymbol{\xi}$$

is defined and continuous in the variables t, τ, $\boldsymbol{\xi}$ for

$$|t - \tau| < r'. \qquad |\tau - t_0| < \sigma', \qquad |\boldsymbol{\xi} - \mathbf{x}_0| < \sigma' \tag{36}$$

and has continuous partial derivatives with respect to $\xi^1, \ldots, \xi^n$,while the mixed partial derivatives

$$\frac{\partial^2 \varphi^i(t; \tau, \boldsymbol{\xi})}{\partial t\, \partial \xi^j}$$

are continuous and independent of the order of differentiation. In addition, the partial derivatives

$$\varphi_k^j(t) = \frac{\partial}{\partial \xi^k} \varphi^i(t; t_0, \boldsymbol{\xi})|_{\boldsymbol{\xi}=\mathbf{x}_0}$$

satisfy the system of equations

$$\dot{\varphi}_k^i(t) = \sum_{j=1}^{n} f_j^i(t, \boldsymbol{\varphi}(t; t_0, \mathbf{x}_0))\varphi_k^j(t) \tag{37}$$

under the initial conditions

$$\varphi_k^i(t_0) = \delta_k^i, \tag{38}$$

where δ_k^i is Kronecker's symbol. The system (37) is called the *system of variational equations* with respect to the initial values.

Proof. We shall utilize the transformation of variables described in proposition (C) in connection with equation (34), i.e., we shall assume that the parameter $\boldsymbol{\mu}$ is absent in equation (4). By Theorems 13 and 14 there exist positive numbers r' and ρ' such that the solution $\mathbf{y} = \boldsymbol{\psi}(s, \tau, \boldsymbol{\xi})$ is defined [see (31)] for

$$|s| < r', \qquad |\tau - t_0|^2 + |\boldsymbol{\xi} - \mathbf{x}_0|^2 < (\rho')^2 \tag{39}$$

and in the same domain has continuous derivatives with respect to the variables $\xi^1, \ldots, \xi^n$, since the right-hand side of equation (29) is differentiable with respect to these variables. In addition, the mixed derivative $\partial^2 \boldsymbol{\psi}/\partial s\, \partial \xi^j$ is continuous and does not depend on the order of differentiation. For a given number ρ' it is possible to select a number σ' so small that the inequality (39) follows from inequalities (36) for the same values of τ and $\boldsymbol{\xi}$. It follows from (32) that when the conditions (36) are satisfied, the solution (35) is defined, continuous, and differentiable with respect to $\xi^1, \ldots, \xi^n$, and also has a continuous mixed derivative $\partial^2 \boldsymbol{\varphi}/\partial t\, \partial \xi^j$ which does not depend on the order of differentiation.

Further, we write down the fact that the functions $\varphi^1(t; \tau, \xi), \ldots, \varphi^n(t; \tau, \xi)$ satisfy the system (33):

$$\frac{\partial}{\partial t} \varphi^i(t; \tau, \xi) = f^i(t, \varphi^1(t; \tau, \xi), \ldots, \varphi^n(t; \tau, \xi)), \qquad i = 1, \ldots, n. \tag{40}$$

By what has been proved above, the functions $\varphi^j(t; \tau, \xi)$ are differentiable with respect to $\xi^1, \ldots, \xi^n$, and the mixed derivative $\partial^2 \varphi^i(t; \tau, \xi)/\partial t\, \partial \xi^k$ is continuous and does not depend on the order of differentiation, so that if we differentiate (40) with respect to ξ^k and set $\tau = t_0$, $\xi = \mathbf{x}_0$, we obtain (37). If we differentiate the initial conditions

$$\varphi^i(t_0; t_0, \xi) = \xi^i$$

with respect to ξ^k, we obtain the initial conditions (38). Thus Theorem 15 is proved.

We remark that to form the system (37) of variational equations it is necessary to know only one solution, $\boldsymbol{\varphi}(t) = \boldsymbol{\varphi}(t; t_0, \mathbf{x}_0)$ of equation (34); knowing this solution, however, does not enable us to calculate the derivatives $\varphi^i_k(t) = (\partial/\partial \xi^k)\varphi^i(t; t_0, \xi)|_{\xi = \mathbf{x}_0}$. Thus the calculation of these derivatives, knowing one solution $\mathbf{x} = \boldsymbol{\varphi}(t)$, is reduced to the simpler problem of the solution of the linear system of equations (37).

23. First integrals. We shall introduce here the concept of a first integral and solve a boundary-value problem for linear partial differential equations.

First integrals. Let

$$\dot{x}^i = f^i(x^1, \ldots, x^n), \qquad i = 1, \ldots, n, \tag{1}$$

be a normal autonomous system of equations whose right-hand sides, together with their partial derivatives, are defined and continuous in some domain Δ of the space of the variables $x^1, \ldots, x^n$, and let

$$\dot{\mathbf{x}} = \mathbf{f}(\mathbf{x}) \tag{2}$$

be the vector notation for this system.

(A) A function

$$u(x^1, \ldots, x^n) = u(\mathbf{x}),$$

which is defined and continuous, together with its partial derivatives, in a certain domain G contained in Δ, is called a *first integral* of the system (1) if the substitution into (1) of an arbitrary solution $\mathbf{x} = \boldsymbol{\varphi}(t)$ of (2) leads to an expression which is independent of t; that is, the function $u(\boldsymbol{\varphi}(t))$ depends only on the choice of the solution $\boldsymbol{\varphi}(t)$, and not on t. Any first integral $u(\mathbf{x})$ of system (1) then satisfies the condition

$$\sum_{i=1}^{n} \frac{\partial u(\mathbf{x})}{\partial x^i} f^i(\mathbf{x}) = 0, \tag{3}$$

and, conversely, any function $u(\mathbf{x})$ which satisfies condition (3) is a first integral of the system (1).

We shall prove that a first integral $u(\mathbf{x})$ of system (1) satisfies (3). Let $\boldsymbol{\xi}$ be an arbitrary point of G and let $\mathbf{x} = \boldsymbol{\varphi}(t, \boldsymbol{\xi})$ be a solution of equation (2) with initial conditions $0, \boldsymbol{\xi}$. We have

$$0 = \frac{d}{dt} u(\boldsymbol{\varphi}(t, \boldsymbol{\xi}))|_{t=0} = \sum_{i=1}^{n} \frac{\partial u(\boldsymbol{\xi})}{\partial \xi^i} f^i(\boldsymbol{\xi});$$

since $\boldsymbol{\xi}$ is an arbitrary point of G, relation (3) is fulfilled in G.

Let us now assume that the relation (3) is fulfilled for the function $u(\mathbf{x})$, and let $\mathbf{x} = \boldsymbol{\varphi}(t)$ be an arbitrary solution of (2). Substituting $\mathbf{x} = \boldsymbol{\varphi}(t)$ into $u(\mathbf{x})$, we obtain a certain function

$$v(t) = u(\boldsymbol{\varphi}(t)).$$

By differentiating this function with respect to t, we obtain

$$\frac{dv(t)}{dt} = \sum_{i=1}^{n} \frac{\partial u(\boldsymbol{\varphi}(t))}{\partial x^i} f^i(\boldsymbol{\varphi}(t)) = 0.$$

Thus $u(\boldsymbol{\varphi}(t))$ does not depend on t.

In what follows, a study of the first integrals of the system (1) will be carried out purely locally in a certain neighborhood of the point $\mathbf{a}$ of Δ which is not a state of equilibrium of the system (1):

$$\mathbf{f}(\mathbf{a}) \neq 0. \tag{4}$$

(B) The first integrals

$$u_1(\mathbf{x}), \ldots, u_k(\mathbf{x})$$

of system (1), defined in a certain neighborhood of $\mathbf{a}$ [see (4)], are called *independent at the point* $\mathbf{a}$, or simply *independent*, if the functional matrix

$$\left(\frac{\partial u_i(\mathbf{a})}{\partial x^j}\right), \qquad i = 1, \ldots, k, \qquad j = 1, \ldots, n,$$

is of rank k. Thus the number of independent first integrals of system (1) cannot exceed $n - 1$, and if

$$u_1(\mathbf{x}), \ldots, u_{n-1}(\mathbf{x})$$

are independent first integrals, then any other first integral $u(\mathbf{x})$ can be written in the form

$$u(\mathbf{x}) = U(u_1(\mathbf{x}), \ldots, u_{n-1}(\mathbf{x})), \tag{5}$$

where $U(u_1, \ldots, u_{n-1})$ is a certain function of $n - 1$ variables and (5) is an identity in $\mathbf{x}$ in a certain neighborhood of $\mathbf{a}$.

Let us prove proposition (B). Let

$$u_1(\mathbf{x}), \ldots, u_n(\mathbf{x}) \tag{6}$$

be a set of n first integrals of the system (1). By (A) we have

$$\sum_{i=1}^{n} \frac{\partial u_j(\mathbf{x})}{\partial x^i} f^i(\mathbf{x}) = 0, \qquad j = 1, \ldots, n.$$

Since the vector $\mathbf{f}(\mathbf{x})$ is different from zero at the point $\mathbf{a}$, then it is also different from zero in some neighborhood of $\mathbf{a}$. Consequently, in that neighborhood the identity

$$\operatorname{Det}\left(\frac{\partial u_j(\mathbf{x})}{\partial x^i}\right) = 0 \tag{7}$$

is valid. Thus the set of n first integrals cannot be independent. Let us now assume that the first $n - 1$ of the first integrals of (6) are independent; then by a known theorem of analysis it follows from (7) that the function $u_n(\mathbf{x})$ may be expressed in terms of the remaining functions $u_1(\mathbf{x}), \ldots, u_{n-1}(\mathbf{x})$, that is, identity (5) is valid for $u(\mathbf{x}) = u_n(\mathbf{x})$. Thus proposition (B) is proved.

(C) There exist $n - 1$ independent first integrals of the system (1) in some neighborhood of the point $\mathbf{a}$ [see (4)].

Let us prove this. Since the vector $\mathbf{f}(\mathbf{a})$ is different from zero, at least one of its components is different from zero. We shall assume that

$$f^n(\mathbf{a}) \neq 0.$$

Let $\boldsymbol{\xi} = (\xi^1, \ldots, \xi^{n-1}, a^n)$ be a point close to the point $\mathbf{a}$, and let $\mathbf{x} = \boldsymbol{\varphi}(t, \boldsymbol{\xi})$ be a solution of (2) with the initial conditions $0, \boldsymbol{\xi}$. This solution can be written in the coordinate form

$$x^i = \varphi^i(t, \xi^1, \ldots, \xi^{n-1}), \qquad i = 1, \ldots, n. \tag{8}$$

We shall regard (8) as a system of equations in the unknowns

$$\xi^1, \ldots, \xi^{n-1}, t. \tag{9}$$

For $x^i = a^i$, $i = 1, \ldots, n$, this system of equations has the obvious solution $\xi^1 = a^1, \ldots, \xi^{n-1} = a^{n-1}$, $t = 0$, and the functional determinant of the system (8) is different from zero at this point, since

$$\frac{\partial \varphi^i(0, a^1, \ldots, a^{n-1})}{\partial \xi^j} = \delta^i_j, \qquad i = 1, \ldots, n, \qquad j = 1, \ldots, n-1, \tag{10}$$

[see (38) of §22], and

$$f^n(\mathbf{a}) \neq 0.$$

Thus there exists a neighborhood G of the point $\mathbf{a}$ such that for $\mathbf{x}$ belonging to G, the system (8) is solvable for the unknowns (9), and the solution may be written in the form

$$\xi^1 = u_1(\mathbf{x}), \ \ldots, \ \xi^{n-1} = u_{n-1}(\mathbf{x}), \qquad t = v(\mathbf{x}). \tag{11}$$

We shall show that the functions

$$u_1(\mathbf{x}), \ldots, u_{n-1}(\mathbf{x}) \tag{12}$$

in these relations are first integrals of the system (1) and, moreover, are independent at $\mathbf{a}$. Since the functional determinant of system (8) has been found [see (10)], it follows that the functional matrix

$$\left(\frac{\partial u_i(\mathbf{a})}{\partial x^j}\right), \qquad i, j = 1, \ldots, n-1,$$

is a unit matrix, and therefore the functions (12) are independent. We shall show that they are first integrals of the system (1). Since the system (11) is the inverse of the system (8), the functions (12) satisfy the identities

$$u_i(\boldsymbol{\varphi}(t, \boldsymbol{\xi})) = \xi^i, \qquad i = 1, \ldots, n-1. \tag{13}$$

Now let $\mathbf{x} = \boldsymbol{\varphi}(t)$ be some solution of equation (2) passing through the domain G. Let t_0, $\mathbf{x}_0$ be its initial values, with $\mathbf{x}_0$ belonging to G. Since the system (8) is solvable when $\mathbf{x} = \mathbf{x}_0$, there exists a solution $\mathbf{x} = \boldsymbol{\varphi}(t, \boldsymbol{\xi}_0)$ passing through $\mathbf{x}_0$, so that the solution $\boldsymbol{\varphi}(t)$ can be written in the form

$$\boldsymbol{\varphi}(t) = \boldsymbol{\varphi}(t + c, \boldsymbol{\xi}_0),$$

where c is a constant [see (B) of §15]. Thus, substituting $\mathbf{x} = \boldsymbol{\varphi}(t)$ into the function $u_i(\mathbf{x})$, we obtain by (13)

$$u_i(\boldsymbol{\varphi}(t)) = u_i(\boldsymbol{\varphi}(t + c, \boldsymbol{\xi}_0)) = \xi^i_0, \qquad i = 1, \ldots, n-1,$$

and proposition (C) is proved.

If certain first integrals of system (1) are unknown to us, then the solution of system (1) can be facilitated in the same way. This case may be formulated precisely in the following proposition.

(D) Let

$$u_{k+1}(\mathbf{x}), \ldots, u_n(\mathbf{x}) \tag{14}$$

be a system of $n - k$ first integrals, independent at the point $\mathbf{a}$ [see (B)], of the autonomous system (1). By using the functions (14), the order of (1) can be decreased by $n - k$, i.e., it can be replaced by an autonomous system of order k; in particular, when we deal with a maximal number $n - 1$ of independent first integrals, the autonomous system (1) can be reduced to the first order and therefore [see (B) of §2] can be solved by quadratures.

We shall prove proposition (D). Since the first integrals (14) are independent, the functional matrix

$$\left(\frac{\partial u_i(\mathbf{a})}{\partial x^j}\right), \qquad i = k+1, \ldots, n, \qquad j = 1, \ldots, n,$$

contains a square matrix of order $n - k$ whose determinant is different from zero. To be definite, we shall assume that the determinant of the matrix

$$\left(\frac{\partial u_i(\mathbf{a})}{\partial x^j}\right), \qquad i, j = k+1, \ldots, n,$$

is different from zero. We can now introduce in the neighborhood of the point $\mathbf{a}$ the new coordinates

$$y^1, \ldots, y^n \tag{15}$$

in the place of the previous coordinates

$$x^1, \ldots, x^n$$

by setting

$$\begin{gathered} y^1 = x^1, \ \ldots, \ y^k = x^k, \\ y^{k+1} = u_{k+1}(\mathbf{x}), \ldots, y^n = u_n(\mathbf{x}). \end{gathered} \tag{16}$$

The new coordinates $y^1, \ldots, y^n$ are, in fact, introduced by these formulas, since the functional determinant of the system (16) is different from zero in the neighborhood of $\mathbf{a}$. In the new system of variables (15) the system (1) takes the form

$$\dot{y}^i = g^i(y^1, \ldots, y^n), \qquad i = 1, \ldots, n. \tag{17}$$

But since every function (14) satisfies (3), we have

$$g^{k+1}(\mathbf{y}) = 0, \quad \ldots, \quad g^n(\mathbf{y}) = 0,$$

and therefore the system (17) is actually an autonomous system of order k.

Linear first-order partial differential equations. The relation (3) can be considered as a *partial differential equation* in the unknown function $u(\mathbf{x})$ with variables $x^1, \ldots, x^n$. Propositions (C) and (B) have shown that for $f(\mathbf{a}) \neq 0$ there exist $n - 1$ independent solutions of this equation in the neighborhood of the point $\mathbf{a}$, and that, having $n - 1$ independent solutions, we can obtain any other solution of this equation with the aid of formula (5). In this connection it is evident that every function given by (5) is a solution of (3), and therefore (3) can be regarded as *solved.* In other words, it has been shown that if we know how to solve (1), we also know how to solve (3). It is possible, however, to approach the solution of (3) from another point of view, that is, it is possible to pose and to solve a *boundary-value problem* for equation (3) and even for an equation of more general form than (3).

(E) Let

$$\sum_{i=1}^{n} f^i(\mathbf{x}) \frac{\partial u}{\partial x^i} = F(\mathbf{x}, u) \tag{18}$$

be a partial differential equation in the unknown function $u(\mathbf{x})$, where $F(\mathbf{x}, u)$ is a certain prescribed function having continuous first-order partial derivatives with respect to all its arguments. Further, let

$$\mathbf{x} = \boldsymbol{\xi}(t^1, \ldots, t^{n-1}) \tag{19}$$

be the vector form of a given $(n - 1)$-dimensional surface which passes through the point $\mathbf{a}$ for $t^1 = \cdots = t^{n-1} = 0$, so that

$$\boldsymbol{\xi}(0, \ldots, 0) = \mathbf{a}.$$

We shall assume that the surface (19) is differentiable and is not tangent to the vector $\mathbf{f}(\mathbf{a})$ at $\mathbf{a}$ so that the vectors

$$\frac{\partial \boldsymbol{\xi}(0, \ldots, 0)}{\partial t^1}, \ldots, \frac{\partial \boldsymbol{\xi}(0, \ldots, 0)}{\partial t^{n-1}}, \quad f(\mathbf{a}) \tag{20}$$

are linearly independent. Finally, let

$$u_0(t^1, \ldots, t^{n-1}) \tag{21}$$

be a certain function defined on the surface (19). Then, in the neighborhood of $\mathbf{a}$, there exists (and uniquely, moreover) a solution $u(\mathbf{x})$ of equation

(18) which coincides on the surface (19) with the given function (21), so that

$$u(\xi(t^1, \dots, t^{n-1})) = u_0(t^1, \dots, t^{n-1})$$

To find the solution $u(\mathbf{x})$ we use the trajectories of the system (1), which start on the surface (19). These trajectories are called the *characteristics* of equation (18).

Let us prove proposition (E). We introduce new coordinates into the neighborhood of $\mathbf{a}$ of the phase space of the system (1) in place of the coordinates $x^1, \dots, x^n$. Let $\mathbf{x} = \boldsymbol{\varphi}(t, t^1, \dots, t^{n-1})$ be a solution of equation (2) which starts at the point $\boldsymbol{\xi}(t^1, \dots, t^{n-1})$ of the surface (19), so that its initial values are 0, $\boldsymbol{\xi}(t^1, \dots, t^{n-1})$. We then have the system of relations

$$x^i = \varphi^i(t, t^1, \dots, t^{n-1}), \qquad i = 1, \dots, n. \tag{22}$$

If we consider as unknowns the variables

$$t, t^1, \dots, t^{n-1}, \tag{23}$$

then for $\mathbf{x} = \mathbf{a}$ this system has the obvious solution

$$t = t^1 = \cdots = t^{n-1} = 0,$$

and its functional determinant does not vanish at this point, a fact which follows from the linear independence of vectors (20) [see §22, formula (38)]. Thus the system (22) allows us to introduce new coordinates (23) into some neighborhood of $\mathbf{a}$ in place of the coordinates $x^1, \dots, x^n$. In these new coordinates the form of (18) is particularly simple, and the boundary-value problem stated in proposition (E) can be easily solved. Let $u(\mathbf{x})$ be a certain function defined in the neighborhood of the point $\mathbf{a}$. Let us substitute into this function the variables (23) in place of the variables $x^1, \dots, x^n$ according to (22); we then obtain the function

$$v(t, t^1, \dots, t^{n-1}) = u(\boldsymbol{\varphi}(t, t^1, \dots, t^{n-1})).$$

We have

$$\frac{\partial v(t, t^1, \dots, t^{n-1})}{\partial t} = \sum_{i=1}^{n} \frac{\partial u(\mathbf{x})}{\partial x^i} f^i(\mathbf{x}),$$

where $\mathbf{x} = \boldsymbol{\varphi}(t, t^1, \dots, t^{n-1})$. Thus, in terms of the variables (23), equation (18) will have the form

$$\frac{\partial v(t, t^1, \dots, t^{n-1})}{\partial t} = F(\boldsymbol{\varphi}(t, t^1, \dots, t^{n-1}), v(t, t^1, \dots, t^{n-1})). \tag{24}$$

Since the surface (19) in the coordinates (23) is given by the equation $t = 0$, we must find a solution of equation (24) which reduces to the given function $u_0(t^1, \ldots, t^{n-1})$ for $t = 0$. To find such a solution, it is necessary to solve (24) by regarding it as an ordinary differential equation with the independent variable t and the variables $t^1, \ldots, t^{n-1}$ as parameters. Moreover, we must find solutions with the initial values

$$0,\ u_0(t^1, \ldots, t^{n-1}).$$

The function $v(t, t^1, \ldots, t^{n-1})$, obtained by virtue of Theorem 14, has continuous derivatives with respect to all variables. Thus the boundary-value problem posed in (E) is solved.

Note. Let

$$\dot{x}^i = f^i(t, x^1, \ldots, x^n) \tag{25}$$

be a nonautonomous system of differential equations. In order to introduce the concept of a first integral of this system, we shall transform it into an autonomous system by introducing the auxiliary unknown function

$$x^{n+1} = t.$$

Then system (25) augmented by the equation

$$\dot{x}^{n+1} = 1$$

will be autonomous; its first integrals may be taken as first integrals of the system (25).

Example

Let

$$H = H(x^1, \ldots, x^n; y^1, \ldots, y^n) = H(\mathbf{x}, \mathbf{y}) \tag{26}$$

be a function of two systems of variables. The system of ordinary differential equations

$$\begin{aligned} \dot{x}^i &= \frac{\partial}{\partial y^i} H(\mathbf{x}, \mathbf{y}), \\ \dot{y}^i &= -\frac{\partial}{\partial x^i} H(\mathbf{x}, \mathbf{y}), \qquad i = 1, \ldots, n, \end{aligned} \tag{27}$$

is called a *Hamiltonian system,* and the function $H(\mathbf{x}, \mathbf{y})$ is the Hamiltonian function of this system. We see directly that the function (26) is a first integral of (27).

24. Behavior of the trajectories on large time intervals. In proving Theorem 2 (see §21) for the given initial values t_0, $\mathbf{x}_0$, a positive number r was found such that a solution $\mathbf{x} = \boldsymbol{\varphi}(t)$ with these initial values exists on the interval $|t - t_0| < r$. Actually, the maximal interval of existence $m_1 < t < m_2$ [see §3, (A)] of the solution $\mathbf{x} = \boldsymbol{\varphi}(t)$ can be larger than the interval $|t - t_0| < r$. However, the question of why the maximal interval of existence $m_1 < t < m_2$ can be bounded from one side or the other, or even bounded in general, has not been discussed up to this point, except in the case of Theorem 3, which dealt with the linear equation, where the problem was solved completely. In this section we shall answer further the question of why the maximal interval of existence can be bounded from one side or the other.

(A) Let

$$\dot{x}^i = f(t, x^1, \ldots, x^n), \qquad i = 1, \ldots, n,$$

be a normal system of differential equations whose right-hand sides, together with their partial derivatives

$$\frac{\partial f^i}{\partial x^j},$$

are defined and continuous in some domain Γ of the space R of the variables $t, x^1, \ldots, x^n$, and let

$$\dot{\mathbf{x}} = \mathbf{f}(t, \mathbf{x}) \tag{1}$$

be the vector notation of this system. In addition, let E be a certain closed bounded set in Γ, and let

$$\mathbf{x} = \boldsymbol{\varphi}(t) \tag{2}$$

be a certain solution of equation (1) with maximal interval of existence $m_1 < t < m_2$ [see §3, (A)]. Thus, if the number m_2 is less than $+\infty$, there exists a positive number ϵ_2 such that for $t > m_2 - \epsilon_2$ the point $(t, \boldsymbol{\varphi}(t))$ is outside the set E. In exactly the same way, if the number m_1 is greater than $-\infty$, there exists a positive number ϵ_1 such that for $t < m_1 + \epsilon_1$ the point $(t, \boldsymbol{\varphi}(t))$ is outside the set E.

To prove proposition (A), we shall use the bound on the number r given in proposition (D) of §21. We shall consider only the case $m_2 < +\infty$ since the case $m_1 > -\infty$ is handled in the same way. We shall introduce a Euclidean metric into R. Since the set E and the complement of the set Γ are closed and E is bounded, the distance ρ between E and the complement of Γ is positive. This means that if the distance between the point $(t_0, \mathbf{x}_0)$ of E and the point $(t, \mathbf{x})$ of R is less than ρ, then $(t, \mathbf{x})$ must belong to the domain Γ. Now let E^* be the set of all points of R whose distance to the set E does not exceed the number $\rho/2$. Then E^* is contained in Γ, so that the right-hand sides of the system (1) and their derivatives are defined on the set E^*. Let us choose two positive numbers q and a, such that

$$q^2 + a^2 < \left(\frac{\rho}{2}\right)^2. \tag{3}$$

Let $(t_0, \mathbf{x}_0)$ be a certain point of the set E and Π be the set of all points $(t, \mathbf{x})$ which satisfy the inequalities

$$|t - t_0| \leq q, \qquad |\mathbf{x} - \mathbf{x}_0| \leq a.$$

It is obvious from (3) that Π is contained in E^*. Since E^* is closed and bounded, it follows that for any point $(t, \mathbf{x})$ of E^*, the inequalities

$$|\mathbf{f}(t, \mathbf{x})| \leq M, \qquad \left|\frac{\partial f^i(t, \mathbf{x})}{\partial x^j}\right| \leq K, \qquad i, j = 1, \ldots, n,$$

are fulfilled, where M and K are certain positive numbers. Thus if the number r satisfies (16), (19), and (22) of §21, the solution $\mathbf{x} = \boldsymbol{\varphi}(t)$ of equation (1) with the initial values $(t_0, \mathbf{x}_0)$ is defined on the interval $|t - t_0| < r$. The only important fact here is that the number r which is found is the same for all points $(t_0, \mathbf{x}_0)$ of the set E. Hence for the number ϵ_2 we may now take r.

Let us assume the opposite, i.e., that for a certain $t_0 > m_2 - r$ the point $(t_0, \boldsymbol{\varphi}(t_0))$ belongs to the set E. Then we can take the values t_0 and $\mathbf{x}_0 = \boldsymbol{\varphi}(t_0)$ as initial values of the solution (2). By the bound given in proposition (D) of §21, the solution (2) is defined on the interval $|t - t_0| < r$, which obviously falls outside the limits of the interval $m_1 < t < m_2$. But this contradicts the fact that $m_1 < t < m_2$ is the maximal interval of existence of the solution (2). With this contradiction, proposition (A) is proved.

(B) Let

$$\dot{x}^i = f^i(t, x^1, \ldots, x^n), \qquad i = 1, \ldots, n, \tag{4}$$

be a normal system of equations, whose right-hand sides are defined and continuous, together with their partial derivatives $\partial f^i/\partial x^j$, in some domain Γ of the space R of the variables $t, x^1, \ldots, x^n$. Here Γ has a special form, i.e., it consists of all points of the form $(t, x^1, \ldots, x^n)$, where t is an arbitrary number and the point $(x^1, \ldots, x^n)$ belongs to a certain well-defined domain Δ of the space S of the variables $x^1, \ldots, x^n$. In the particular case when system (4) is autonomous, we can take Δ to be an arbitrary domain in which the functions f^i and $\partial f^i/\partial x^j$ are defined and continuous, and then form Γ on the basis of this domain Δ. Let

$$\dot{\mathbf{x}} = \mathbf{f}(t, \mathbf{x}) \tag{5}$$

be the vector representation of the system (4), and let F be a certain closed, bounded set in Δ and $\mathbf{x} = \boldsymbol{\varphi}(t)$ a certain solution of (5) with maximal interval of existence $m_1 < t < m_2$. Then if the number m_2 is less than $+\infty$, there exists a positive number ϵ_2 such that for $t > m_2 - \epsilon_2$ the point $\boldsymbol{\varphi}(t)$ lies outside the set F. In exactly the same way, if the number m_1 is greater than $-\infty$, there exists a positive number ϵ_1 such that for $t < m_1 + \epsilon_1$ the point $\boldsymbol{\varphi}(t)$ lies outside F.

As in the proof of (A), we shall confine ourselves to the case $m_2 < +\infty$. In order to reduce the proof of (B) to that of (A), we select in a suitable way a closed, bounded set E. Let m be an arbitrary number which satisfies the inequality $m < m_2$; we shall define E as the set of all points $(t, \mathbf{x})$, where $m \leq t \leq m_2$ and $\mathbf{x}$ belongs to the set F. By (A), there exists a positive number ϵ_2 such that for $t > m_2 - \epsilon_2$, the point $(t, \boldsymbol{\varphi}(t))$ does not belong to E. We can assume here that $\epsilon_2 < m_2 - m$, that is, $m_2 - \epsilon_2 > m$. Since the number t satisfies the inequalities $m_2 - \epsilon_2 < t < m_2$, so that the inequalities $m \leq t \leq m_2$ are satisfied, the point $(t, \boldsymbol{\varphi}(t))$ need not belong to E because $\boldsymbol{\varphi}(t)$ does not belong to F.

Example

To illustrate the results of this section, we shall consider a first-order autonomous equation

$$\dot{x} = \frac{1}{f(x)}, \tag{6}$$

where $f(x)$ is a polynomial all of whose roots are real and simple; let $a_1, a_2, \ldots, a_n$ be their enumeration in increasing order. The phase space of equation (6) is a straight line P, all of whose points can serve as the domain Δ with the exception of the points $a_1, a_2, \ldots, a_n$ where the right-hand side of (6) becomes infinite. If we set

$$F(x) = \int_0^x f(\xi)\, d\xi,$$

then the set of all solutions of (6) is described by the relation

$$F(x) = t + c.$$

Since in an autonomous equation a constant time shift does not change the trajectories, the set of all trajectories of (6), together with the description of the motion of $x(t)$ along them, is given by the relation $F(x) = t$. Let us study the motion of $x(t)$ along the interval $a_1 < x < a_2$. Since $f(x)$ does not change sign on the interval $a_1 < x < a_2$, we have $F(a_1) \neq$

$F(a_2)$. To be definite, we shall assume that

$$m_1 = F(a_1) < m_2 = F(a_2).$$

It is easy to see that as t traverses the interval $m_1 < t < m_2$, the point $x(t)$ traverses the interval $a_1 < x < a_2$. Hence it is obvious that $m_1 < t < m_2$ is the maximal interval of existence of the corresponding solution. Here both endpoints of this interval are finite; on the basis of proposition (B), this is explained by the fact that in approaching the endpoints of the interval $m_1 < t < m_2$, the point $x(t)$ approaches the boundary of Δ.

25. Global theorems of continuity and differentiability. Theorems 13, 14, and 15, which were proved in §22, establish certain properties of the solutions, i.e., continuity and differentiability with respect to the parameters and the initial values. All these theorems, however, were of a *local character*, that is, they referred to sufficiently small intervals of time. In this section we shall prove the so-called *integral theorems* or *global theorems* of continuity and differentiability of a solution with respect to the parameters and the initial conditions. The term "integral," as applied to Theorems 16, 17, and 18 to be proved here, has no relation to the operation of integration, but means that continuity and differentiability are established "integrally," i.e., on some "large" interval of time.

The proof of the integral theorem of continuity (Theorem 16) differs from that of Theorem 13 and will be presented without reference to Theorem 13. The integral theorems of differentiability (Theorems 17 and 18) are proved in exactly the same way as the corresponding local Theorems 14 and 15, so that their proofs will not be written out in detail.

We shall consider the normal system of equations

$$\dot{x}^i = f^i(t, x^1, \ldots, x^n, \mu^1, \ldots, \mu^l), \qquad i = 1, \ldots, n, \tag{1}$$

whose right-hand sides depend on the parameters $\mu^1, \ldots, \mu^l$. System (1) may be written in vector form

$$\dot{\mathbf{x}} = \mathbf{f}(t, \mathbf{x}, \boldsymbol{\mu}), \tag{2}$$

and from now on we shall assume that the right-hand sides of (1) are defined and continuous, together with their partial derivatives

$$\frac{\partial f^i(t, \mathbf{x}, \boldsymbol{\mu})}{\partial x^j} = f^i_j(t, \mathbf{x}, \boldsymbol{\mu}), \tag{3}$$

in a certain domain $\widetilde{\Gamma}$ of the space $\widetilde{R}$ of the variables

$$t, x^1, \ldots, x^n, \mu^1, \ldots, \mu^l.$$

Let

$$\mathbf{x} = \boldsymbol{\varphi}(t, \boldsymbol{\mu}) \tag{4}$$

be the solution of equation (2) with initial values t_0, $\mathbf{x}_0$. It will be proved that if the solution $\mathbf{x} = \boldsymbol{\varphi}(t, \boldsymbol{\mu}_0)$ is defined on the interval $r_1 \leq t \leq r_2$, then when $\boldsymbol{\mu}$ is sufficiently close to $\boldsymbol{\mu}_0$, the solution (4) is defined on the same interval and the function (4) is continuous in $(t, \boldsymbol{\mu})$. In addition, under the hypothesis that the partial derivatives $\partial\mathbf{f}/\partial\mu^k$ exist and are continuous in $\widetilde{\Gamma}$, it will be proved that (4) is differentiable with respect to the parameters $\mu^1, \ldots, \mu^l$ for $r_1 \leq t \leq r_2$ and for $\boldsymbol{\mu}$ sufficiently close to $\boldsymbol{\mu}_0$. It is from these theorems and the use of the construction (C) of §22 that we shall prove the corresponding integral theorem of continuity and differentiability with respect to the initial conditions.

It is now clear that the content of this section is very similar to that of §22; the main difference is that we now assume that one solution $\boldsymbol{\varphi}(t, \boldsymbol{\mu}_0)$ is given on a specific interval $r_1 \leq t \leq r_2$, and the existence, continuity, and differentiability of the solution $\boldsymbol{\varphi}(t, \boldsymbol{\mu})$ (for $\boldsymbol{\mu}$ close to $\boldsymbol{\mu}_0$) is established over the *entire* given interval $r_1 \leq t \leq r_2$. The situation is the same for the dependence on the initial conditions.

Continuous dependence of the solutions on the parameters. We shall prove first the following proposition, which plays an auxiliary role in the proof of the integral theorem of continuity.

(A) Let $u(t)$ be a continuous function of t on the interval $t_0 \leq t \leq t_1$; if $u(t)$ satisfies the integral inequality

$$u(t) \leq \int_{t_0}^{t} (\alpha u(\tau) + \beta)\, d\tau, \qquad \alpha > 0, \qquad \beta > 0, \tag{5}$$

on this interval, then the estimate,

$$u(t) \leq \frac{\beta}{\alpha} (e^{\alpha(t-t_0)} - 1) \tag{6}$$

is valid.

For the proof we set

$$v(t) = \int_{t_0}^{t} (\alpha u(\tau) + \beta)\, d\tau, \tag{7}$$

so that

$$\dot{v}(t) = \alpha u(t) + \beta.$$

From the last equality we have

$$u(t) = \frac{1}{\alpha} (\dot{v}(t) - \beta).$$

Then, from the original inequality we have

$$\frac{1}{\alpha} (\dot{v}(t) - \beta) \leq v(t),$$

or, what is the same thing,

$$\dot{v}(t) - \alpha v(t) \leq \beta.$$

If we multiply this last inequality by $e^{-\alpha t}$, we obtain

$$\dot{v}(t)e^{-\alpha t} - v(t)\alpha e^{-\alpha t} = \frac{d}{dt}\left(v(t)e^{-\alpha t}\right) \leq \beta e^{-\alpha t}.$$

Integrating the last inequality from t_0 to t and using the fact that $v(t_0) = 0$, which follows from (7), we find that

$$v(t)e^{-\alpha t} \leq \frac{\beta}{\alpha}\left(e^{-\alpha t_0} - e^{-\alpha t}\right),$$

or

$$v(t) \leq \frac{\beta}{\alpha}\left(e^{\alpha(t-t_0)} - 1\right). \tag{8}$$

From (7), (5), and (8), we obtain the required inequality (6).

THEOREM 16. Let $(t_0, \mathbf{x}_0, \boldsymbol{\mu}_0)$ be a certain point of the domain $\tilde{\Gamma}$ and let

$$\mathbf{x} = \boldsymbol{\varphi}(t, \boldsymbol{\mu}) \tag{9}$$

be a solution of equation (2) which satisfies the initial condition

$$\boldsymbol{\varphi}(t_0, \boldsymbol{\mu}) = \mathbf{x}_0.$$

If the solution

$$\mathbf{x} = \boldsymbol{\varphi}(t, \boldsymbol{\mu}_0)$$

is defined on the interval $r_1 \leq t \leq r_2$, then there exists a positive number ρ such that for $|\boldsymbol{\mu} - \boldsymbol{\mu}_0| < \rho$ the solution (9) is defined on the same interval $r_1 \leq t \leq r_2$ and the function $\boldsymbol{\varphi}(t, \boldsymbol{\mu})$ is continuous in the variables $t, \boldsymbol{\mu}$ for $|\boldsymbol{\mu} - \boldsymbol{\mu}_0| < \rho$, $r_1 \leq t \leq r_2$.

Proof. As the number t traverses the interval $r_1 \leq t \leq r_2$, the point $(t, \boldsymbol{\varphi}(t, \boldsymbol{\mu}_0), \boldsymbol{\mu}_0)$ describes some curve Q in the space $\tilde{R}$. We shall construct a certain closed neighborhood $\tilde{\Pi}$ of the curve Q. Let a and b be two positive numbers. We shall denote by $\tilde{\Pi}$ the set of all points $(t, \mathbf{x}, \boldsymbol{\mu})$ of $\tilde{R}$ which satisfy the conditions $r_1 \leq t \leq r_2$, $|\mathbf{x} - \boldsymbol{\varphi}(t, \boldsymbol{\mu}_0)| \leq a$, $|\boldsymbol{\mu} - \boldsymbol{\mu}_0| \leq b$. From the fact that the solution $\boldsymbol{\varphi}(t, \boldsymbol{\mu}_0)$ is continuous and passes through $\tilde{\Gamma}$ it follows that there exist positive numbers a and b such that the set $\tilde{\Pi}$ is contained in $\tilde{\Gamma}$. Henceforth we shall assume that a and b satisfy this condition. Since the derivatives (3) are continuous on the set $\tilde{\Pi}$ and are therefore bounded in modulus by some number K on $\tilde{\Pi}$, it follows from the inequality (6) of §21 that for any two points $(t, \mathbf{x}_1, \boldsymbol{\mu})$, $(t, \mathbf{x}_2, \boldsymbol{\mu})$ of $\tilde{\Pi}$ we have the relation

$$|\mathbf{f}(t, \mathbf{x}_2, \boldsymbol{\mu}) - \mathbf{f}(t, \mathbf{x}_1, \boldsymbol{\mu})| \leq n^2K|\mathbf{x}_2 - \mathbf{x}_1|. \tag{10}$$

Further, from the uniform continuity of the function $\mathbf{f}(t, \mathbf{x}, \boldsymbol{\mu})$ on the set $\tilde{\Pi}$ it follows that there exists a positive monotonic function $\beta(\epsilon)$ of ϵ which tends to zero with $\epsilon > 0$, such that for any two points $(t, \mathbf{x}, \boldsymbol{\mu}_0)$, $(t, \mathbf{x}, \boldsymbol{\mu})$ of $\tilde{\Pi}$ the relation

$$|\mathbf{f}(t, \mathbf{x}, \boldsymbol{\mu}) - \mathbf{f}(t, \mathbf{x}, \boldsymbol{\mu}_0)| < \beta(|\boldsymbol{\mu} - \boldsymbol{\mu}_0|) \tag{11}$$

is satisfied.

Now let $\mathbf{x} = \boldsymbol{\varphi}(t, \boldsymbol{\mu})$, $|\boldsymbol{\mu} - \boldsymbol{\mu}_0| \leq b$, be the solution of (2) with initial values $(t_0, \mathbf{x}_0)$. We shall assume that this solution is defined on the interval $t_1 \leq t \leq t_2$, where $r_1 \leq t_1 < t_0 < t_2 \leq r_2$, and is completely contained in the set $\tilde{\Pi}$, that is, for $t_1 < t < t_2$ the point $(t, \boldsymbol{\varphi}(t, \boldsymbol{\mu}), \boldsymbol{\mu})$ lies in $\tilde{\Pi}$. We shall estimate the difference $|\boldsymbol{\varphi}(t, \boldsymbol{\mu}) - \boldsymbol{\varphi}(t, \boldsymbol{\mu}_0)|$ on the interval. We shall carry out calculations only on the interval $t_0 \leq t \leq t_2$, since the calculations on the interval $t_1 \leq t \leq t_0$ are similar.

Let us write equation (2) in the form of an integral [see §21, (A)] for both values of the parameters $\boldsymbol{\mu}$ and $\boldsymbol{\mu}_0$, and subtract the second integral from the first; we then obtain

$$\boldsymbol{\varphi}(t, \boldsymbol{\mu}) - \boldsymbol{\varphi}(t, \boldsymbol{\mu}_0) = \int_{t_0}^{t} [\mathbf{f}(\tau, \boldsymbol{\varphi}(\tau, \boldsymbol{\mu}), \boldsymbol{\mu}) - \mathbf{f}(\tau, \boldsymbol{\varphi}(\tau, \boldsymbol{\mu}_0), \boldsymbol{\mu}_0)]\, d\tau, \qquad t_0 \leq t \leq t_2 .$$

Let us estimate the difference on the right under the integral sign. We have

$$|\mathbf{f}(\tau, \boldsymbol{\varphi}(\tau, \boldsymbol{\mu}), \boldsymbol{\mu}) - \mathbf{f}(\tau, \boldsymbol{\varphi}(\tau, \boldsymbol{\mu}_0), \boldsymbol{\mu}_0)| \leq |\mathbf{f}(\tau, \boldsymbol{\varphi}(\tau, \boldsymbol{\mu}), \boldsymbol{\mu}) - \mathbf{f}(\tau, \boldsymbol{\varphi}(\tau, \boldsymbol{\mu}_0), \boldsymbol{\mu})| + |\mathbf{f}(\tau, \boldsymbol{\varphi}(\tau, \boldsymbol{\mu}_0), \boldsymbol{\mu}) - \mathbf{f}(\tau, \boldsymbol{\varphi}(\tau, \boldsymbol{\mu}_0), \boldsymbol{\mu}_0)|.$$

The first term on the right-hand side can be estimated by means of the inequality (10) and the second by the inequality (11). Combining these two estimates, we obtain

$$|\boldsymbol{\varphi}(t, \boldsymbol{\mu}) - \boldsymbol{\varphi}(t, \boldsymbol{\mu}_0)| \leq \int_{t_0}^{t} [n^2K \cdot |\boldsymbol{\varphi}(\tau, \boldsymbol{\mu}) - \boldsymbol{\varphi}(\tau, \boldsymbol{\mu}_0)| + \beta(|\boldsymbol{\mu} - \boldsymbol{\mu}_0|)]\, d\tau.$$

If we set $u(t) = |\boldsymbol{\varphi}(t, \boldsymbol{\mu}) - \boldsymbol{\varphi}(t, \boldsymbol{\mu}_0)|$, it follows from proposition (A) that, for $t_0 \leq t \leq t_2 \leq r_2$,

$$|\boldsymbol{\varphi}(t, \boldsymbol{\mu}) - \boldsymbol{\varphi}(t, \boldsymbol{\mu}_0)| \leq \frac{\beta(|\boldsymbol{\mu} - \boldsymbol{\mu}_0|)}{n^2K} (e^{n^2K(t-t_0)} - 1)$$
$$< \frac{\beta(|\boldsymbol{\mu} - \boldsymbol{\mu}_0|)}{n^2K} (e^{n^2K(r_2-r_1)} - 1) = C\beta(|\boldsymbol{\mu} - \boldsymbol{\mu}_0|).$$

By making a similar estimate on the interval $t_1 \leq t \leq t_0$, we obtain

$$|\boldsymbol{\varphi}(t, \boldsymbol{\mu}) - \boldsymbol{\varphi}(t, \boldsymbol{\mu}_0)| \leq C\beta(|\boldsymbol{\mu} - \boldsymbol{\mu}_0|). \tag{12}$$

Thus the estimate (12) is valid over the entire interval $t_1 \leq t \leq t_2$.

Let ρ be a positive number which satisfies the inequalities

$$\rho \leq b, \qquad C\beta(\rho) < a, \tag{13}$$

and $\boldsymbol{\mu}_1$ a fixed value of $\boldsymbol{\mu}$, for which

$$|\boldsymbol{\mu}_1 - \boldsymbol{\mu}_0| \leq \rho.$$

We shall show that the solution $\boldsymbol{\varphi}(t, \boldsymbol{\mu}_1)$ is defined on the entire interval $r_1 \leq t \leq r_2$.

Let $m_1 < t < m_2$ be the maximal interval of existence of the solution $\boldsymbol{\varphi}(t, \boldsymbol{\mu}_1)$ [see §3, (A)]. We shall show that $m_2 > r_2$ (the inequality $m_1 < r_1$ is proved analogously). For this we shall denote by E the set of all points $(t, \mathbf{x}, \boldsymbol{\mu})$ of $\widetilde{\Pi}$ for which $\boldsymbol{\mu} = \boldsymbol{\mu}_1$. Let us assume that $m_2 \leq r_2$. It then follows from (A) of §24 that the point $(t, \boldsymbol{\varphi}(t, \boldsymbol{\mu}_1), \boldsymbol{\mu}_1)$ must leave the closed set E as $t \to m_2$. This is possible only by violating the inequality $|\boldsymbol{\varphi}(t, \boldsymbol{\mu}_1) - \boldsymbol{\varphi}(t, \boldsymbol{\mu}_0)| \leq a$, since the inequality $t \leq r_2$ holds by hypothesis. Since the inequality $|\boldsymbol{\varphi}(t, \boldsymbol{\mu}_1) - \boldsymbol{\varphi}(t, \boldsymbol{\mu}_0)| \leq a$ is violated for some value of t, there exists a value of t, which we shall denote by t_2, such that

$$|\boldsymbol{\varphi}(t_2, \boldsymbol{\mu}_1) - \boldsymbol{\varphi}(t_2, \boldsymbol{\mu}_0)| = a,$$
$$|\boldsymbol{\varphi}(t, \boldsymbol{\mu}_1) - \boldsymbol{\varphi}(t, \boldsymbol{\mu}_0)| < a \qquad \text{for} \qquad t_0 < t < t_2.$$

This is impossible, however, because of the estimate (12) [see (13)].

Thus the solution $\boldsymbol{\varphi}(t, \boldsymbol{\mu})$ is defined on the entire interval $r_1 \leq t \leq r_2$ for $|\boldsymbol{\mu} - \boldsymbol{\mu}_0| \leq \rho$. Thus the first assertion in Theorem 16 is proved; we shall prove the second assertion, i.e., the continuity of the function $\boldsymbol{\varphi}(t, \boldsymbol{\mu})$ for $r_1 \leq t \leq r_2$, $|\boldsymbol{\mu} - \boldsymbol{\mu}_0| \leq \rho$. We shall prove the continuity of $\boldsymbol{\varphi}(t, \boldsymbol{\mu})$ at the point $(t_0^*, \boldsymbol{\mu}_0^*)$, where $r_1 \leq t_0^* \leq r_2$, $|\boldsymbol{\mu}_0^* - \boldsymbol{\mu}_0| \leq \rho$. Since the solution $\boldsymbol{\varphi}(t, \boldsymbol{\mu}_0^*)$ satisfies the hypotheses of Theorem 16, it follows from what we have already proved that there exist for it a positive number ρ^*, analogous to ρ, and a function $C^*\beta^*(\epsilon)$, analogous to $C\beta(\epsilon)$. Now let $\boldsymbol{\mu}$ be a value of the parameter satisfying the condition $|\boldsymbol{\mu} - \boldsymbol{\mu}_0^*| \leq \rho^*$ and t a number satisfying the condition $r_1 \leq t \leq r_2$. Let us estimate the difference $\boldsymbol{\varphi}(t, \boldsymbol{\mu}) - \boldsymbol{\varphi}(t_0^*, \boldsymbol{\mu}_0^*)$. We have

$$|\boldsymbol{\varphi}(t, \boldsymbol{\mu}) - \boldsymbol{\varphi}(t_0^*, \boldsymbol{\mu}_0^*)| \leq |\boldsymbol{\varphi}(t, \boldsymbol{\mu}) - \boldsymbol{\varphi}(t, \boldsymbol{\mu}_0^*)| + |\boldsymbol{\varphi}(t, \boldsymbol{\mu}_0^*) - \boldsymbol{\varphi}(t_0^*, \boldsymbol{\mu}_0^*)|.$$

The second term on the right-hand side tends to zero as $t \to t_0^*$ because of the continuity of $\boldsymbol{\varphi}(t, \boldsymbol{\mu}_0^*)$ in t, while the first tends to zero as $\boldsymbol{\mu} \to \boldsymbol{\mu}_0^*$

(uniformly with respect to t) because of the existence of $C^*\beta^*(\epsilon)$. This proves Theorem 16 completely.

Differentiability of the solution with respect to the parameters.

THEOREM 17. Let us assume that the partial derivatives

$$e_k^i(t, \mathbf{x}, \boldsymbol{\mu}) = \frac{\partial}{\partial \mu^k} f^i(t, \mathbf{x}, \boldsymbol{\mu}), \qquad i = 1, \ldots, n, \quad k = 1, \ldots, l,$$

of the right-hand sides of the system (1) exist and are continuous in the domain $\widetilde{\Gamma}$. Let $(t_0, \mathbf{x}_0, \boldsymbol{\mu}_0)$ be a certain point of $\widetilde{\Gamma}$ and

$$\mathbf{x} = \boldsymbol{\varphi}(t, \boldsymbol{\mu}) \tag{14}$$

a solution of equation (2) which satisfies the initial condition

$$\boldsymbol{\varphi}(t_0, \boldsymbol{\mu}) = \mathbf{x}_0. \tag{15}$$

If the solution

$$\mathbf{x} = \boldsymbol{\varphi}(t, \boldsymbol{\mu}_0)$$

is defined on the interval $r_1 \leq t \leq r_2$, then there exists a positive number ρ' such that for $|\boldsymbol{\mu} - \boldsymbol{\mu}_0| < \rho'$ the solution (14) is defined on the interval $r_1 \leq t \leq r_2$ and has continuous partial derivatives

$$\frac{\partial \varphi(t, \boldsymbol{\mu})}{\partial \mu^k}; \tag{16}$$

the mixed partial derivatives

$$\frac{\partial^2 \boldsymbol{\varphi}(t, \boldsymbol{\mu})}{\partial t \, \partial \mu^k}$$

are also continuous and do not depend on the order of differentiation. In addition, the partial derivatives (16) satisfy on the interval $r_1 \leq t \leq r_2$ the system of variational equations

$$\frac{\partial}{\partial t}\left(\frac{\partial \varphi^i(t, \boldsymbol{\mu})}{\partial \mu^k}\right) = \sum_{j=1}^{n} f_j^i(t, \boldsymbol{\varphi}(t, \boldsymbol{\mu}), \boldsymbol{\mu}) \frac{\partial \varphi^j(t, \boldsymbol{\mu})}{\partial \mu^k} + e_k^i(t, \boldsymbol{\varphi}(t, \boldsymbol{\mu}), \boldsymbol{\mu})$$

under the initial condition

$$\frac{\partial \varphi^i(t_0, \boldsymbol{\mu})}{\partial \mu^k} = 0.$$

Proof. The proof of Theorem 17 repeats almost verbatim the proof of Theorem 14 with the following modifications. The reference to Theorem 13 is replaced by a reference to Theorem 16. In addition, the solution $\boldsymbol{\varphi}(t, \boldsymbol{\mu})$

is considered on the *entire* interval $r_1 \leq t \leq r_2$, and not on the interval $|t - t_0| < r$; the same is true of the functions (20) of §22. Analogous modifications must be made in order to prove proposition (B) below and Theorem 18.

We shall now make one simple deduction from Theorem 17.

(B) If all partial derivatives of the right-hand sides of system (1) with respect to the variables $x^1, \ldots, x^n, \mu^1, \ldots, \mu^l$ up to order m inclusive exist and are continuous, then the functions $\varphi^i(t, \boldsymbol{\mu})$, $i = 1, \ldots, n$, which comprise a solution of (1) and satisfy the initial conditions (15), also have, for $|\boldsymbol{\mu} - \boldsymbol{\mu}_0| < \rho'$, $r_1 \leq t \leq r_2$ (where ρ' is a sufficiently small positive number), continuous partial derivatives up to order m inclusive with respect to the parameters $\mu^1, \ldots, \mu^l$.

[See the proof of proposition (B) of §22.]

Continuous dependence and differentiability of the solutions in terms of the initial values. By means of the construction given in (B) of §22, we can obtain the following result from Theorems 16 and 17.

THEOREM 18. Let

$$\dot{x}^i = f^i(t, x^1, \ldots, x^n), \qquad i = 1, \ldots, n,$$

be a normal system of differential equations whose right-hand sides are defined and continuous, together with their partial derivatives

$$f^i_j(t, \mathbf{x}) = \frac{\partial}{\partial x^j} f^i(t, \mathbf{x}),$$

in some domain Γ of the space of the variables $t, x^1, \ldots, x^n$. In addition, let

$$\dot{\mathbf{x}} = \mathbf{f}(t, \mathbf{x}) \tag{17}$$

be the vector representation of this system. If $(t_0, \mathbf{x}_0)$ is an arbitrary point of Γ and if the solution

$$\mathbf{x} = \boldsymbol{\varphi}(t)$$

of equation (17) with the initial values t_0, $\mathbf{x}_0$ is defined on the interval $r_1 \leq t \leq r_2$, then there exists a positive number ρ' such that for

$$r_1 \leq t \leq r_2, \qquad |\tau - t_0| < \sigma', \qquad |\boldsymbol{\xi} - \mathbf{x}_0| < \sigma',$$

the solution

$$\mathbf{x} = \boldsymbol{\varphi}(t, \tau, \boldsymbol{\xi})$$

of equation (17) with initial values τ, $\boldsymbol{\xi}$ is defined and continuous in the variables $t, \tau, \boldsymbol{\xi}$ and has continuous partial derivatives with respect to the components $\xi^1, \ldots, \xi^n$ of the vector ξ, while the mixed partial derivatives

$$\frac{\partial^2 \varphi^i(t, \tau, \xi)}{\partial t\, \partial \xi^j}$$

are continuous and do not depend on the order of differentiation. In addition, the functions

$$\varphi_k^i(t) = \frac{\partial}{\partial \xi^k} \varphi^i(t, t_0, \xi)|_{\xi = \mathbf{x}_0}$$

satisfy the linear system of equations (the variational equations)

$$\dot{\varphi}_k^i(t) = \sum_{j=1}^{n} f_j^i(t, \varphi(t))\varphi_k^j(t)$$

under the initial conditions

$$\varphi_k^i(t_0) = \delta_k^i.$$

(See the proof of Theorem 15.)

(C) When the conditions of Theorem 18 are satisfied for any positive number ϵ, a positive number δ can be found such that for $|\tau - t_0| < \delta$, $|\xi - \mathbf{x}_0| < \delta$ the inequality

$$|\varphi(t, \tau, \xi) - \varphi(t, t_0, \mathbf{x}_0)| < \epsilon \tag{18}$$

is valid for any t in the interval $r_1 \leq t \leq r_2$.

To prove this inequality, we note that for $0 < \sigma'' < \sigma'$ the function $\varphi(t, \tau, \xi)$, which is continuous by Theorem 18, is uniformly continuous on the closed set defined by the inequalities

$$|\tau - t_0| \leq \sigma'', \qquad |\xi - \mathbf{x}_0| \leq \sigma'', \qquad r_1 \leq t \leq r_2.$$

Thus for any positive ϵ, a positive δ can be found such that for $|\tau - t_0| < \delta$, $|\xi - \mathbf{x}_0| < \delta$, $|t_1 - t_2| < \delta$, the inequality

$$|\varphi(t_2, \tau, \xi) - \varphi(t_1, t_0, \mathbf{x}_0)| < \epsilon$$

is valid. Inequality (18) follows from this inequality for $t_1 = t_2 = t$.

CHAPTER 5

STABILITY

The performance of numerous mechanical, electrical, and other types of devices (machines, instruments, etc.) is described by systems of ordinary differential equations. A system of ordinary differential equations always has an infinite number of solutions, and in order to find a certain definite solution it is necessary to specify its initial values. However, devices which are used in practice usually operate under completely well-defined conditions, and in their performance it is impossible, at first glance at any rate, to discover an infinite number of operational conditions which correspond to the various solutions of the system of equations. This can be explained either by the fact that the initial values of the solution are chosen at the start of the operation in a certain well-defined way, or by the fact that the initial values lose their effect during continued operation of the device, and the device itself stabilizes its operation at a *stationary* solution. We have already encountered the latter phenomenon when we analyzed the performance of electrical circuits. We shall give one more example. A clock runs with a completely prescribed amplitude of its pendulum, although when the clock is started, the pendulum can deviate either slightly or greatly from the vertical position. If in starting the clock the pendulum is not deflected sufficiently, then it will stop after a few oscillations. If the deflection is great enough, then after a short time the oscillation amplitude of the pendulum becomes quite well-defined, and the clock will run with this amplitude for an indefinite period of time, if not forever. Thus the system of equations which describes the performance of the clock has two stationary solutions: a state of equilibrium which corresponds to the rest position, and a periodic solution which corresponds to what we think of as the normal performance of the clock. Every other solution—and there are undoubtedly an infinite number of these solutions—approaches one of these two stationary solutions very rapidly, and after a short time becomes practically indistinguishable from it. Each of the two stationary solutions noted appears to be in a certain sense *stable*. This means that if we take a solution which is not stationary, but which deviates very little from a stationary solution at time zero, then the non-stationary solution approaches the stationary solution. This is a definition of the stability of a solution, though not formulated in precise terms. From this example it is seen that the phase space of the system of equations which describes the performance of the clock decomposes into two attractive *domains*. If the initial value is taken in one of the domains, then the solution will tend toward the state of equilibrium; if the initial value is taken in the other domain, then the solution will tend toward the periodic solution.

From what we have seen, it is clear that in order to understand completely the performance of any device we must analyze the phase space of the system of equations which describes the performance of the device. Here it is most important to know all *stable solutions* of this system of equations.

From the integral theorem on the continuous dependence of a solution on its initial values (see §25), we already know that if a definite *finite* interval of time is set, then for a sufficiently small deviation of the initial values, the solution will undergo only a small deviation on the entire time interval; this property of the solution, however, does not necessarily mean stability. Where stability is concerned, the deviation on an *infinitely large* time interval must be small only if the deviation from the initial values is small.

The present chapter is devoted basically to the problem of stability of states of equilibrium and of periodic solutions.

We also include two important applications to engineering problems: Vyshnegradskiy's work on the performance of a steam engine with Watt's governor and Andronov's work on the operation of a vacuum-tube oscillator generating nondamping electrical oscillations. The first of these studies has been the basic principle in the theory of automatic control, and the second, in the theory of nonlinear oscillations.

In §30 we shall study the behavior of trajectories close to the equilibrium states of a second-order autonomous system, a problem which does not pertain entirely to the stability problem. This section is somewhat more difficult than the average level of the book. Still more difficult in content is the last section of this chapter (§31).

26. Lyapunov's theorem. We shall describe here the concept of stability and sufficient stability conditions in connection with the equilibrium state of an autonomous system (see §15).

The stability of an equilibrium state. Let

$$\dot{x}^i = f^i(x^1, \ldots, x^n), \qquad i = 1, \ldots, n, \tag{1}$$

be a normal autonomous system and let

$$\dot{\mathbf{x}} = \mathbf{f}(\mathbf{x}) \tag{2}$$

describe the system in vector notation. We shall assume that the functions

$$f^i(x^1, \ldots, x^n), \qquad i = 1, \ldots, n, \tag{3}$$

are defined and have continuous first-order partial derivatives in a certain domain Δ of the space of variables $x^1, \ldots, x^n$. Later, in establishing stability criteria, we shall strengthen the differentiability requirements by assuming that the functions (3) have continuous second-order partial derivatives in Δ.

Without giving the formal definition of Lyapunov stability, we shall first try to convey the idea of stability. The equilibrium state $\mathbf{a} = (a^1, \ldots, a^n)$ of equation (2) should be considered *stable* if any solution of equation (2) starting at $t = 0$ from a point sufficiently close to $\mathbf{a}$ remains in the neighborhood of $\mathbf{a}$ during subsequent variation (i.e., for $t > 0$). The physical sense of stability is clear. A physical object (for example, some machine) whose motion is controlled by equation (2) can be in the equilibrium state $\mathbf{a}$ only if this state of equilibrium is stable, since a trivial deviation from the state of equilibrium caused by a random impulse can force the object far out of the state of equilibrium.

Below, we shall denote by $\boldsymbol{\varphi}(t, \boldsymbol{\xi})$ the solution of equation (2) with initial values $t = 0$, $\mathbf{x} = \boldsymbol{\xi}$, so that $\boldsymbol{\varphi}(t, \boldsymbol{\xi})$ is a vector function of the scalar variable t and the vector variable $\boldsymbol{\xi}$, which satisfies the condition

$$\boldsymbol{\varphi}(0, \boldsymbol{\xi}) = \boldsymbol{\xi}. \tag{4}$$

Definition. The equilibrium state $\mathbf{a}$ of equation (2) is called *Lyapunov stable* if (1) there exists a positive number ρ so small that for $|\boldsymbol{\xi} - \mathbf{a}| < \rho$ the solution $\boldsymbol{\varphi}(t, \boldsymbol{\xi})$ of (2) is defined for all positive values of t; (2) for any positive number ϵ there exists a positive number $\delta < \rho$ such that for $|\boldsymbol{\xi} - \mathbf{a}| < \delta$ we have $|\boldsymbol{\varphi}(t, \boldsymbol{\xi}) - \mathbf{a}| < \epsilon$ for all $t > 0$. An equilibrium state $\mathbf{a}$ of equation (2) which is Lyapunov stable is called *asymptotically stable* if (3) there exists a small positive number $\sigma < \rho$ such that for $|\boldsymbol{\xi} - \mathbf{a}| < \sigma$ we have

$$\lim_{t \to +\infty} |\boldsymbol{\varphi}(t, \boldsymbol{\xi}) - \mathbf{a}| = 0.$$

We shall first give sufficient conditions for the stability of the equilibrium state for a linear homogeneous system with constant coefficients.

(A) Let

$$\dot{\mathbf{x}} = A\mathbf{x} \tag{5}$$

be a linear homogeneous equation with constant coefficients written in vector form. Its solution with initial values 0, $\boldsymbol{\xi}$ will be denoted by $\boldsymbol{\psi}(t, \boldsymbol{\xi})$. If all eigenvalues of the matrix A have negative real parts, then there exist positive numbers α and r such that the inequality

$$|\boldsymbol{\psi}(t, \boldsymbol{\xi})| \leq r|\boldsymbol{\xi}|e^{-\alpha t}, \qquad t \geq 0, \tag{6}$$

holds. It follows directly from (6) that the equilibrium state $\mathbf{x} = 0$ of equation (5) is Lyapunov stable and asymptotically stable.

Let us prove inequality (6). Let us set

$$A = (a^i_j), \qquad L(p) = (a^i_j - p\delta^i_j).$$

Then, by using the differentiation symbol p (see §7), equation (5) can be written in a scalar form as the system

$$\sum_{j=1}^{n} L^i_j(p)x^j = 0, \qquad i = 1, \ldots, n. \tag{7}$$

Let $M^j_i(p)$ be the minor of the element $L^i_j(p)$ of the matrix $L(p)$ taken with the appropriate sign, so that the identity

$$\sum_{i=1}^{n} M^k_i(p)L^i_j(p) = \delta^k_j D(p)$$

holds, where $D(p)$ is the determinant of $L(p)$. By multiplying (7) by the polynomial $M^k_i(p)$ and summing with respect to i, we obtain

$$0 = \sum_{i=1}^{n}\sum_{j=1}^{n} M^k_i(p)L^i_j(p)x^j = \sum_{j=1}^{n} \delta^k_j D(p)x^j = D(p)x^k.$$

Thus, if

$$\mathbf{x} = (x^1, \ldots, x^n)$$

is some solution of equation (5), then every function x^i satisfies the differential equation

$$D(p)x^i = 0.$$

Since, by hypothesis, all roots of the polynomial $D(p)$ have negative real parts [see §9, (A)], for the function x^i the inequality

$$|x^i| \leq Re^{-\alpha t}, \qquad i = 1, \ldots, n, \quad t \geq 0,$$

holds, where R and α are positive numbers which do not depend on the number i. From this inequality follows the inequality

$$|\mathbf{x}| \leq \sqrt{n}\, Re^{-\alpha t}, \qquad t \geq 0,$$

which has already been proved [see §11, (B)] under more general assumptions; here this proof will be carried out again.

Let $\mathbf{e}_i$ be the unit coordinate vector corresponding to i, so that

$$\mathbf{e}_i = (0, \ldots, 1, \ldots, 0),$$

where the ith place contains the number 1. Further, let $\psi_i(t)$ be the solution of (5) with initial value $\mathbf{e}_i$, so that

$$\psi_i(0) = \mathbf{e}_i, \qquad i = 1, \dots, n.$$

Then the solution $\psi(t, \xi)$ of (5) with the initial value

$$\xi = (\xi^1, \dots, \xi^n)$$

can obviously be written in the form

$$\psi(t, \xi) = \sum_{i=1}^{n} \xi^i \psi_i(t). \tag{8}$$

Since the inequality

$$|\psi_i(t)| \leq \sqrt{n}\, Re^{-\alpha t}, \qquad t \geq 0,$$

holds for every solution $\psi_i(t)$, it follows that inequality (6) holds for the solution $\psi(t, \xi)$.

Lyapunov stability of the equilibrium state $\mathbf{x} = 0$ follows directly from (6). Actually, if ϵ is a given positive number, then it is sufficient to take for δ the number ϵ/r. Asymptotic stability also follows from (6).

The Lyapunov function. In establishing a criterion for the equilibrium state of the *nonlinear* system (1), the so-called *differentiation with respect to a system of equations* is used; this type of differentiation finds other applications in addition to the proof of Lyapunov's theorem.

(B) Let

$$F(x^1, \dots, x^n) = F(\mathbf{x})$$

be some differentiable function of variables $x^1, \dots, x^n$ defined in a domain Δ. Its t-derivative with respect to the system of equations (1) is defined at the point $\mathbf{x} = (x^1, \dots, x^n)$ in the following manner. Let $\varphi(t)$ be a solution of equation (2) which at some value $t = t_0$ satisfies the initial condition

$$\varphi(t_0) = \mathbf{x}.$$

The derivative

$$\dot{F}_{(1)}(\mathbf{x})$$

with respect to system (1) is defined by the formula

$$\dot{F}_{(1)}(\mathbf{x}) = \frac{d}{dt} F(\varphi(t))|_{t=t_0},$$

or by means of the total derivative formula

$$\dot{F}_{(1)}(\mathbf{x}) = \sum_{i=1}^{n} \frac{\partial F(\mathbf{x})}{\partial x^i} f^i(\mathbf{x}). \tag{9}$$

From formula (9) it is evident that the derivative $\dot{F}_{(1)}(\mathbf{x})$ does not depend on the solution $\boldsymbol{\varphi}(t)$ but is determined uniquely by the choice of the point $\mathbf{x}$.

We shall now prove a property of an autonomous system.

(C) The solution of the autonomous system (2) with initial values 0, $\boldsymbol{\xi}$ will be denoted as before by $\boldsymbol{\varphi}(t, \boldsymbol{\xi})$. The function $\boldsymbol{\varphi}(t, \boldsymbol{\xi})$ satisfies the identity

$$\boldsymbol{\varphi}(t, \boldsymbol{\varphi}(s, \boldsymbol{\xi})) = \boldsymbol{\varphi}(s + t, \boldsymbol{\xi}). \tag{10}$$

We shall prove formula (10). Let us set

$$\boldsymbol{\eta} = \boldsymbol{\varphi}(s, \boldsymbol{\xi}), \tag{11}$$

where s is a fixed number, and consider the solution

$$\boldsymbol{\varphi}_1(t) = \boldsymbol{\varphi}(t, \boldsymbol{\eta})$$

of equation (2). Since $\boldsymbol{\varphi}(t, \boldsymbol{\xi})$ is a solution of (2) and since this equation is autonomous [see §15, (A)], the function $\boldsymbol{\varphi}_2(t)$, which is determined by the relation

$$\boldsymbol{\varphi}_2(t) = \boldsymbol{\varphi}(t + s, \boldsymbol{\xi}),$$

is also a solution. We thus have two solutions, $\boldsymbol{\varphi}_1(t)$ and $\boldsymbol{\varphi}_2(t)$, of equation (2). Furthermore,

$$\boldsymbol{\varphi}_1(0) = \boldsymbol{\varphi}(0, \boldsymbol{\eta}) = \boldsymbol{\eta}$$

[see (4)], and

$$\boldsymbol{\varphi}_2(0) = \boldsymbol{\varphi}(s, \boldsymbol{\xi}) = \boldsymbol{\eta}$$

[see (11)]. Therefore, since the solutions $\boldsymbol{\varphi}_1(t)$ and $\boldsymbol{\varphi}_2(t)$ have common initial values, they coincide and relation (10) holds.

In the proof of Lyapunov's theorem the basic role is played by a certain positive definite quadratic form called the *Lyapunov function*. First we shall note certain properties of positive definite quadratic forms [see (D)], and then we shall construct the Lyapunov function itself [see (E)].

(D) Let

$$\mathbf{x} = (x^1, \ldots, x^n) \tag{12}$$

be a variable vector in an n-dimensional space. By a *quadratic form* of the vector $\mathbf{x}$ we shall mean a function $W(\mathbf{x})$ determined by the formula

$$W(\mathbf{x}) = \sum_{i,j=1}^{n} w_{ij} x^i x^j,$$

where $w_{ij} = w_{ji}$ are real numbers. The quadratic form $W(\mathbf{x})$ is called *positive definite* if for $\mathbf{x} \neq 0$ we have

$$W(\mathbf{x}) > 0.$$

We find that for any positive definite quadratic form $W(\mathbf{x})$, it is always possible to choose two positive numbers μ and ν such that the inequality

$$\mu|\mathbf{x}|^2 \leq W(\mathbf{x}) \leq \nu|\mathbf{x}|^2 \tag{13}$$

holds for an arbitrary vector $\mathbf{x}$. From this it follows that for an arbitrary $\mathbf{x}$ [see (12)] the inequality

$$|x^i| \leq \sqrt{\frac{1}{\mu} W(\mathbf{x})} \tag{14}$$

holds.

We shall prove the existence of the numbers μ and ν. For this we shall consider the values of the function $W(\boldsymbol{\xi})$ when the vector $\boldsymbol{\xi}$ belongs to the unit sphere, i.e., when $\boldsymbol{\xi}$ satisfies the condition

$$|\boldsymbol{\xi}| = 1. \tag{15}$$

Since the sphere (15) is a closed bounded set and the function $W(\boldsymbol{\xi})$ is continuous, it attains its minimum μ and its maximum ν on the sphere (15). Since all vectors of the sphere (15) are nonzero, the numbers μ and ν are positive. Let $\mathbf{x}$ be an arbitrary vector; then we have $\mathbf{x} = \lambda\boldsymbol{\xi}$, where the vector $\boldsymbol{\xi}$ belongs to the sphere (15) and therefore the inequalities

$$\mu \leq W(\boldsymbol{\xi}) \leq \nu$$

hold for the vector $\boldsymbol{\xi}$. By multiplying this last inequality by λ^2, we obtain the inequality (13).

We shall now proceed to the construction of the Lyapunov function.

(E) Let

$$\dot{x}^i = \sum_{j=1}^{n} a^i_j x^j, \qquad i = 1, \ldots, n, \tag{16}$$

be a linear homogeneous system of equations with constant coefficients, where all eigenvalues of the matrix $A = (a^i_j)$ have negative real parts. Then there exists a positive definite quadratic form $W(\mathbf{x})$ whose derivative with respect to the system (16) [see (B)] satisfies the inequality

$$\dot{W}_{(16)}(\mathbf{x}) \leq -\beta W(\mathbf{x}), \tag{17}$$

where $\mathbf{x}$ is an arbitrary vector and β is a positive number independent of $\mathbf{x}$.

Let us construct the form $W(\mathbf{x})$. We shall assume that the system (16) is the scalar form of the vector equation (5). The solution of (5) with initial values 0, ξ will be denoted by $\psi(t, \xi)$ as in proposition (A); we then have

$$\psi(t, \xi) = \sum_{i=1}^{n} \xi^i \psi_i(t) \tag{18}$$

[see (8)]. We now form

$$W(\xi) = \int_0^\infty |\psi(\tau, \xi)|^2 \, d\tau. \tag{19}$$

It follows from (18) that

$$W(\xi) = \sum_{i\ j=1}^{n} \xi^i \xi^j \int_0^\infty (\psi_i(\tau), \psi_j(\tau)) \, d\tau. \tag{20}$$

Since every function $\psi_i(t)$ satisfies (6), each improper integral on the right-hand side of (20) converges, so that $W(\mathbf{x})$ is a quadratic form for the vector ξ. This quadratic form is positive definite, since for $\xi \neq 0$ the integrand of (19) is positive, so that $W(\xi) > 0$. Let us now calculate the derivative $\dot{W}'_{(16)}(\xi)$ of the function $W(\xi)$ with respect to the system (16). To do this, according to the hypotheses of (B), we draw the solution $\psi(t, \xi)$ through the point ξ and then calculate the derivative of the function $W[\psi(t, \xi)]$ at $t = 0$. We remark first that, as a consequence of (C),

$$\psi(\tau, \psi(t, \xi)) = \psi(\tau + t, \xi),$$

so that

$$\begin{aligned} W(\psi(t, \xi)) &= \int_0^\infty |\psi(\tau, \psi(t, \xi))|^2 \, d\tau \\ &= \int_0^\infty |\psi(t + \tau, \xi)|^2 \, d\tau = \int_0^\infty |\psi(\tau, \xi)|^2 \, d\tau. \end{aligned}$$

Thus we have

$$\begin{aligned} \dot{W}_{(16)}(\xi) &= \frac{d}{dt} W(\psi(t, \xi)) \Big|_{t=0} = \frac{d}{dt} \int_t^\infty |\psi(\tau, \xi)|^2 \, d\tau \Big|_{t=0} \\ &= -|\psi(t, \xi)|^2|_{t=0} = -|\xi|^2. \end{aligned}$$

We have thus obtained the equality

$$\dot{W}_{(16)}(\xi) = -|\xi|^2,$$

but, by the second of the inequalities (13), we have

$$-|\xi|^2 \leq -\frac{1}{\nu} W(\xi),$$

and therefore we obtain

$$\dot{W}_{(16)}(\xi) \leq -\frac{1}{\nu} W(\xi).$$

Thus inequality (17) is proved.

Lyapunov's theorem. Let us proceed, finally, to the formulation and proof of Lyapunov's theorem.

Let

$$\mathbf{a} = (a^1, \ldots, a^n)$$

be an equilibrium state of the autonomous system (1). We shall set

$$x^i = a^i + \Delta x^i, \qquad i = 1, 2, \ldots, n, \tag{21}$$

and take as new unknown functions the quantities

$$\Delta x^1, \ldots, \Delta x^n. \tag{22}$$

Substituting (21) into the system (1) and expanding the right-hand sides into Taylor series in the variables (22), we obtain

$$\Delta \dot{x}^i = f^i(\mathbf{a}) + \sum_{j=1}^{n} \frac{\partial f^i(\mathbf{a})}{\partial x^j} \Delta x^j + R^i, \qquad i = 1, \ldots, n, \tag{23}$$

where R^i is an infinitesimal of the second order with respect to the unknowns (22). Since $\mathbf{a}$ is an equilibrium state of the system (1),

$$f^i(\mathbf{a}) = 0;$$

in addition, if we set

$$a^i_j = \frac{\partial f^i(\mathbf{a})}{\partial x^j}, \tag{24}$$

we can write (23) in the form

$$\Delta \dot{x}^i = \sum_{j=1}^{n} a^i_j \Delta x^j + R^i, \qquad i = 1, \ldots, n. \tag{25}$$

THEOREM 19. If all eigenvalues of the matrix $A = (a^i_j)$ [see (24)] have negative real parts, then the equilibrium state $\mathbf{a}$ of the system (1) is asymptotically stable; more precisely, there exists a positive number σ so small that for $|\xi - \mathbf{a}| < \sigma$, the inequality

$$|\varphi(t, \xi) - \mathbf{a}| \leq r|\xi - \mathbf{a}|e^{-\alpha t}, \tag{26}$$

holds, where r and α are positive numbers which do not depend on ξ.

Proof. We shall assume that the equilibrium state $\mathbf{a}$ of (1) coincides with the origin, i.e., that $\mathbf{a} = 0$. This can always be attained by parallel translation of axes; the matrix A is invariant under this transformation. Assuming that $\mathbf{a} = 0$, we have

$$\Delta x^i = x^i,$$

so that system (25) may be written in the form

$$\dot{x}^i = \sum_{j=1}^{n} a^i_j x^j + R^i, \qquad i = 1, \ldots, n, \tag{27}$$

where

$$R^i = \frac{1}{2} \sum_{j,k=1}^{n} \frac{\partial^2 f^i(\theta \mathbf{x})}{\partial x^j \, \partial x^k} \, x^j x^k.$$

Now let $W(\mathbf{x})$ be a Lyapunov function [see (E)] for the linear system

$$\dot{x}^i = \sum_{j=1}^{n} a^i_j x^j, \qquad i = 1, \ldots, n, \tag{28}$$

which can be obtained from (27) by linearization, that is, by discarding the remainder terms R^i. If we calculate the derivative $\dot{W}_{(27)}(\mathbf{x})$ of the function $W(\mathbf{x})$ with respect to the system (27), we have

$$\begin{aligned}\dot{W}_{(27)}(x) &= \sum_{i,j=1}^{n} \frac{\partial W(\mathbf{x})}{\partial x^i} \, a^i_j x^j + \sum_{i=1}^{n} \frac{\partial W(\mathbf{x})}{\partial x^i} \, R^i \\ &= \dot{W}_{(28)}(\mathbf{x}) + \sum_{i=1}^{n} \frac{\partial W(\mathbf{x})}{\partial x^i} \, R^i.\end{aligned}$$

Since the function $W(\mathbf{x})$ satisfies (17), we have

$$\dot{W}_{(27)}(\mathbf{x}) \leq -\beta W(\mathbf{x}) + \sum_{i=1}^{n} \frac{\partial W(\mathbf{x})}{\partial x^i} \, R^i.$$

Let us now choose a positive number b so small that for

$$W(\mathbf{x}) \leq b \tag{29}$$

the vector $\mathbf{x}$ will belong to the domain Δ [such a number exists by virtue of (13)]. Since the second derivatives $\partial^2 f^i(\theta\mathbf{x})/\partial x^j \, \partial x^k$, being continuous functions, are bounded in the ellipsoid (29), they are also bounded in the ellipsoid

$$|R^i| \leq k|\mathbf{x}|^2 \leq \frac{k}{\mu} \, W(\mathbf{x}),$$

where k is a certain constant. Furthermore, since $\partial W(\mathbf{x})/\partial x^i$ is a linear form in $x^1, \ldots, x^n$, then

$$\left|\frac{\partial W(\mathbf{x})}{\partial x^i}\right| \leq l\sqrt{W(\mathbf{x})},$$

where l is a certain constant [see (14)]. Thus there exists a positive number q such that for $W(\mathbf{x}) \leq b$ we have

$$\sum_{i=1}^{n} \frac{\partial W(\mathbf{x})}{\partial x^i} R^i \leq qW(\mathbf{x})^{3/2}.$$

If we now choose a positive number c in such a way that

$$c \leq b, \quad q\sqrt{c} \leq \frac{\beta}{2},$$

we have

$$\dot{W}_{(27)}(\mathbf{x}) \leq -\frac{\beta}{2} W(\mathbf{x})$$

whenever the inequality

$$W(\mathbf{x}) \leq c \tag{30}$$

is satisfied. If we set $\alpha = \beta/4$, we obtain the inequality

$$\dot{W}_{(27)}(\mathbf{x}) \leq -2\alpha W(\mathbf{x}),$$

which is valid if inequality (30) holds for $\mathbf{x}$.

Let $\boldsymbol{\xi}$ be an interior point of the ellipsoid (30), i.e., a point satisfying the inequality

$$W(\boldsymbol{\xi}) < c. \tag{31}$$

We shall denote the solution of (27) with the initial values 0, $\boldsymbol{\xi}$ as before by $\boldsymbol{\varphi}(t, \boldsymbol{\xi})$, and we set

$$w(t) = W(\boldsymbol{\varphi}(t, \boldsymbol{\xi})).$$

The function $w(t)$ is defined for all those values of $t \geq 0$ for which the solution $\boldsymbol{\varphi}(t, \boldsymbol{\xi})$ is defined, and by (B) it satisfies the condition

$$\dot{w}(t) \leq -2\alpha w(t) \tag{32}$$

whenever

$$w(t) \leq c \tag{33}$$

holds. If the solution $\boldsymbol{\varphi}(t, \boldsymbol{\xi})$ does not exist for all positive values t, then the point $\mathbf{x} = \boldsymbol{\varphi}(t, \boldsymbol{\xi})$ must necessarily leave the ellipsoid (30) with increasing t [see §24, (B)]. Let us assume that the point $\mathbf{x} = \boldsymbol{\varphi}(t, \boldsymbol{\xi})$ leaves this ellipsoid, and let $t' > 0$ be the first value of t for which the point crosses the boundary. Then on the interval $0 \leq t \leq t'$ the point $\boldsymbol{\varphi}(t, \boldsymbol{\xi})$ belongs to the ellipsoid (30) so that (32) holds and $\dot{w}(t)$ is not positive. Consequently we have a contradiction: $c = w(t') \leq w(0) < c$.

Thus the solution $\boldsymbol{\varphi}(t, \boldsymbol{\xi})$, as well as the function $w(t)$, is defined for all positive values of t, and inequality (32) holds for all these values. If $\boldsymbol{\xi} \neq 0$, then $w(t) > 0$, and we can carry out the following calculations, beginning with inequality (32):

$$\frac{\dot{w}(t)}{w(t)} \leq -2\alpha; \qquad \int_0^t \frac{\dot{w}(t)}{w(t)}\, dt \leq -2\alpha t, \qquad \text{for} \qquad t \geq 0;$$

$$\ln w(t) - \ln w(0) \leq -2\alpha t.$$

The last inequality gives

$$W\big(\boldsymbol{\varphi}(t, \boldsymbol{\xi})\big) \leq W(\boldsymbol{\xi}) e^{-2\alpha t}.$$

Combining this inequality with (13) we obtain

$$|\boldsymbol{\varphi}(t, \boldsymbol{\xi})|^2 \leq \frac{\nu}{\mu} |\boldsymbol{\xi}|^2 e^{-2\alpha t}, \qquad t \geq 0, \tag{34}$$

which is valid whenever (31) holds for $\boldsymbol{\xi}$.

Inequality (31) follows from the second of the inequalities (13), together with the relation

$$|\boldsymbol{\xi}| < \sigma = \sqrt{\frac{c}{\nu}} \cdot \tag{35}$$

Thus if (35) is valid, then inequality (34) is also valid, and if we take the square root of (34), we obtain the inequality

$$|\boldsymbol{\varphi}(t, \boldsymbol{\xi})| \leq \sqrt{\frac{\nu}{\mu}} |\boldsymbol{\xi}| e^{-\alpha t}, \qquad t \geq 0,$$

which coincides with (26), where $r = \sqrt{\nu/\mu}$ and $\mathbf{a} = 0$. Thus Theorem 19 is proved.

The following proposition (F) describes a case which is in a certain sense opposite to that considered in Theorem 19.

(F) The equilibrium state $\mathbf{a}$ of equation (2) will be called *completely unstable* if there exists a positive number σ such that any solution $\boldsymbol{\varphi}(t, \boldsymbol{\xi})$ of equation (2) starting at the point $\boldsymbol{\xi} \neq \mathbf{a}$ of the sphere $|\boldsymbol{\xi} - \mathbf{a}| < \sigma$ leaves this sphere and does not return to it. That is, there exists a positive number $T = T(\boldsymbol{\xi})$ such that for $t = T$ the solution $\boldsymbol{\varphi}(t, \boldsymbol{\xi})$ is defined, and, for all values $t > T$ for which this solution is defined, it satisfies the inequality $|\varphi(t, \xi) - \mathbf{a}| \geq \sigma$. It turns out that if all eigenvalues of the matrix $(\partial f^i(\mathbf{a})/\partial x^j)$ have positive real parts, then the equilibrium state $\mathbf{a}$ of equation (2) is completely unstable.

To prove proposition (F) we shall use certain results established in the proof of Theorem 19; here, as before, we shall assume that $\mathbf{a} = 0$. In order to prove this, let us consider along with equation (2), for which all eigenvalues of the matrix $(\partial f^i(\mathbf{a})/\partial x^j)$ have positive real parts by hypothesis, the equation

$$\dot{\mathbf{x}} = -f(\mathbf{x}), \tag{36}$$

for which the point 0 obviously satisfies the conditions of Theorem 19. By virtue of the construction given in the proof of Theorem 19, a Lyapunov function $W(\mathbf{x})$ exists for equation (36) and satisfies the inequality

$$\dot{W}_{(36)}(\mathbf{x}) \leq -2\alpha W(\mathbf{x})$$

under condition (30). Writing out explicitly the left-hand side of this inequality [see (9)], we obtain

$$\dot{W}_{(36)}(\mathbf{x}) = \sum_{i=1}^{n} \frac{\partial W(\mathbf{x})}{\partial x^i} \left(-f^i(\mathbf{x})\right) \leq -2\alpha W(\mathbf{x})$$

or

$$\dot{W}_{(1)}(\mathbf{x}) \geq 2\alpha W(\mathbf{x}).$$

This inequality is automatically valid whenever (30) holds. Now let ξ be some interior point of the ellipsoid (30) [see (31)]. Let us set

$$w(t) = W(\varphi(t, \xi)).$$

The inequality

$$\dot{w}(t) \geq 2\alpha w(t) \tag{37}$$

holds for the function $w(t)$ whenever the inequality

$$w(t) \leq c$$

is valid. Since $\xi \neq 0$, then $w(t) > 0$ and the following calculations can be carried out, starting from inequality (37):

$$\frac{\dot{w}(t)}{w(t)} \geq 2\alpha; \qquad \int_0^t \frac{\dot{w}(t)}{w(t)}\, dt \geq 2\alpha t, \qquad \text{for} \qquad t \geq 0;$$

$$w(t) \geq w(0)e^{2\alpha t}; \qquad W(\varphi(t, \xi)) \geq W(\xi)e^{2\alpha t}.$$

From the last inequality it follows that, as t increases, the point $\mathbf{x} =$

$\boldsymbol{\varphi}(t, \boldsymbol{\xi})$ goes out to the boundary of the ellipsoid (30) and hence leaves the interior. We shall show that it then cannot return to the interior of the ellipsoid (30). If we assume the opposite, then we could find a positive value t' such that $w(t') = c$ and, for all sufficiently small positive values of Δt, the inequality $w(t' + \Delta t) < c$ would hold. From the last two relations it would follow that $\dot{w}(t') \leq 0$, thus contradicting (37) which holds for $t = t'$, since $w(t') = c$. Thus we have proved that the trajectory $\mathbf{x} = \boldsymbol{\varphi}(t, \boldsymbol{\xi})$, where $\boldsymbol{\xi} \neq 0$ is an interior point of the ellipsoid (30), must leave the ellipsoid (30) and cannot return to it. From the second of the inequalities (13) and from the inequalities (35) we obtain (31), so that the sphere (35) is contained in the ellipsoid (30). What we have proved, in view of this fact, implies the validity of the assertion (F).

Example

As a supplement to (A) we shall show that if the matrix A has an eigenvalue λ with a positive real part, then the equilibrium state $\mathbf{x} = 0$ of equation (5) is no longer Lyapunov stable. Actually, by (A) of §14, the solution of (5) is a vector function $\mathbf{x} = c\mathbf{h}e^{\lambda t}$, where c is an arbitrary real constant and $\mathbf{h}$ is an eigenvector of the matrix A with the eigenvalue λ. If λ is a real number, then for sufficiently small c the solution starts at the point $c\mathbf{h}$, which is arbitrarily close to the equilibrium state $\mathbf{x} = 0$, but whose modulus becomes arbitrarily large with time. If λ is a complex number, then the solution $c(\mathbf{h}e^{\lambda t} + \overline{\mathbf{h}}e^{\overline{\lambda} t})$ of equation (5) has the same property.

27. The centrifugal governor and the analysis of Vyshnegradskiy. In modern technology, thanks to the abundance of automatic control devices, a large role is played by the *theory of automatic control*. One of the most important problems arising for the designer of an automatic control device is that of the *operational stability* of a machine-control system. This problem can be solved in many cases on the basis of Lyapunov's theorem (see §26).

The oldest existing automatic control system is the system of the steam engine and Watt's centrifugal governor. The centrifugal governor, which at the end of the 18th and the first half of the 19th centuries handled its task quite adequately, began to operate unreliably because of design changes in the middle of the 19th century. A number of theoreticians and engineers sought a way out of the crisis that had arisen. The problem was solved with complete clarity and simplicity by the outstanding Russian engineer Vyshnegradskiy, one of the originators of the theory of automatic control. Vyshnegradskiy's work, *Direct-action governors* (1876), was one of the first studies in the theory of machine control which sought to

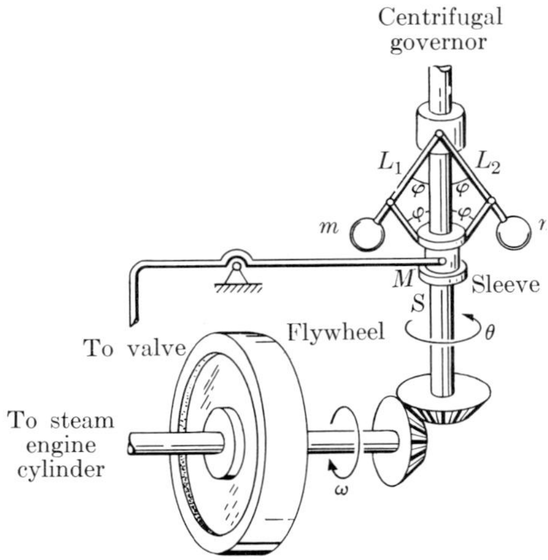

FIGURE 46

answer questions raised by industrial practice. In the present section, Vyshnegradskiy's study is presented in a simplified form.

The centrifugal governor (Fig. 46) is a vertical spindle or rod S which can rotate about its vertical axis, at the upper end of which are attached on hinges two identical arms L_1 and L_2 with identical weights at their ends. The arms L_1 and L_2 are joined together by supplementary links so that they can only deviate from the vertical position simultaneously by the same angle φ, and they lie in a common vertical plane which contains the spindle S. When the arms L_1 and L_2 deviate from their vertical position by an angle φ, they set in motion with the aid of the links a special sleeve M, which is fitted on the spindle S so that the distance from this sleeve to the upper end of S is proportional to cos φ. The length of the vertical arms L_1 and L_2 will be taken as unity, and the mass of each of the weights fixed on their ends will be denoted by m. If the spindle S rotates with angular velocity θ and the arms L_1 an' L_2 are inclined from the vertical position by an angle φ, then each of the weights is subject to the centrifugal force

$$m\theta^2 \sin \varphi. \tag{1}$$

Simultaneously, a gravitational force equal to

$$mg \tag{2}$$

acts on each weight. Since the forces acting upon the weights in the direction of the arm L_i are balanced by the reaction of arm L_i, then, in order to calculate the force acting on the weights, it is necessary to resolve both of

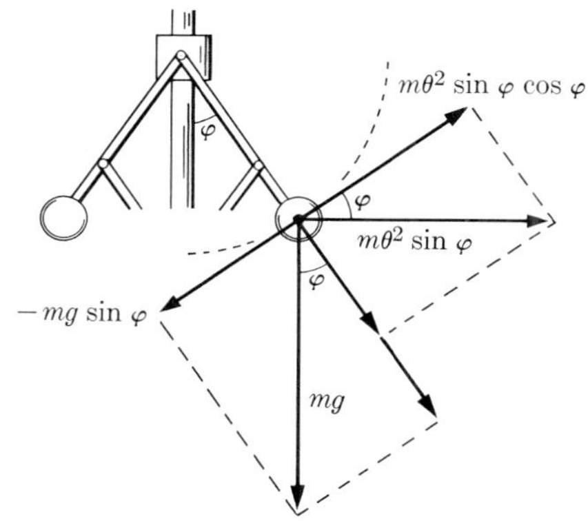

FIGURE 47

these forces in the direction of the axes, the first of which is parallel to the arm and the second perpendicular to the arm, i.e., in the direction of increasing φ. It is immediately seen (Fig. 47) that the component of the force (1) in the direction of increasing φ is equal to

$$m\theta^2 \sin\varphi \cos\varphi, \tag{3}$$

and the component of the gravitational force (2) in the same direction is equal to

$$-mg \sin\varphi. \tag{4}$$

Thus the resultant of the two forces (3) and (4) is given by the formula

$$m\theta^2 \sin\varphi \cos\varphi - mg \sin\varphi. \tag{5}$$

The simplified version of the performance of a centrifugal governor is that for a given angular velocity θ, the arms L_1 and L_2 deviate by an angle φ under the action of forces (1) and (2); the angle φ may be determined from the equality

$$m\theta^2 \sin\varphi \cos\varphi - mg \sin\varphi = 0, \tag{6}$$

i.e., by equating the force (5) to zero. The relation (6) determines the angle φ as a single-valued monotonic increasing function of the velocity θ; in this sense, Watt's governor can be considered as a measure of the velocity of rotation. This is the so-called *static* study of a governor. In reality, we have here a *dynamic* phenomenon. Under the action of the force (5), the mass m performs a motion described by a differential equation. In addition to the force (5), the mass m is acted upon during its motion by a frictional force in the hinge joints. This force depends in a rather complex manner on the motion which takes place. Simplifying substantially the complexity here, we shall assume that the frictional force is proportional to the velocity $\dot{\varphi}$ of the motion of the mass m and that its sign is opposite to that of $\dot{\varphi}$, i.e., it has the value

$$-b\dot{\varphi},$$

where b is a constant. Thus, if we take φ as the coordinate which determines the position of the mass m, then we obtain for φ the differential equation

$$m\ddot{\varphi} = m\theta^2 \sin\varphi \cos\varphi - mg \sin\varphi - b\dot{\varphi}. \tag{7}$$

[The calculation of the force (5) is carried out here under the assumption that θ and φ are constants. For variable θ and φ, additional forces enter, which, however, are balanced by the reactions of the arms and of the hinges which force the arms to move in a plane. Thus equation (7) turns out to be valid.]

The steam engine is represented by a flywheel with moment of inertia J, which is set into rotary motion by the force of the steam and is capable of performing useful work (for example, hoisting a cage from a mine). The differential equation of a steam engine can thus be written in the form

$$J\dot{\omega} = P_1 - P, \tag{8}$$

where ω is the angular velocity of the rotation of the flywheel, P_1 is the moment due to the action of the steam, and P is the moment due to the weight of the cage acting on the flywheel. The moment P_1 depends on how much the valve is opened to admit steam into the cylinders of the steam engine, and the moment P depends on the load of the cage.

The centrifugal governor is connected to a steam engine in order to maintain a uniformity of operation. It "measures" the speed of rotation of the flywheel and, if it is too great, decreases the supply of steam; if it is too small, it increases the supply of steam. To this end, the flywheel of the steam engine is connected by a set of gears with the spindle of the governor (Fig. 46) so that between the angular velocities ω and θ, there exists a constant relation

$$\theta = n\omega, \tag{9}$$

where n is the so-called *transmission ratio*. This is the *effect of the engine on the governor*, as a result of which the flywheel speed is measured. On the other hand, the sleeve M of the governor is connected with the valve which admits the steam, so that

$$P_1 = F_1 + k(\cos\varphi - \cos\varphi^*), \tag{10}$$

where φ^* is a certain "mean" value of φ near which the regulated value φ must be maintained, F_1 is the value of the force due to the action P_1 of the steam at $\varphi = \varphi^*$, and $k > 0$ is a constant factor of proportionality.

As is obvious from (10), the *reverse action of the governor on the steam engine* is realized in such a way that as the angle φ increases, the supply of steam (together with the force due to the action P_1 of the steam) decreases. As a result of the interaction described between the engine and the governor, the latter, it would seem, completely accomplishes its task, increasing the steam supply when the flywheel speed decreases and decreasing the steam supply when the flywheel speed increases. In this connection, it is natural to expect that the rotational speed of the flywheel will be stabilized. This was in fact observed in steam engines built before the middle of the nineteenth century. In order to explain the reasons for the breakdown in performance of the governor in steam engines, which began to be observed after the middle of the 19th century, it is necessary to make a detailed study of the *dynamics* of the performance of the engine-governor system and of its stability, which is what was done by Vyshnegradskiy.

As is evident from relations (7) to (10), the engine-governor system is described by the two differential equations

$$\begin{aligned} m\ddot{\varphi} &= mn^2\omega^2 \sin\varphi\cos\varphi - mg\sin\varphi - b\dot{\varphi}, \\ J\dot{\omega} &= k\cos\varphi - F, \end{aligned} \tag{11}$$

where $F = P - F_1 + k\cos\varphi^*$ is a quantity which depends on the load. The first of these equations is of second order. To reduce this system to a normal form we shall introduce a new variable ψ by setting

$$\psi = \dot{\varphi},$$

so that system (11) may be written in the normal form

$$\begin{aligned} \dot{\varphi} &= \psi, \\ \dot{\psi} &= n^2\omega^2 \sin\varphi\cos\varphi - g\sin\varphi - \frac{b}{m}\psi \\ \dot{\omega} &= \frac{k}{J}\cos\varphi - \frac{F}{J}. \end{aligned} \tag{12}$$

The satisfactory performance of a steam engine requires that the angular speed ω of the flywheel remain constant both for a fixed load P, i.e., for a constant F, and for a stationary steam supply valve. Because of the latter requirement the angle φ remains unchanged. Thus it is a matter of seeking a solution of the system (12) of the form

$$\varphi = \varphi_0, \qquad \psi = 0, \qquad \omega = \omega_0,$$

i.e., finding the state of equilibrium of this system. The problem is that of finding the state of equilibrium of system (12) and then studying its stability.

By equating to zero the right-hand sides of (12) and solving the resulting equations, we find the coordinates of the equilibrium state

$$\begin{aligned} \psi_0 &= 0, \\ \cos\varphi_0 &= \frac{F}{k}, \\ n^2\omega_0^2 &= \frac{g}{\cos\varphi_0}. \end{aligned} \tag{13}$$

Let us set

$$\varphi = \varphi_0 + \Delta\varphi, \qquad \psi = \psi_0 + \Delta\psi, \qquad \omega = \omega_0 + \Delta\omega.$$

As a result of such substitution and linearization of equations (12), we obtain the system

$$\begin{aligned} \Delta\dot{\varphi} &= \Delta\psi, \\ \Delta\dot{\psi} &= n^2\omega_0^2\cos 2\varphi_0\,\Delta\varphi + n^2\omega_0\sin 2\varphi_0\,\Delta\omega - g\cos\varphi_0\,\Delta\varphi - \frac{b}{m}\,\Delta\psi, \\ \Delta\dot{\omega} &= -\frac{k}{J}\sin\varphi_0\,\Delta\varphi. \end{aligned}$$

Substituting into the second of these equations the value of $n^2\omega_0^2$ given in (13), we obtain, after a simple calculation,

$$\Delta\dot{\psi} = -\frac{g\sin^2\varphi_0}{\cos\varphi_0}\,\Delta\varphi - \frac{b}{m}\,\Delta\psi + \frac{2g\sin\varphi_0}{\omega_0}\,\Delta\omega.$$

The characteristic polynomial of the linear system of equations obtained for $\Delta\varphi$, $\Delta\psi$, $\Delta\omega$ is

$$D(p) = \begin{vmatrix} -p & 1 & 0 \\ -\dfrac{g\sin^2\varphi_0}{\cos\varphi_0} & -\dfrac{b}{m} - p & \dfrac{2g\sin\varphi_0}{\omega_0} \\ -\dfrac{k}{J}\sin\varphi_0 & 0 & -p \end{vmatrix},$$

or, after calculating the determinant and multiplying it by -1,

$$-D(p) = p^3 + \frac{b}{m}\,p^2 + \frac{g\sin^2\varphi_0}{\cos\varphi_0}\,p + \frac{2kg\sin^2\varphi_0}{J\omega_0}.$$

All coefficients of this polynomial are positive, and therefore the necessary and sufficient condition for its stability is (by Theorem 6) that the inequalities

$$\frac{b}{m} \cdot \frac{g \sin^2 \varphi_0}{\cos \varphi_0} > 1 \cdot \frac{2kg \sin^2 \varphi_0}{J\omega_0}$$

or

$$\frac{bJ}{m} > \frac{2k \cos \varphi_0}{\omega_0} = \frac{2F}{\omega_0} \tag{14}$$

[see (13)] are satisfied. Relation (14) represents, by Lyapunov's theorem (Theorem 19), the *sufficient condition for the stability* of the engine-governor system.

In order to clarify the meaning of the right-hand side of the last inequality, we shall introduce the concept of *nonuniformity of performance* of a steam engine, which plays an important role in engineering. From (13) it is evident that changing the value $F = P - F_1 + k \cos \varphi^*$ (i.e., changing the load P) alters the stable speed ω_0. The quantity $d\omega_0/dP$ characterizes the rate of change of ω_0 when the load P is changed. Its absolute value $\nu = |d\omega_0/dP|$(as we shall soon see, the derivative $d\omega_0/dP$ is negative) is called the *nonuniformity of performance* of the steam engine. By (13) we have

$$F\omega_0^2 = \text{const},$$

and therefore by differentiating, we obtain

$$\frac{d\omega_0}{dF} = -\frac{\omega_0}{2F}.$$

Thus

$$\nu = \frac{\omega_0}{2F},$$

and the stability condition (14) may be rewritten in the final form

$$\frac{bJ}{m} \cdot \nu > 1. \tag{15}$$

From formula (15) Vyshnegradskiy made the following deductions:

1. An increase of the mass m of the balls has a harmful effect on the stability.

2. A decrease of the coefficient of friction b has a harmful effect on stability.

3. A decrease of the moment of inertia J of the flywheel has a harmful effect on stability.

4. A decrease of the nonuniformity ν has a harmful effect on stability.

In order to make his conclusions accessible to engineers and to attract attention to the more important results, Vyshnegradskiy formulated his famous "theses" at the end of his work.

First thesis. The cataract (friction) is an essential element of a sensitive and correctly operating governor, or briefly, "without a cataract, there can be no governor."

Second thesis. Astatic governors (i.e., governors with zero nonuniformity) should not be used even with the cataract, or briefly, "without nonuniformity, there can be no governor."

The breakdown in the performance of governors in the middle of the 19th century is explained by the fact that, due to the development of engineering, all four quantities appearing in (15) were subjected to changes which served to diminish the stability. Specifically, because of the increasing weight of the valve (in connection with the growth of engine power), heavier balls were being used. Improved machining of the surfaces of the engine parts led to a reduction in friction. The increase of the operating speed of the engines made it necessary to decrease the moment of inertia J of the flywheel. Finally, the tendency to decrease the dependence of speed on load led to the reduction in nonuniformity of performance.

Having explained the unfavorable effects of all the factors indicated, Vyshnegradskiy recommended in his theses an artificial increase of friction (by means of a special device, the cataract) and an increase of the nonuniformity of performance (by changing the numbers n and k, which depend on the design of the engine).

28. Limit cycles. In this section we shall define, and to a certain degree study, the concept of *limit cycle* which was introduced by the French mathematician Poincaré; we shall also give a criterion for establishing the existence of a limit cycle. It should be noted that at the present time the concept of limit cycle plays a most important role both in the theory of ordinary differential equations and in its applications in engineering.

We shall investigate the normal autonomous system (see §15) of equations

$$\dot{x}^i = f^i(x^1, \ldots, x^n), \qquad i = 1, \ldots, n, \tag{1}$$

whose right-hand sides are defined and have continuous partial derivatives $\partial f^i/\partial x^j$ in a certain domain Δ of the phase space R of the variables $x^1, \ldots, x^n$. We shall also use the vector form of this system

$$\dot{\mathbf{x}} = \mathbf{f}(\mathbf{x}). \tag{2}$$

The most essential considerations of this section will be concerned with the case $n = 2$. In order to emphasize that this case is two-dimensional, we shall speak about the *phase plane* P of system (1) and not about its phase space R. In the study of phase planes, intuitive geometrical constructions will play an essential role. The case when the domain Δ coincides with the entire phase plane P is by no means trivial, and, for the sake of simplicity, our entire attention can be concentrated on it.

The limit cycle and the behavior of trajectories in its vicinity. By a *limit cycle* of equation (2) ($n = 2$) we shall mean an isolated periodic solution of this equation. More precisely, let $\mathbf{x} = \varphi(t)$ be a periodic solution of (2) and let K be the closed curve in the plane P described by this solution. The solution $\mathbf{x} = \varphi(t)$ (and also the trajectory K) is assumed to be an *isolated* periodic solution and is called a *limit cycle* if there exists a positive number ρ such that for any point ξ in the plane P, whose distance from K is positive and smaller than ρ, the solution of (2) passing through the point ξ is not periodic.

This means geometrically that there can be no other closed trajectories of this equation in the vicinity of the closed trajectory K. The question of how the trajectories of equation (2) behave in the neighborhood of a limiting cycle K is answered by the following theorem.

THEOREM 20. Let $\mathbf{x} = \varphi(t)$ be a limit cycle of equation (2) ($n = 2$) and K a closed trajectory described by this solution in the plane P. The closed curve, as is known, separates the plane into two domains, an *interior* and an *exterior;* since the trajectories of (2) cannot intersect, each trajectory distinct from K is interior or exterior to the trajectory K. It is found that for exterior as well as for interior trajectories there are two mutually exclusive possibilities of behavior in the neighborhood of K. That is, all interior trajectories starting in a neighborhood of K spiral around K, either as $t \to +\infty$ [Fig. 48(a)] or as $t \to -\infty$ [Fig. 48(b)]. The same is true also for exterior trajectories [Figs. 48(a) and (b)].

If all trajectories (both interior and exterior) start in the neighborhood of K and spiral toward K as $t \to +\infty$, then the limit cycle is called *stable* [Fig. 48(a)]. If all trajectories starting in the neighborhood of K spiral away from K as $t \to -\infty$, then the limit cycle is called *completely unstable* [Fig. 48(b)]. In the two other cases (i.e., if the interior trajectories spiral away from K as $t \to -\infty$ and the exterior trajectories spiral toward K as $t \to +\infty$) the limiting cycle K is called *semistable* [Fig. 48(c)].

Proof. To the proof of Theorem 20 we preface a proposition.

(A) Let $\mathbf{x} = \varphi(t, \xi)$ be a solution of equation (2) ($n = 2$) with initial values 0, ξ, and let L be a curve defined in the plane P by an equation in vector parametric form

$$\mathbf{x} = \psi(u), \tag{3}$$

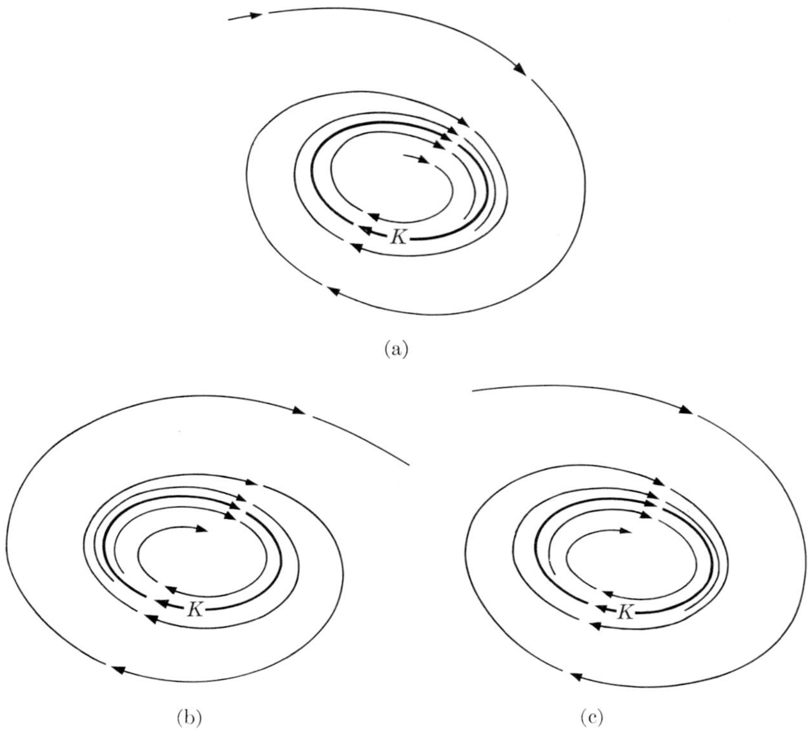

FIGURE 48

where the vector function $\boldsymbol{\psi}(u)$ of the parameter u has a continuous derivative which is not zero. If the trajectory $\mathbf{x} = \boldsymbol{\varphi}(t, \boldsymbol{\xi}_0)$ and the curve L intersect at the point

$$\boldsymbol{\varphi}(t_0, \boldsymbol{\xi}_0) = \boldsymbol{\psi}(u_0) = \mathbf{b} \tag{4}$$

but are not tangent there, i.e., the vectors $\mathbf{f}(\mathbf{b})$ and $\boldsymbol{\psi}'(u_0)$ are linearly independent, then for sufficiently small $|\boldsymbol{\xi} - \boldsymbol{\xi}_0|$, the trajectory $\boldsymbol{\varphi}(t, \boldsymbol{\xi})$ and the curve L intersect at the point

$$\boldsymbol{\varphi}(t(\boldsymbol{\xi}), \boldsymbol{\xi}) = \boldsymbol{\psi}(u(\boldsymbol{\xi})),$$

where the values $t(\boldsymbol{\xi}) - t_0$ and $u(\boldsymbol{\xi}) - u_0$ are small and the functions $t(\boldsymbol{\xi})$ and $u(\boldsymbol{\xi})$ have continuous derivatives with respect to the components of the vector $\boldsymbol{\xi}$. From this it follows, in particular, that every trajectory passing sufficiently close to $\mathbf{b}$ intersects the line L.

For the proof let us consider the vector equation

$$\boldsymbol{\varphi}(t, \boldsymbol{\xi}) - \boldsymbol{\psi}(u) = 0 \tag{5}$$

determining the desired point of intersection, in which the variables t and u are unknown functions of the vector $\boldsymbol{\xi}$. By hypothesis, equation (5) has the solution $t = t_0$, $u = u_0$ at $\boldsymbol{\xi} = \boldsymbol{\xi}_0$ [see (4)]. In order to prove the existence and differentiability of the solution of (5) we write this equation in the scalar form

$$\varphi^1(t, \boldsymbol{\xi}) - \psi^1(u) = 0,$$

$$\varphi^2(t, \boldsymbol{\xi}) - \psi^2(u) = 0.$$

For $t = t_0$, $u = u_0$, $\boldsymbol{\xi} = \boldsymbol{\xi}_0$, the functional determinant of this system is obviously

$$\begin{vmatrix} \dfrac{\partial \varphi^1}{\partial t} & \dfrac{\partial \varphi^2}{\partial t} \\ -\dfrac{d\psi^1}{du} & -\dfrac{d\psi^2}{du} \end{vmatrix} = - \begin{vmatrix} f^1(\mathbf{b}) & f^2(\mathbf{b}) \\ \psi^{1\prime}(u_0) & \psi^{2\prime}(u_0) \end{vmatrix},$$

which does not vanish since by hypothesis the vectors $\mathbf{f}(\mathbf{b})$ and $\boldsymbol{\psi}'(u_0)$ are linearly independent. Hence, by the well-known implicit function theorem, and also by Theorem 18 on the differentiability with respect to the initial values, the validity of proposition (A) follows.

The proof of Theorem 20 is based on a consideration of the *succession function* of equation (2) in the neighborhood of the closed trajectory K. In the construction of this function we shall use only the fact that K is a closed trajectory, and not the fact that it is a limit cycle. The period of the trajectory K will be denoted by T. Let L be a curve which is defined on the plane P by equation (3) and which intersects K at a unique point $\mathbf{b} = \boldsymbol{\psi}(u_0)$ and is not tangent to it at this point. For example, we can assume that L is a straight-line segment with the interior point $\mathbf{b}$. Let $\mathbf{p}$ be a variable point of the curve L. Since for $\mathbf{p} = \mathbf{b}$ the trajectory $\boldsymbol{\varphi}(t, \mathbf{p})$ intersects L at the point $\mathbf{b}$ for $t = T$, it follows from proposition (A) that, for $\mathbf{p}$ close to $\mathbf{b}$, the trajectory $\boldsymbol{\varphi}(t, \mathbf{p})$ intersects L at the point $\mathbf{q}$ for t close to T and that the point $\mathbf{q}$ is the *successor* of the point $\mathbf{p}$ (Fig. 49). This means that on the segment $\mathbf{pq}$ of the trajectory $\boldsymbol{\varphi}(t, \mathbf{p})$ there are no points of intersection with the curve L which are distinct from $\mathbf{p}$ and $\mathbf{q}$ (see Theorem 18). If

$$\mathbf{p} = \boldsymbol{\psi}(u), \qquad \mathbf{q} = \boldsymbol{\psi}(v),$$

then by proposition (A) the quantity v is a function of u, that is, $v = \chi(u)$,

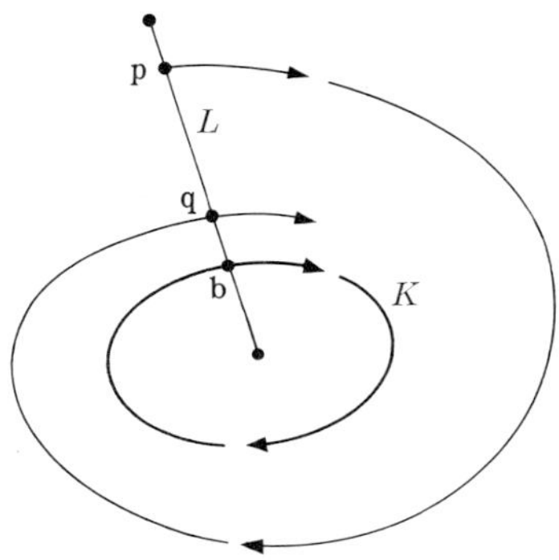

FIGURE 49

where $\chi'(u)$ is continuous. The function $\chi(u)$ is called the *succession function.* It is obvious that

$$\chi(u_0) = u_0. \tag{6}$$

Furthermore, since the trajectory $\boldsymbol{\varphi}(t, \mathbf{p})$ cannot intersect K for $\mathbf{p} \neq \mathbf{b}$, the points $\mathbf{p}$ and $\mathbf{q}$ lie on the same side of K, so that the numbers

$$u - u_0, \quad \chi(u) - u_0 \tag{7}$$

have the same sign.

We shall now show that the coordinate u of the point $\mathbf{p}$ is a single-valued differentiable function of the coordinate v of the point $\mathbf{q}$, i.e., that the relation $v = \chi(u)$ has an inverse $u = \chi^{-1}(v)$. To do this we may consider, instead of equation (2), the equation

$$\dot{\mathbf{x}} = -\mathbf{f}(\mathbf{x}), \tag{8}$$

which differs from (2) only by the sign of the right-hand side. By equation (8), the point $\mathbf{p}$ is the *successor* of the point $\mathbf{q}$, so that, by what we have already proved, $u = \chi^{-1}(v)$, where $\chi^{-1}(v)$ is the succession function for (8) in the neighborhood of its periodic trajectory K.

Because the quantities (7) have the same sign, the derivative of the function $\chi(u)$ at the point u_0 is nonnegative [see (6)], and since $\chi(u)$ has a differentiable inverse, the equality $\chi'(u_0) = 0$ cannot occur. Thus the inequality $\chi'(u_0) > 0$ holds, and therefore

$$\chi'(u) > 0$$

for all values of u sufficiently close to u_0. Hence it follows that the succession function is *monotone increasing.*

It is obvious that to every solution of the equation

$$\chi(u) = u \tag{9}$$

corresponds a periodic solution $\varphi[t, \psi(u)]$ of equation (2), and since K is a limit cycle, there exists a positive number ϵ such that on the interval $|u - u_0| < \epsilon$, equation (9) has only one solution $u = u_0$ [see (6)]. We shall restrict our consideration of the function $\chi(u)$ to that interval.

The points $\psi(u)$ for $u < u_0$ and $u > u_0$ lie on opposite sides of the curve K. To be specific, we shall assume that for $u > u_0$ they lie *outside* the curve K, and we shall study the behavior of exterior trajectories by means of the properties of the function $\chi(u)$ in the interval $u_0 < u < u_0 + \epsilon$. Since equation (9) has no solutions on this interval, one of the inequalities

$$\chi(u) < u, \tag{10}$$

$$\chi(u) > u \tag{11}$$

must hold on this entire interval. The first of these inequalities means that outside K the point **q** is closer to **b** than **p** on the curve L (Fig. 49), and the second inequality means the opposite. Inequality (11) is obviously equivalent to the inequality

$$\chi^{-1}(v) < v. \tag{12}$$

We shall analyze the case of inequality (10). Let u_1 be an arbitrary point of the interval $u_0 < u < u_0 + \epsilon$, and let us set

$$u_{i+1} = \chi(u_i), \qquad i = 1, 2, \ldots. \tag{13}$$

By (10) the sequence

$$u_1, u_2, \ldots, u_i, \ldots$$

is monotone decreasing, each of its terms being larger than u_0. Let u^* be the limit of this sequence; we shall show that $u^* = u_0$. Passing to the limit in equation (13) as $i \to \infty$, we obtain $\chi(u^*) = u^*$, and since $u = u_0$ is a unique solution of equation (9), then $u^* = u_0$. The points $\psi(u_1)$, $\psi(u_2), \ldots, \psi(u_i), \ldots$ are successive points of intersection of the trajectory $\varphi[t, \psi(u_1)]$ with the line L, and because they converge to the point **b** for sufficiently small ϵ, this trajectory spirals toward K as $t \to \infty$ (see Theorem 18). This is true for the trajectory which starts at any point $\psi(u_1)$, $u_0 < u_1 < u_0 + \epsilon$; thus, spiral trajectories of this form fill the entire exterior half-neighborhood of K.

If (12) is satisfied for the succession function of equation (8), then by applying to equation (8) the assertion just proved, we conclude that the exterior half-neighborhood of the trajectory K is filled with spiral trajectories of equation (8) which spiral toward it as $t \to +\infty$. Since the trajectories of equations (2) and (8) coincide geometrically and proceed only in opposite directions, then from what was proved for equation (8) we may conclude in the case of equation (2) that whenever (11) is satisfied, the exterior half-neighborhood of K is filled with trajectories which spiral toward K as $t \to -\infty$. The behavior of interior trajectories may be studied in exactly the same manner. Thus Theorem 20 is proved.

Note. In order to combine in one formulation the relation between the behavior of the function $\chi(u)$ in the neighborhood of u_0 and the behavior both of exterior and interior trajectories, we shall consider the inequalities

$$
\begin{aligned}
|\chi(u) - u_0| &< |u - u_0|, \\
|\chi(u) - u_0| &> |u - u_0|.
\end{aligned}
\tag{14}
$$

If the first of these inequalities holds in a half-neighborhood of the curve K (exterior or interior), then the point $\mathbf{q}$ lies closer to $\mathbf{b}$ than to $\mathbf{p}$ on the line L, so that the trajectories spiral toward K in this half-neighborhood as $t \to +\infty$. However, if in this half-neighborhood the second of inequalities (14) holds, then the trajectories spiral toward K in this half-neighborhood as $t \to -\infty$.

The geometrical study of the succession function $\chi(u)$ has a certain attractiveness. We shall represent it in the form of a graph of the equation

$$
v = \chi(u) \tag{15}
$$

in the uv-plane, assuming here for convenience that $u_0 > 0$. In order to obtain a solution of equation (9), we shall consider, along with the curve (15), the bisector of the first quadrant

$$
v = u \tag{16}
$$

(Fig. 50). To find all solutions of equation (9) it is necessary to find all the points of intersection of the curves (15) and (16). In order that the closed curve K be a limit cycle, it is necessary and sufficient that the point (u_0, u_0) be an isolated point of intersection of the curves (15) and (16). If these curves are not tangent at the point (u_0, u_0), that is, if $\chi'(u_0) \neq 1$, then their point of intersection (u_0, u_0) is necessarily isolated. In this case K is called a "rough" limit cycle. Whenever $\chi'(u_0) < 1$ (Fig. 50) in both half-neighborhoods, the first of inequalities (14) is obviously fulfilled, so that the limit cycle K is stable. Whenever $\chi'(u_0) > 1$ (Fig. 51), the second of the inequalities (14) is fulfilled so that the limit cycle K is completely unstable.

If the curves (15) and (16) are tangent at the point (u_0, u_0) but the curve (15) passes from one side of the bisector (16) to the other, then the

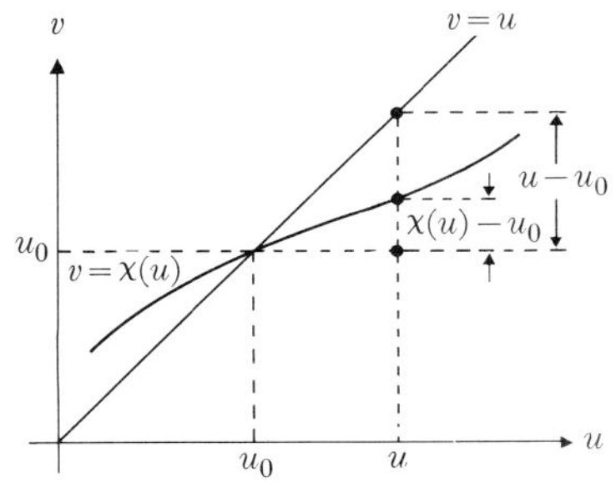

FIGURE 50

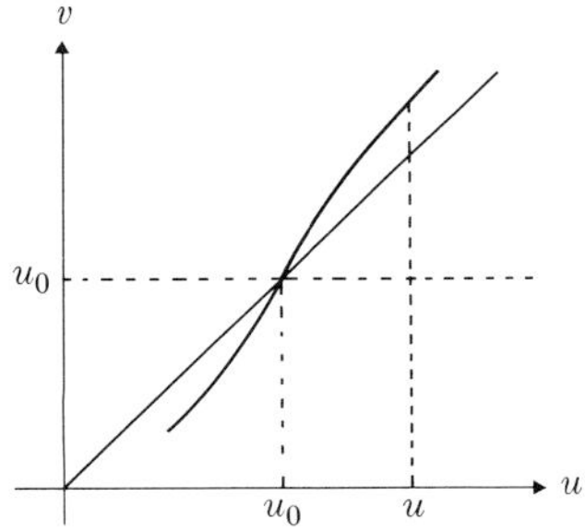

FIGURE 51

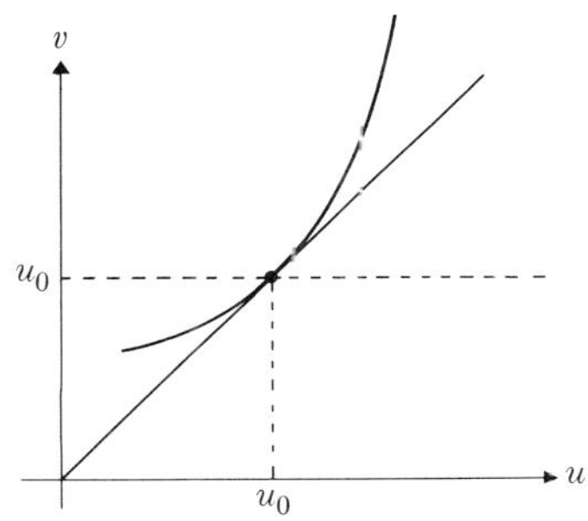

FIGURE 52

limit cycle K is either stable or completely unstable. However, if (15) is tangent to the bisector (16) and lies on one side of it (Fig. 52), then the corresponding limit cycle is semistable.

Criterion for the existence of a limit cycle.

(B) Let $\boldsymbol{\varphi}(t)$ be a certain solution of equation (2) (n is arbitrary) defined for all values $t \geq t_0$ and remaining in a closed bounded set F of the domain Δ for these values of t. The point $\mathbf{p}$ of the space R is called an *ω-limit point* of the solution $\boldsymbol{\varphi}(t)$ if there exists an unbounded increasing sequence of values (larger than t_0)

$$t_1, t_2, \ldots, t_k, \ldots, \qquad \lim_{k\to\infty} t_k = \infty,$$

such that

$$\lim_{k\to\infty} \boldsymbol{\varphi}(t_k) = \mathbf{p}.$$

The set Ω of all ω-limit points of the solution $\boldsymbol{\varphi}(t)$ is called the *ω-limit set.* Thus the set Ω is nonempty, closed, and bounded. The set Ω also consists of entire trajectories; that is, if the point $\boldsymbol{\xi}$ belongs to Ω, then the solution $\boldsymbol{\varphi}(t, \boldsymbol{\xi})$ with initial values $(0, \boldsymbol{\xi})$ is defined for all values of t, and the entire trajectory $\boldsymbol{\varphi}(t, \boldsymbol{\xi})$ is contained in the set Ω. It is obvious that the ω-limit set of the trajectory $\boldsymbol{\varphi}(t, \boldsymbol{\xi})$ is entirely contained in Ω.

We shall prove proposition (B). From the fact that the set F is closed and bounded it follows that the set Ω (which is obviously contained in F) is nonempty and bounded. We shall show that the set Ω is closed. Let

$$\mathbf{p}_1, \mathbf{p}_2, \ldots, \mathbf{p}_k, \ldots$$

be a certain sequence of points of the set Ω which converges to some point $\mathbf{p}$ of the set F; we shall prove that $\mathbf{p}$ belongs to Ω. Let $\epsilon_1, \epsilon_2, \ldots, \epsilon_k, \ldots$ and $s_1, s_2, \ldots, s_k, \ldots$ be two sequences of positive numbers such that

$$\lim_{k\to\infty} \epsilon_k = 0; \qquad \lim_{k\to\infty} s_k = \infty.$$

Since the point $\mathbf{p}_k$ belongs to Ω, there exists a value $t_k \geq s_k$ such that the distance between $\mathbf{p}_k$ and $\boldsymbol{\varphi}(t_k)$ is smaller than ϵ_k. For these values

$$t_1, t_2, \ldots, t_k, \ldots, \qquad \lim_{k\to\infty} t_k = \infty$$

we obtain

$$\lim_{k\to\infty} \boldsymbol{\varphi}(t_k) = \mathbf{p},$$

which means that $\mathbf{p}$ is in Ω.

We shall now show that the set Ω consists of complete trajectories. Let $\boldsymbol{\xi}$ be an arbitrary point of the set Ω and $\boldsymbol{\varphi}(t, \boldsymbol{\xi})$ a solution with initial values $(0, \boldsymbol{\xi})$. Moreover, let T be a value of t (which can be negative) for which the solution $\boldsymbol{\varphi}(t, \boldsymbol{\xi})$ is defined, so that the point $\boldsymbol{\varphi}(T, \boldsymbol{\xi})$ exists. Since the point $\boldsymbol{\xi}$ belongs to Ω, there exists an unbounded increasing sequence

$$t_1, t_2, \ldots, t_k, \ldots, \qquad \lim_{k\to\infty} t_k = \infty,$$

such that

$$\lim_{k\to\infty} \boldsymbol{\varphi}(t_k) = \boldsymbol{\xi}. \tag{17}$$

Since the solution $\boldsymbol{\varphi}(t)$ is defined for all sufficiently large values of t, then for a given T (starting from some k) the points

$$\boldsymbol{\varphi}(t_k + T) = \boldsymbol{\varphi}(T, \boldsymbol{\varphi}(t_k))$$

are defined [see §26, (C)]. From (17) and Theorem 18 we have

$$\lim_{k\to\infty} \boldsymbol{\varphi}(t_k + T) = \lim_{k\to\infty} \boldsymbol{\varphi}(T, \boldsymbol{\varphi}(t_k)) = \boldsymbol{\varphi}(T, \boldsymbol{\xi}),$$

from which it follows that the point $\boldsymbol{\varphi}(t, \boldsymbol{\xi})$ belongs to the set Ω and con-

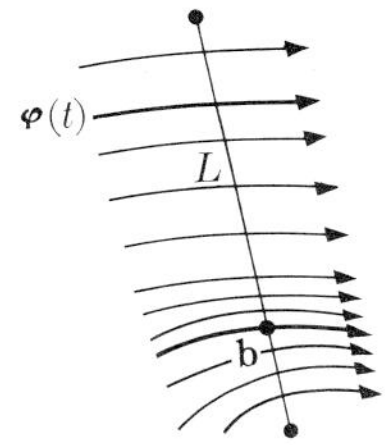

FIGURE 53

sequently to the set F. Thus the trajectory $\varphi(t, \xi)$ cannot leave the set F either for increasing t or for decreasing t, and therefore by proposition (B) of §24 it is defined for all values of t. Thus proposition (B) is proved.

Let us consider some special cases of the ω-limit set. If the solution $\varphi(t)$ [see (B)] is an equilibrium state, i.e., if $\varphi(t) \equiv \mathbf{x}_0$, then the ω-limit set of the solution $\varphi(t)$ obviously consists of one point $\mathbf{x}_0$. If $\varphi(t)$ is a periodic solution describing a closed trajectory K, then the ω-limit set of $\varphi(t)$ clearly coincides with K. Finally, if K is a periodic solution and $\varphi(t)$ is a trajectory which spirals toward this solution as $t \to +\infty$, then K is the ω-limit set of $\varphi(t)$.

We shall now prove a theorem which makes it possible to establish the existence of a periodic solution in certain cases. Whenever the right-hand sides of (1) are analytic, this periodic solution will be either a limit cycle or will be contained inside the family of periodic trajectories (see Example 3).

THEOREM 21. *Let $\varphi(t)$ be a solution of equation (2) ($n = 2$) defined for all values $t \geq t_0$ and remaining for these values of t in a closed, bounded set F which is contained in Δ, and let Ω be the ω-limit set of the solution $\varphi(t)$. If the set Ω does not contain any equilibrium states, then it consists of one closed trajectory K. Here two cases are possible: (1) $\varphi(t)$ is a periodic solution and K the trajectory described by it, and (2) the trajectory described by the solution $\varphi(t)$ as $t \to +\infty$ spirals toward the trajectory K.*

Proof. If $\varphi(t)$ is a periodic solution, then the set Ω consists of a unique periodic trajectory K described by the solution $\varphi(t)$, and the conclusion of the theorem is obvious [case (1)]. Let us assume that $\varphi(t)$ is not periodic and let $\mathbf{b}$ be an arbitrary point of the set Ω. We shall draw a straight-line segment L through the point $\mathbf{b}$ so that it is not collinear with the phase velocity vector $\mathbf{f}(\mathbf{b})$ which starts from the point $\mathbf{b}$ [the vector $\mathbf{f}(\mathbf{b})$ can not equal zero since the point $\mathbf{b}$ of the set Ω is by hypothesis not a state of equilibrium]. We choose a segment so short that all trajectories passing through a point of this segment will intersect it (without being tangent) in the same direction as the trajectory passing through $\mathbf{b}$ (Fig. 53). Since

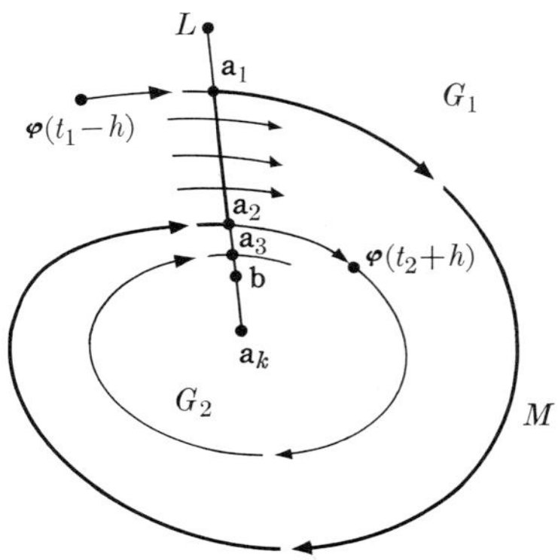

FIGURE 54

$\mathbf{b}$ is the ω-limit point for the trajectory $\boldsymbol{\varphi}(t)$ and the latter is not closed, this trajectory obviously must intersect the segment L an infinite number of times and at distinct points [see (A)]. Let $\mathbf{a}_1 = \boldsymbol{\varphi}(t_1)$ and $\mathbf{a}_2 = \boldsymbol{\varphi}(t_2)$ be two points of intersection of the trajectory $\boldsymbol{\varphi}(t)$ with the segment L following each other in time ($t_1 < t_2$). That segment of the trajectory $\boldsymbol{\varphi}(t)$ with $t_1 \leq t \leq t_2$ will be denoted by M. Together with the segment $\overline{\mathbf{a}_1\mathbf{a}_2}$ it forms a closed curve Q, which divides the plane into two domains G_1 and G_2. Let h be a small positive number. Geometrically it is obvious (Fig. 54) that the points $\boldsymbol{\varphi}(t_1 - h)$ and $\boldsymbol{\varphi}(t_2 + h)$ lie on different sides of the curve Q; we shall assume that the first belongs to the domain G_1 and the second to the domain G_2. All trajectories going from the domain G_1 into the domain G_2 pass through the segment $\overline{\mathbf{a}_1\mathbf{a}_2}$. Thus, no trajectory can leave the domain G_2 through this segment, nor can any trajectory enter or leave the domain G_2 through the curve M, because M is a piece of a trajectory and trajectories cannot intersect. Since a portion M of the trajectory $\boldsymbol{\varphi}(t)$ intersects L only at its endpoints, these endpoints of L lie on opposite sides of the curve Q. We denote by $\mathbf{a}$ that endpoint of L which is located in the domain G_2. The whole trajectory $\boldsymbol{\varphi}(t)$, starting from $t > t_2 + h$, runs in the domain G_2 and cannot intersect the segment $\overline{\mathbf{a}_1\mathbf{a}_2}$; therefore the point $\mathbf{b}$ does not belong to the segment $\overline{\mathbf{a}_1\mathbf{a}_2}$ [see (A)] so that it must lie on $\overline{\mathbf{a}_1\mathbf{a}_2}$. Now if $\mathbf{a}_3 = \boldsymbol{\varphi}(t_3)$ is that point of intersection of the trajectory with segment L following $\mathbf{a}_2$ (with respect to time), then it is evident from analogous considerations that it lies on the segment (Fig. 54). Denoting by

$$\mathbf{a}_4 = \boldsymbol{\varphi}(t_4), \quad \ldots, \quad \mathbf{a}_k = \boldsymbol{\varphi}(t_k), \ldots$$

the points of intersection following each other (in time) of the trajectory $\boldsymbol{\varphi}(t)$ with L, we can verify that they form on the segment L a monotonic sequence of points directed from $\mathbf{a}_1$ to $\mathbf{b}$. We shall show that the limit $\mathbf{b}'$ of the sequence $\mathbf{a}_1, \mathbf{a}_2, \ldots, \mathbf{a}_k, \ldots$ coincides with $\mathbf{b}$.

In order to show this we shall first prove that the sequence $t_1, t_2, \ldots, t_k, \ldots$ increases without bound. Let us assume that $\lim_{k\to\infty} t_k = \tau < +\infty$. Then $\boldsymbol{\varphi}(\tau) = \mathbf{b}'$ and $\mathbf{f}(\mathbf{b}') = \boldsymbol{\varphi}'(\tau) = \lim_{k\to\infty} (\boldsymbol{\varphi}(\tau) - \boldsymbol{\varphi}(t_k))/(\tau - t_k)$, which is impossible since the vector $\boldsymbol{\varphi}(\tau) - \boldsymbol{\varphi}(t_k)$ is directed along the segment L, and the vector $\mathbf{f}(\mathbf{b}')$ is not collinear with this segment. Thus the relation $\lim_{k\to\infty} t_k = +\infty$ must be fulfilled, so that the whole trajectory $\boldsymbol{\varphi}(t)$ intersects L only at the points $\mathbf{a}_1, \mathbf{a}_2, \ldots, \mathbf{a}_k, \ldots$ for $t \geq t_1$. Consequently, this trajectory has only one ω-limit point $\mathbf{b}'$ on the segment L [see (A)], so that $\mathbf{b}' = \mathbf{b}$. We remark that only the fact that *the point* $\mathbf{b}$ *itself* is not an equilibrium state has been used in the proof.

We shall now show that the trajectory $\boldsymbol{\varphi}(t)$ cannot enter the ω-limit set for any other trajectory $\boldsymbol{\psi}(t)$. Let us assume the opposite. Then every point of the trajectory $\boldsymbol{\varphi}(t)$ is an ω-limit point for $\boldsymbol{\psi}(t)$ [see (B)]; in particular, the point $\mathbf{a}_1$ will be such a point. Since the point $\mathbf{a}_1$ is not an equilibrium state, then, by what has been proved above, the successive points of intersection

$$\mathbf{b}_1, \mathbf{b}_2, \ldots, \mathbf{b}_k, \ldots$$

of the trajectory $\boldsymbol{\psi}(t)$ with L form a monotonic sequence which converges to $\mathbf{a}_1$, and other ω-limit points of trajectory $\boldsymbol{\psi}(t)$ on segment L do not exist. But this is contradicted by the fact that all points $\mathbf{a}_2, \mathbf{a}_3, \ldots$ located on the trajectory $\boldsymbol{\varphi}(t)$ are ω-limit points of the trajectory $\boldsymbol{\psi}(t)$.

Thus we have proved that *an open trajectory, among whose ω-limit points there are no states of equilibrium, cannot itself be an ω-limit trajectory.*

Since the trajectory K is contained in the ω-limit set Ω of the trajectory $\boldsymbol{\varphi}(t)$ and this set is closed [see (B)], all ω-limit points of K are contained in Ω and therefore are not states of equilibrium. Thus the proposition proved above can be applied to the trajectory K, so that K must be closed. From the entire construction it is evident that the trajectory $\boldsymbol{\varphi}(t)$ spirals toward K, and therefore the set Ω consists only of the closed trajectory K which passes through the point $\mathbf{b}$.

Thus Theorem 21 is proved.

Examples

1. We shall give an example of a system of equations of the form (1) ($n = 2$) which has periodic solutions of a different type, in particular, limit cycles of different forms. Initially we shall define it in polar coordinates φ and ρ, and then we shall transform it into rectangular coordinates x and y. Remembering the subsequent transformation into

rectangular coordinates, we shall define it in the form

$$\dot{\varphi} = 1; \qquad \dot{\rho} = \rho g(\rho^2), \tag{18}$$

where $g(u)$ is a continuously differentiable function defined for all its nonnegative values. In considering it in polar coordinates we shall use only positive values for ρ.

The set of all positive values of ρ for which $g(\rho^2) = 0$ will be denoted by N, and its complement in the set of positive numbers by D. To every number u_0 from N corresponds, obviously, the solution

$$\varphi = t, \qquad \rho = u_0$$

of equation (18). The corresponding trajectory K_{u_0} is closed; it is a circle in the plane P with center at the origin and with radius u_0. Since N is closed in the set of all positive numbers, D is open and consists of a finite or countable number of intervals which do not intersect each other in pairs. Let $u_1 < \rho < u_2$ be one of the finite intervals. Then the closed trajectories K_{u_1} and K_{u_2} bound an annulus Q in the plane P. For all numbers ρ of the interval $u_1 < \rho < u_2$ the function $g(\rho^2)$ does not change sign, so that on the entire interval one of the inequalities

$$g(\rho^2) < 0, \qquad g(\rho^2) > 0 \tag{19}$$

holds. Let

$$\varphi = t, \qquad \rho = \rho(t, u) \tag{20}$$

be a solution of (18) with initial values $t = 0$, $\varphi = 0$, and $\rho = u$, where $u_1 < u < u_2$. By what we proved in Example 1 of §15, the function $\rho(t, u)$ is defined for all values of t; as $t \to \infty$ it approaches one end of the interval $u_1 < \rho < u_2$, and as $t \to -\infty$ it approaches the other end. Hence it follows that the trajectory (20) spirals toward the circumferences K_{u_1} as $t \to \infty$ and K_{u_2} as $t \to -\infty$. That is, if the first of the inequalities (19) holds, then the trajectory (20) is a spiral which winds toward K_{u_1} as $t \to \infty$ and away from K_{u_2} as $t \to -\infty$ (Fig. 55). If the second of the inequalities (19) holds, then the solution (20) is a spiral which winds away from K_{u_1} as $t \to -\infty$ and toward K_{u_2} as $t \to +\infty$ (Fig. 56). Thus the annulus Q is filled by the same type of spirals of one of two forms depending on which of the inequalities (19) is fulfilled on the interval $u_1 < \rho < u_2$. If the set N is bounded and u^* is its least upper bound, then on the infinite interval $u^* < \rho < +\infty$, the trajectory (20) spirals in one direction on K_{u^*}, and, in the other direction, it recedes to infinity.

If the point u_0 of the set N is an isolated point, then the closed trajectory K_{u_0} is a limit cycle, whose form depends on the type of spiral filling the

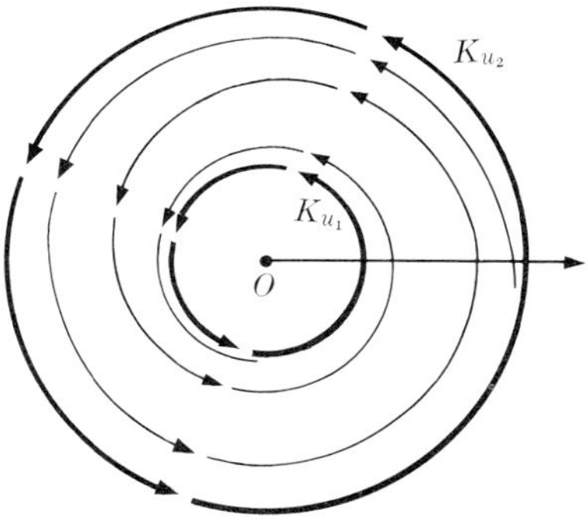

FIGURE 55

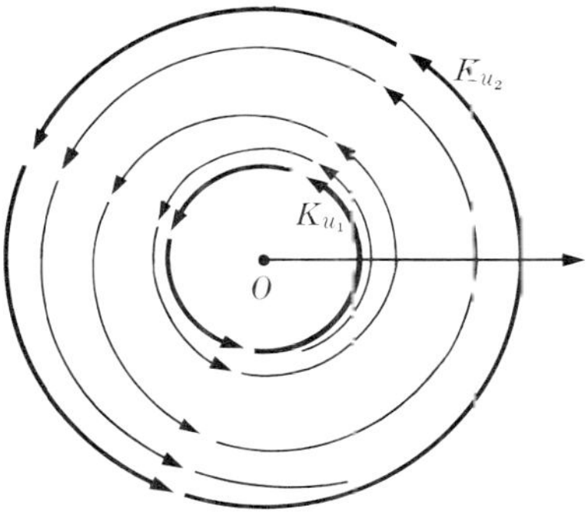

FIGURE 56

annuli adjacent to the trajectory K_{u_0}. If the point u_0 of the set N is not an isolated point, then the periodic solution K_{u_0} is not a limit cycle. If, in addition, the entire interval with center u_0 is contained in N, then the periodic solution K_{u_0} is contained in a whole family of periodic solutions forming a set of concentric circles with a common center at the origin. If the number u_0 forms the endpoint of an entire segment of numbers of the set N and is, at the same time, the endpoint of an interval of D from the other side, then the trajectory K_{u_0} is extremal in the family of closed trajectories which are contiguous to K_{u_0} from one side, while from the other side the family of trajectories spirals toward K_{u_0}. It is possible, however, for the contiguity of closed trajectories to a periodic solution K_{u_0} to be much more complex. Such cases can be easily imagined; for example, N can be the perfect set of Cantor.

We shall now write system (18) in rectangular coordinates by setting

$$x = \rho \cos \varphi, \qquad y = \rho \sin \varphi. \tag{21}$$

By differentiating (21), we obtain

$$\begin{aligned} \dot{x} &= \dot{\rho} \cos \varphi - \rho\dot{\varphi} \sin \varphi = \rho g(\rho^2) \cdot \frac{x}{\rho} - \rho \cdot \frac{y}{\rho} = xg(x^2 + y^2) - y; \\ \dot{y} &= \dot{\rho} \sin \varphi + \rho\dot{\varphi} \cos \varphi = \rho g(\rho^2) \cdot \frac{y}{\rho} + \rho \cdot \frac{x}{\rho} = yg(x^2 + y^2) + x. \end{aligned} \tag{22}$$

Thus, system (18) may be written in the form of rectangular coordinates:

$$\dot{x} = xg(x^2 + y^2) - y; \qquad \dot{y} = yg(x^2 + y^2) + x. \tag{23}$$

(Here g can be, for example, an arbitrary polynomial.) System (23) has a state of equilibrium at the origin.

2. Let

$$\dot{x}^1 = f^1(x^1, x^2, \mu), \qquad \dot{x}^2 = f^2(x^1, x^2, \mu)$$

be a normal autonomous second-order system whose right-hand sides depend on the numerical parameter μ and have continuous first-order partial derivatives with respect to x_1, x_2, μ. In addition, let

$$\dot{\mathbf{x}} = \mathbf{f}(\mathbf{x}, \mu) \tag{24}$$

be the vector form of this system. The solution of equation (24) with initial values 0, $\boldsymbol{\xi}$ will be denoted by $\boldsymbol{\varphi}(t, \boldsymbol{\xi}, \mu)$; let us assume that $\boldsymbol{\varphi}(t, \boldsymbol{\xi}_0, \mu_0)$ is a periodic solution of equation (24) at $\mu = \mu_0$ with period T. We shall answer the question of what happens to this solution when the parameter μ varies in the neighborhood of μ_0.

We shall represent the solutions of equation (24) in the same plane P independently of the value of the parameter μ. Let K be a closed trajectory corresponding to the solution $\boldsymbol{\varphi}(t, \boldsymbol{\xi}_0, \mu_0)$ and L a smooth curve defined in the plane P by means of the parametric vector equation

$$\mathbf{x} = \boldsymbol{\psi}(u),$$

which intersects the trajectory K at the unique point

$$\boldsymbol{\xi}_0 = \boldsymbol{\varphi}(0, \boldsymbol{\xi}_0, \mu_0) = \boldsymbol{\varphi}(T, \boldsymbol{\xi}_0, \mu_0) = \boldsymbol{\psi}(u_0) \tag{25}$$

and is not tangent to it. We consider the vector equation

$$\boldsymbol{\varphi}(t, \boldsymbol{\psi}(u), \mu) - \boldsymbol{\psi}(v) = 0, \tag{26}$$

in which μ and u are taken as independent variables and t and v as unknown functions. Let u vary in the neighborhood of u_0 and μ in the neighborhood of μ_0. We shall seek solutions for t close to T and for v close to u_0. For $u = u_0$ and $\mu = \mu_0$ there is the trivial solution of equation (26), $t = T$, $v = u_0$ [see (25)], and the functional determinant corresponding to the system of equations for these values of the variables is distinct from zero, since the vectors $\mathbf{f}(\boldsymbol{\xi}_0, \mu_0)$ and $\boldsymbol{\psi}'(u_0)$ are independent. For $\mu = \mu_0$ equation (26) defines the succession function $v = \chi(u, \mu_0)$ of equation (24) with $\mu = \mu_0$ in the neighborhood of the closed trajectory K. For μ close to μ_0, the function $v = \chi(u, \mu)$ is also defined from equation (26) and can be considered the *succession function* of equation (24) in the neighborhood of the periodic solution K. However, equation (24) need not have a periodic solution for $\mu \neq \mu_0$. To find the periodic solution of equation (24) for μ close to μ_0, we consider the equation

$$\chi(u, \mu) - u = 0 \tag{27}$$

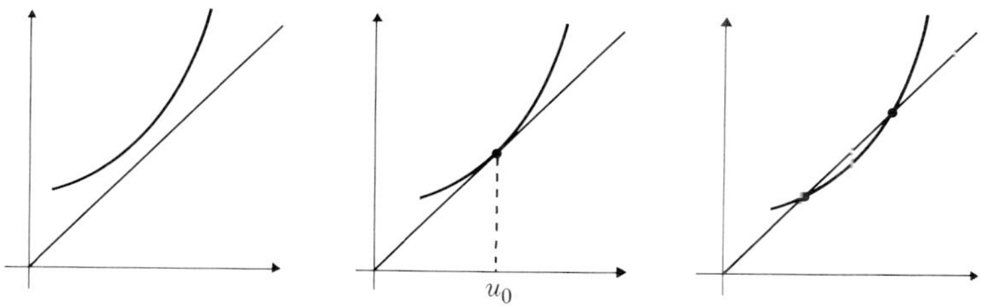

(a) (b) (c)

FIGURE 57

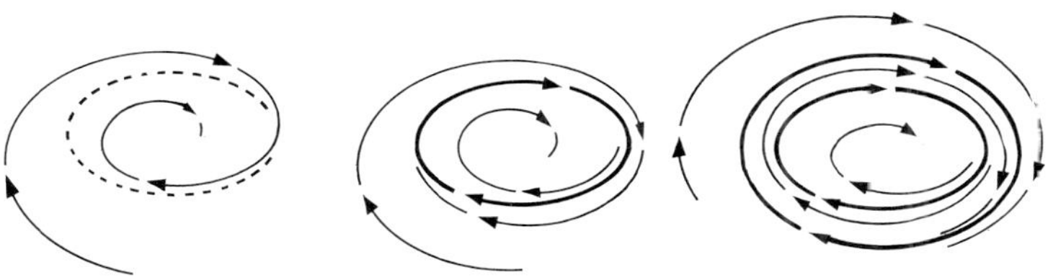

(a) (b) (c)

FIGURE 58

in the unknown function $u(\mu)$ of μ. If the derivative of the left-hand side of (27) with respect to u is distinct from zero at $u = u_0, \mu = \mu_0$, that is, if

$$\frac{\partial}{\partial u} \chi(u_0, \mu_0) \neq 1, \tag{28}$$

then (27) automatically has the differentiable solution $u(\mu)$, so that for μ close to μ_0 equation (24) has a unique periodic solution, which is "smoothly" dependent on μ and which reduces to K for $\mu = \mu_0$. Condition (28) expresses the degree of "roughness" of the cycle K, and the result obtained justifies our use of the word "rough," or "coarse." A "rough" limit cycle does not "disappear" (and remains rough) for small variations of the right-hand sides of the system; it is stable under these variations.

If the curve given by the equation

$$v = \chi(u, \mu) \tag{29}$$

in the uv-plane is tangent for $\mu = \mu_0$ to the bisector

$$v = u \tag{30}$$

at the point (u_0, u_0) with the order of contact equal to unity [Fig. 57(b)], then for $\mu = \mu_0$ the curve (29) lies on one side of the bisector (30), and the limit cycle K is semistable [Fig. 58(b)]. As the parameter μ varies in the neighborhood of μ_0, the most natural behavior of the graph of (29) is for the point of intersection of (29) and (30) to disappear altogether [Fig. 57(a)] for values of μ on one side of μ_0, and for *two* points of intersection of these curves [Fig. 57(b)] to appear for values of μ on the other side, so that (24) has two rough limit cycles close to K [Fig. 58(c)]. Thus, when μ passes through μ_0, we do not have a limiting cycle at first [Fig. 58(a)]; further, at $\mu = \mu_0$, one semistable cycle appears which, for subsequent variation of μ, decomposes into two coarse limit cycles which are close to K. The phenomenon described is called the "generation" of limit cycles of equation (24) by variation of the right-hand side.

3. We point out some important properties of a periodic solution K of (2) whenever the right-hand sides are analytic. Here we shall use without proof the fact that the solution $\boldsymbol{\varphi}(t, \boldsymbol{\xi})$ of equation (2) in this case is an analytic function of t, ξ^1, ξ^2. In the construction of the succession function we shall assume that the curve L is defined by an analytic equation. Under these assumptions the succession function $\chi(u)$ will be analytic, since it is the solution of an analytic equation.

Since a periodic solution of equation (2) corresponds to the zeros of the function $\chi(u) - u$, then, in view of the analyticity of $\chi(u)$, only two mutually exclusive cases are possible: (1) the periodic solution K is a limit cycle in the case when u_0 is an isolated zero of $\chi(u) - u$; (2) the periodic solution K is contained in the family of periodic solutions in the case when $\chi(u) - u$ is identically zero. If any other trajectory spirals toward K, then K is not contained in the family of periodic solutions and consequently it is a limit cycle. Thus, when the right-hand sides are analytic in the second case, the periodic solution K of Theorem 2 is a limit cycle.

29. The vacuum-tube oscillator. Here we shall describe systematically the design of the simplest vacuum-tube oscillator, a device which is a source of periodic (undamped) electrical oscillations. A qualitative mathematical theory of the operation of the generator will be given. The equation describing the operation of a vacuum-tube oscillator is nonlinear, and its limit cycle corresponds to the periodic oscillations generated by the oscillator. One of the first studies of the adequacy of the mathematical concept of a limit cycle and of the physical concept of an undamped oscillation generated by a vacuum-tube oscillator was that of the outstanding Soviet scientist A. A. Andronov. Before Andronov's studies, attempts were made to explain the operation of a vacuum-tube oscillator by means of linear differential equations; such attempts could not give the correct mathematical picture of the generator's performance.

(A) A *triode* (one form of electron tube) is represented by a three-terminal element *aks*. The conventional representation of the triode is shown in Fig. 59. Here a is the plate, k is the cathode, and s is the grid.

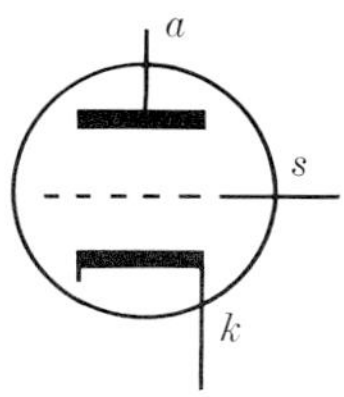

FIGURE 59

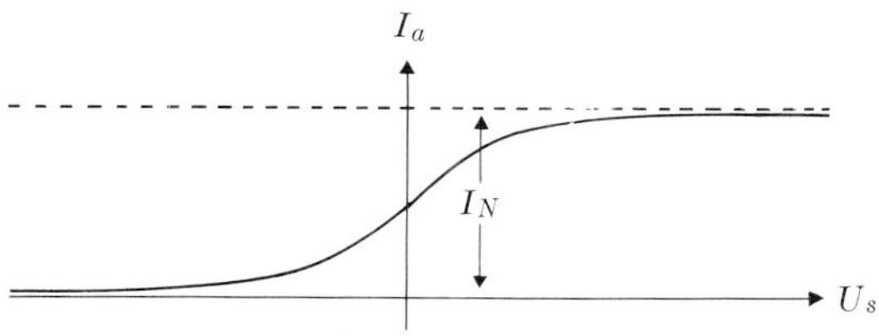

FIG. 60. Characteristic of a triode

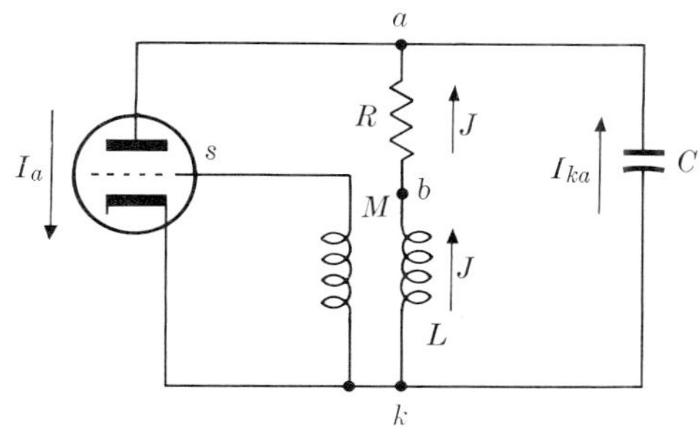

FIGURE 61

Between poles s and k there exists a voltage difference U_s (the grid voltage), but there is no current between the poles s and k; from pole a to pole k the current I_a (plate current) flows through the tube. The law governing the operation of the triode may be expressed by the formula

$$I_a = f(U_s). \tag{1}$$

The function f is called the *characteristic* of the triode. We shall assume that it is monotone increasing and positive, and satisfies the conditions

$$\lim_{U_s \to -\infty} f(U_s) = 0, \qquad \lim_{U_s \to +\infty} f(U_s) = I_N,$$

where I_N is the saturation current of the triode (Fig. 60). Usually it is also assumed that the maximum of the function $f'(U_s)$ is attained at the point $U_s = 0$.

The three-terminal element described in (A) under the name triode in reality includes besides the electron tube also the plate battery, the battery of the grid bias, and the filament battery.

(B) A vacuum-tube oscillator with an oscillatory loop in the plate circuit is diagrammed in Fig. 61. It has four junction points a, k, s, b, and consists of a triode aks [see (A)] with the characteristic $f(U_s)$, a capacitor ak with capacitance C, resistance ab with value R, inductance bk of magnitude

L, and an additional inductance sk of insignificant magnitude. Inductances kb and ks are connected by a negative mutual induction $M(M > 0)$, which effects the so-called feedback coupling in the oscillator. If we denote by J the current which flows through the resistance ba, or, what is the same thing, through the inductance kb, so that

$$J = I_{ba} = I_{kb},$$

hens it turns out that J, as a function of time t, satisfies the following differential equation:

$$L\ddot{J} + R\dot{J} + \frac{J}{C} = \frac{1}{C} f(M\dot{J}). \tag{2}$$

Let us derive equation (2). By Kirchhoff's first law, we have

$$J + I_{ka} = I_a, \tag{3}$$

where I_{ka} is the current flowing through the capacitor ka. In addition, by the properties of the triode we have

$$I_{sk} = 0. \tag{4}$$

Applying Kirchhoff's second law to the oscillatory loop $kbak$ we obtain [see (4)]

$$L\dot{I}_{kb} + RI_{ba} + \frac{1}{C}\int I_{ak}\,dt = 0.$$

Differentiating this relation, we obtain

$$L\ddot{I}_{kb} + R\dot{I}_{ba} + \frac{1}{C} I_{ak} = 0. \tag{5}$$

By virtue of the mutual induction between inductances kb and ks we obtain [see (4) and also §13, (B)]

$$U_s = M\dot{I}_{kb}. \tag{6}$$

Thus equation (2) follows from (1), (3), (5), and (6).

(C) In the phase plane of J and $\dot{J}$ equation (2) has a unique state of equilibrium with coordinates

$$J = f(0), \qquad \dot{J} = 0. \tag{7}$$

This state of equilibrium is asymptotically stable if

$$R > \frac{M}{C} f'(0), \tag{8}$$

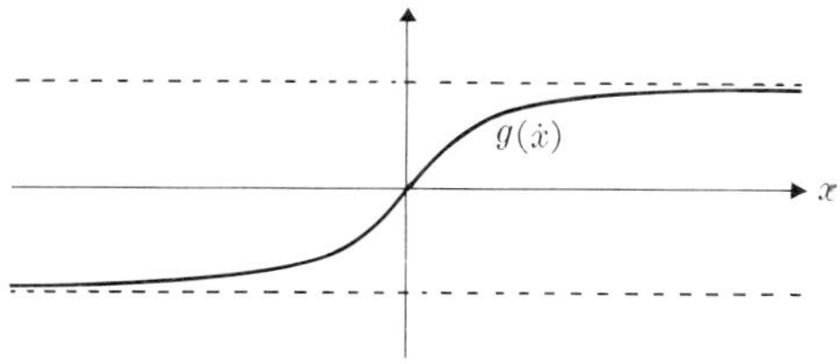

FIGURE 62

and completely unstable [see §26 (F)] if

$$R < \frac{M}{C} f'(0). \tag{9}$$

The point at infinity in the plane of $J, \dot{J}$ is completely unstable in all cases. This means that there exists a circle K in the plane $J, \dot{J}$ so large that any trajectory of (2) beginning at a certain instant enters into this circle and remains in it. When (9) is satisfied, the state of equilibrium (7) is also completely unstable. Thus by Theorem 21 (see §28) the ω-limit set of any trajectory which is not a state of equilibrium (7) is a closed trajectory. Thus, whenever (9) is satisfied, the vacuum-tube oscillator is a source of periodic undamped electric oscillations.

Note: With the appropriate choice of the characteristic f, equation (2) has a *unique* limit cycle, toward which spiral all other trajectories of (2) which are not states of equilibrium of (7). One of the characteristics with this property will be illustrated by an example.

For the proof of proposition (C) we shall replace the unknown function J by a new unknown x by setting

$$J = x + f(0), \tag{10}$$

so that the origin in the $(x, \dot{x})$-plane corresponds to the point (7).

By substituting (10) into (2), we obtain the equation

$$\ddot{x} + \frac{R}{L}\dot{x} + \frac{1}{LC}x = \frac{1}{LC}[f(M\dot{x}) - f(0)]. \tag{11}$$

Let $g(\dot{x})$ denote the function of $\dot{x}$ on the right-hand side of this equation. It is seen immediately that g is bounded and monotone increasing, and vanishes only for the value zero of the argument (Fig. 62). If, in addition we set

$$\frac{R}{L} = 2\delta,$$

$$\frac{1}{LC} = \omega^2,$$

we can write (11) in the form

$$\ddot{x} + 2\,\delta\dot{x} + \omega^2 x = g(\dot{x}).$$

Introducing a new variable $y = \dot{x}$, we obtain from this equation a normal system

$$\begin{aligned} \dot{x} &= y, \\ \dot{y} &= -\omega^2 x - 2\,\delta y + g(y). \end{aligned} \tag{12}$$

To find the state of equilibrium of (12), we equate its right-hand sides to zero:

$$\begin{aligned} y &= 0, \\ -\omega^2 x - 2\,\delta y + g(y) &= 0. \end{aligned}$$

The system obtained has a unique solution

$$x = 0, \qquad y = 0.$$

Thus, the origin is a unique state of equilibrium of (12), and therefore it follows that the point (7) is the unique equilibrium state of equation (2).

We shall now determine the stability conditions of the state of equilibrium (0, 0) of (12) by linearizing this system at the point (0, 0). We obtain the system

$$\begin{aligned} \dot{x} &= y, \\ \dot{y} &= -\omega^2 x - 2\,\delta y + g'(0)y. \end{aligned} \tag{13}$$

A simple calculation gives the characteristic polynomial

$$\lambda^2 + (2\delta - g'(0))\lambda + \omega^2$$

of the linear system (13). In the new notation the conditions (8) and (9) take the form $2\delta > g'(0)$, $2\delta < g'(0)$, respectively. Thus, whenever (8) is fulfilled, the equilibrium state (0, 0) is asymptotically stable [see Theorem 19 and §9(B)], and whenever (9) is fulfilled, it is completely unstable [see §26(F)].

To determine the asymptotic behavior of the trajectories of the system (12) in the phase plane of x, y, we consider the linear system

$$\begin{aligned} \dot{x} &= y, \\ \dot{y} &= -\omega^2 x - 2\,\delta y, \end{aligned} \tag{14}$$

which is obtained from (12) by discarding the term $g(y)$ which is bounded in the entire plane. A simple calculation yields the characteristic polynomial of system (14),

$$\lambda^2 + 2\,\delta\lambda + \omega^2, \tag{15}$$

and, since the numbers 2δ and ω^2 are positive, its roots have negative real parts. Thus, by proposition (E) of §26 for the linear system (14), there exists a Lyapunov function $W(x, y)$ which satisfies the condition

$$\dot{W}_{(14)}(x, y) \leq -\beta W(x, y). \tag{16}$$

Let us now calculate the derivative $\dot{W}_{(12)}(x, y)$ of the function $W(x, y)$ with respect to the system (12):

$$\dot{W}_{(12)}(x, y) = \dot{W}_{(14)}(x, y) + \frac{\partial W(x, y)}{\partial y} g(y). \tag{17}$$

Since the function $g(y)$ is bounded, the inequality

$$\left| \frac{\partial W(x, y)}{\partial y} \cdot g(y) \right| \leq \gamma \sqrt{W(x, y)} \tag{18}$$

holds [see formula (14), §26], where γ is some positive constant.

If we now set

$$c = \frac{2\gamma}{\beta}, \qquad \alpha = \frac{\beta}{4},$$

we obtain from (16), (17), and (18) the inequality

$$\dot{W}_{(12)}(x, y) \leq -2\alpha W(x, y), \qquad \text{whenever} \qquad W(x, y) \geq c^2. \tag{19}$$

The equation

$$W(x, y) = c^2 \tag{20}$$

defines an ellipse in the xy-plane. It follows directly from inequality (19) that at the point (x, y) of the ellipse (20) the function $W(x, y)$ decreases along the trajectory of system (12) which passes through the point (x, y). Thus whenever the trajectories of system (12) intersect the ellipse (20), they must enter it. If

$$x = \varphi(t), \qquad y = \psi(t) \tag{21}$$

is a solution of system (12) which starts at a point (ξ, η) outside the ellipse (20), then, by setting

$$w(t) = W(\varphi(t), \psi(t)),$$

we obtain for the function $w(t)$ the inequality

$$\dot{w}(t) \leq -2\alpha w(t), \tag{22}$$

which is valid whenever

$$w(t) \geq c^2.$$

If we integrate (22), we obtain

$$W(\varphi(t), \psi(t)) \leq W(\xi, \eta)e^{-2\alpha t}.$$

From this it follows that the trajectory (21) must enter the ellipse (20). No trajectory leaves this ellipse, since at its boundary points trajectories can only enter.

Let K be some circle in the xy-plane containing the ellipse (20). It follows from what has been proved that any trajectory of (12) which is distinct from the state of equilibrium (0, 0) necessarily enters the circle K and remains inside. Since the point (0, 0) is completely unstable, it cannot be among the ω-limit points of this trajectory, so that by Theorem 21 (see §28) it either spirals toward a periodic solution or is a periodic solution. Thus proposition (C) is proved.

Example

A. A. Andronov, who first derived the nonlinear equation (2) for the oscillator, considered the case when the characteristic f of the triode was of a particularly simple form, that is, equal to zero for negative values of the argument and equal to a positive constant b for positive values of the argument. Assuming that $f(0) = b/2$ and making the change of variables (10), we arrive at the system (12) in which the function $g(y)$ is defined by

$$g(y) = \begin{cases} -\omega^2 a, & y < 0, \\ \omega^2 a, & y > 0, \end{cases} \tag{23}$$

where $a = b/2$. The system (12) with the discontinuous function $g(y)$ may be written for the upper half-plane ($y > 0$) in the form

$$\begin{aligned} \dot{x} &= y, \\ \dot{y} &= -\omega^2 x - 2\,\delta y + \omega^2 a, \end{aligned} \tag{24}$$

and for the lower half-plane ($y < 0$) it may be written in the form

$$\begin{aligned} \dot{x} &= y, \\ \dot{y} &= -\omega^2 x - 2\,\delta y - \omega^2 a. \end{aligned} \tag{25}$$

We shall assume that the roots of the polynomial (15) are complex. Thus the state of equilibrium (0, 0) of (14) is a stable focus [see §16, (C)]. However, the systems (24) and (25) differ from system (14) only by a translation; their states of equilibrium are located not at the origin as in the system (14), but at the point $(a, 0)$ of the system (24) and at the point $(-a, 0)$ of the system (25). We remark that the spirals of the linear system (14) wind around the state of equilibrium $(0, 0)$ *in a clockwise direction*, and we note further that in the circuit of a half-turn of the spiral the phase point approaches the origin, so that its original distance from the origin is multiplied by a certain number $\lambda < 1$ which does not depend on the initial position of the point [see §16, (C)].

In order to visualize the phase plane of system (12) in the case when the function $g(y)$ is defined by (23), it is necessary to fill the upper half-plane with half-turns of spiral trajectories of system (24) and the lower half-plane with half-turns of spiral trajectories of system (25). However, in crossing the line $y = 0$, it is necessary to pass continuously from certain trajectories to others. On the basis of this description of the phase picture of system (12) [see (23)], we shall look for its closed trajectories.

Let us consider the trajectory of system (12) [see (23)] which starts on the axis of abscissas at the point $\xi > 0$. Since the motion in the phase plane of (12) is clockwise, the trajectory will go from the point chosen into the lower half-plane and consequently will be governed by system (25). After one half-turn of the spiral in the lower half-plane, the phase point again crosses the axis of abscissas at a point with the coordinate

$$-(a + \lambda(a + \xi)). \tag{26}$$

This follows from the fact that after one half-turn of the spiral, the distance of the phase point from the state of equilibrium $(-a, 0)$ is multiplied by λ. The point with coordinate (26) lying on the axis of abscissas will then move by virtue of system (24) and, after one half-turn of the spiral in the upper half-plane, it will arrive at the axis of abscissas at a point with the coordinate

$$a + \lambda(2a + \lambda(a + \xi)). \tag{27}$$

Thus a trajectory starting at a point with coordinate $\xi > 0$ on the positive part of the axis of abscissas will, after a complete turn, again contact the positive part of the axis of abscissas, but this time at the point with coordinate (27), and we obtain a mapping χ of the positive semiaxis of abscissas into itself, defined by the relation

$$\chi(\xi) = a + 2\lambda a + \lambda^2 a + \lambda^2 \xi.$$

The function $\chi(\xi)$ is the succession function for the system (12) [see (23)]. There is only one value of ξ which satisfies the condition

$$\chi(\xi) = \xi,$$

and to this value of ξ corresponds the limit cycle of system (12), which is both rough and stable, since $\chi'(\xi) = \lambda^2 < 1$ (see §28).

30. The states of equilibrium of a second-order autonomous system. In this section we shall classify and study nondegenerate states of equilibrium of the normal autonomous second-order system

$$\begin{aligned} \dot{x} &= f(x, y), \\ \dot{y} &= g(x, y), \end{aligned} \tag{1}$$

where it will be assumed that the right-hand sides are twice continuously differentiable, and, for the purposes of Theorem 23, that they are three times continuously differentiable.

Nondegenerate states of equilibrium. Since the state of equilibrium can always be taken as the origin, we shall assume in what follows that the state of equilibrium of system (1) is the origin. Linearizing system (1) at the point (0, 0), that is, expanding the right-hand sides of system (1) into Taylor series in x and y and discarding second-order terms, we obtain the linear system

$$\begin{aligned} \dot{x} &= a_1^1 x + a_2^1 y, \\ \dot{y} &= a_1^2 x + a_2^2 y. \end{aligned} \tag{2}$$

Let λ and μ be eigenvalues of the matrix (a_j^i). The state of equilibrium (0, 0) of system (1) is called *nondegenerate* if the numbers λ and μ are not equal and if their real parts are not zero. The behavior of trajectories of the linear system (2) was studied in detail in §16. Here it will be shown that for the nondegenerate state of equilibrium the behavior of trajectories in the neighborhood of the state of equilibrium (0, 0) of system (1) essentially coincides with the behavior of trajectories in the neighborhood of the state of equilibrium (0, 0) of system (2).

We shall retain the notation of §16 for the state of equilibrium (0, 0) of (1). If the numbers λ and μ are both real and negative, then the state of equilibrium is called a *stable node.* If the numbers λ and μ are both real and positive, then the state of equilibrium is called an *unstable node.* If λ and μ are complex conjugate and have negative real parts, then the state of equilibrium is called a *stable focus.* If λ and μ are complex conjugate and have positive real parts, then the state of equilibrium is called an *unstable focus.* Finally, if λ and μ are real and are of opposite sign, then the state of equilibrium is called a *saddle.*

The simplest properties of the behavior of trajectories in the neighborhood of the state of equilibrium can be established directly on the basis of the Lyapunov theorem (Theorem 19) and proposition (F) of §26. Thus we obtain the following proposition.

(A) A stable node and a stable focus are asymptotically stable states of equilibrium. An unstable node and an unstable focus are completely unstable states of equilibrium.

This proposition already solves to a considerable degree the problem of the behavior of trajectories in the neighborhood of a node or of a focus. Actually, if it is known that a given state of equilibrium is asymptotically stable, then from the point of view of applications it often does not matter how the trajectories approach it. The same also applies to the completely unstable state of equilibrium. An entirely different role is played by the saddle: if we know the behavior of trajectories in its neighborhood, we can describe the behavior of the trajectories in the entire plane. At the same time, the theorem on the behavior of a trajectory in the neighborhood of a saddle is considerably more difficult to describe than the corresponding theorems concerning the node and the focus.

Let us perform a linear transformation of the coordinates in the phase plane of system (1) in order to give (1) its simplest form.

(B) Expanding the right-hand sides of (1) in Taylor series in x and y up to second-order terms, we obtain

$$\begin{aligned} \dot{x} &= a_1^1 x + a_2^1 y + r(x, y), \\ \dot{y} &= a_1^2 x + a_2^2 y + s(x, y), \end{aligned} \tag{3}$$

where the remainders $r(x, y)$ and $s(x, y)$ vanish together with their first derivatives with respect to x and y at the point $x = 0$, $y = 0$ and can be written in the form

$$\begin{aligned} r(x, y) &= r_{11}x^2 + 2r_{12}xy + r_{22}y^2, \\ s(x, y) &= s_{11}x^2 + 2s_{12}xy + s_{22}y^2, \end{aligned} \tag{4}$$

where the coefficients r_{ij} and s_{ij} of these "quadratic forms" are functions of x and y which are bounded in the neighborhood of the origin. It is found that by performing a real linear transformation of x and y into ξ, η, the system (3) can be reduced to a simple form in which two cases are to be distinguished: (1) If the eigenvalues λ and μ of the matrix (a_j^i) are real and distinct, then the system of equations for ξ and η may be written in the form

$$\dot{\xi} = \lambda\xi + \rho(\xi, \eta), \qquad \dot{\eta} = \mu\eta + \sigma(\xi, \eta). \tag{5}$$

(2) If the eigenvalues of (a_j^i) are complex conjugate, i.e. if they have the form $\mu + i\nu$ and $\mu - i\nu$, then the system of equations for ξ and η may be written in the form

$$\dot{\xi} = \mu\xi - \nu\eta + \rho(\xi, \eta), \qquad \dot{\eta} = \nu\xi + \mu\eta + \sigma(\xi, \eta). \tag{6}$$

In both cases the remainders $\rho(\xi, \eta)$ and $\sigma(\xi, \eta)$ have the same properties which were mentioned above for the functions $r(x, y)$ and $s(x, y)$. In the first case the system takes the form (5) if the directions of the eigenvectors of the matrix (a_j^i) are taken as axes.

To prove proposition (B) it is sufficient to find a linear transformation of the coordinates x and y into the coordinates ξ and η such that (2) assumes a simple form. Such a transformation has already been found [see §14, (F)]. By applying the same transformation to system (3), we obtain system (5) or system (6).

Behavior of trajectories in the neighborhood of a saddle.

THEOREM 22. Let us assume that the state of equilibrium $O = (0, 0)$ of system (1) is a saddle. Let P be a straight line passing through the point O in the direction of an eigenvector of the matrix (a_j^i) with a negative eigenvalue, and let Q be a straight line passing through the point O in the direction of an eigenvector of the matrix (a_j^i) with a positive eigenvalue. Then (Fig. 63) there exist exactly two trajectories U_1 and U_2 of system (1) which tend asymptotically to the point O as $t \to +\infty$. These trajectories, together with the point O, form a continuous differentiable curve U which is tangent to the straight line P at the point O. In the same way there exist exactly two trajectories V_1 and V_2 of system (1) which tend asymptotically to the point O as $t \to -\infty$; these trajectories, together with the point O, form a continuous differentiable curve V which is tangent to the straight line Q at the point O. The remaining trajectories of system (1) passing through the neighborhood of the point O behave, in general, in the same way as in the case of a linear equation (see §16).

The trajectories U_1 and U_2 are called *stable branches* of the saddle O, and the trajectories V_1 and V_2 are called *unstable branches* of the saddle O.

Proof. First of all we shall take the line P as the axis of abscissas and the line Q as the axis of ordinates; then system (1) may be written in the form (5). Adopting again the notation x and y in place of ξ and η, we obtain the system of equations

$$\begin{aligned} \dot{x} &= f(x, y) = \lambda x + r(x, y), \\ \dot{y} &= g(x, y) = \mu y + s(x, y), \end{aligned} \tag{7}$$

where $r(x, y)$ and $s(x, y)$ have the form (4); here $\lambda < 0$, $\mu > 0$. We shall note for what follows that in the proof below we shall use only the following properties of the right-hand sides of (7): they are continuously differentiable with respect to x and y and their functions r_{ij} and s_{ij} [see (4)] are bounded in the neighborhood of the origin.

The proof will be divided into two principal parts: (a) the proof of the existence of the branch U_1 which approaches the point O along the positive part of the axis of abscissas when the coordinate x decreases; and (b) the proof of its uniqueness. The existence and uniqueness of the branch U_2 is proved analogously. To examine the branches V_1 and V_2 it is sufficient

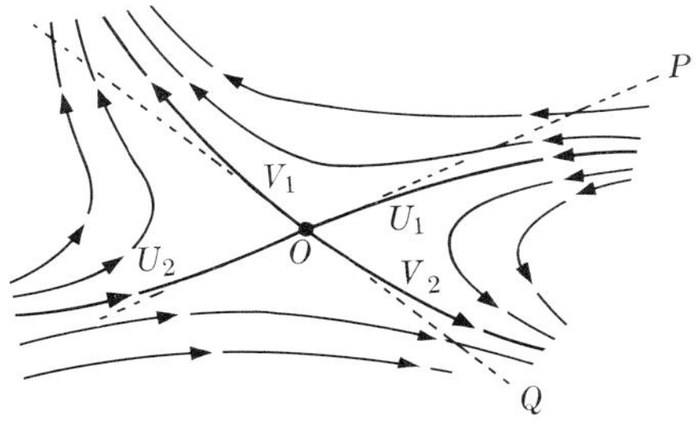

FIGURE 63

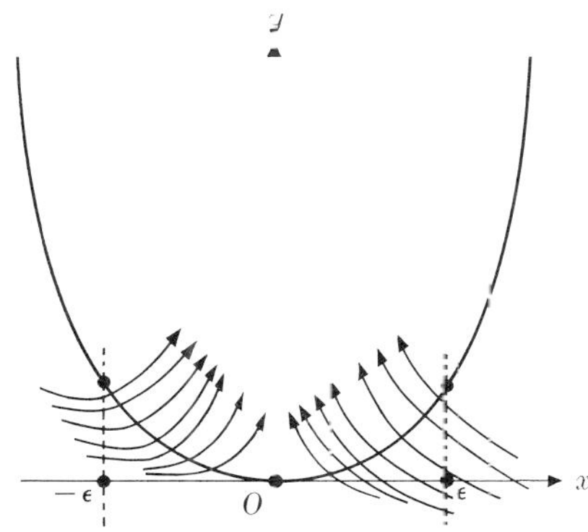

FIGURE 64

to change the sign of the time t; in this connection, stable branches will go over into unstable branches and conversely.

Let us turn to the proof of the existence of the branch U_1. For this we shall set

$$\omega(x, y) = y - ax^2, \qquad \alpha > 0,$$

and we shall consider the parabola in the xy-plane defined by the equation

$$\omega(x, y) = 0. \tag{8}$$

The parabola (8) divides the plane into two parts: a positive part, containing the positive semiaxis of ordinates, and a negative part. The positive domain will be the interior of the parabola. First we shall show that if α is a sufficiently large positive number and x is sufficiently small ($|x| \leq \epsilon$), then all trajectories of system (7) (except the state of equilibrium O) which intersect that part of the parabola (8) for which $|x| \leq \epsilon$ pass from the negative side into the positive, i.e., from the outside to the inside of the parabola (Fig. 64). For this we shall calculate the derivative $\dot{\omega}_{(7)}(x, y)$ of the function $\omega(x, y)$. System (7) applied to the points of the parabola (8) yields

$$\dot{\omega}_{(7)}(x, \alpha x^2) = \dot{y} - 2\alpha x\dot{x} = \alpha(\mu - 2\lambda)x^2 + s_{11}x^2 + \cdots$$

(here the unwritten terms contain x to at least the third power). The number $\mu - 2\lambda$ is positive and the function s_{11} is bounded in the neighborhood of the origin; therefore a number α can be chosen so large that

$$\alpha(\mu - 2\lambda) - |s_{11}| > \delta, \qquad \delta > 0.$$

The omitted terms in the expression for $\dot{\omega}_{(7)}(x, \alpha x^2)$ are at least third-

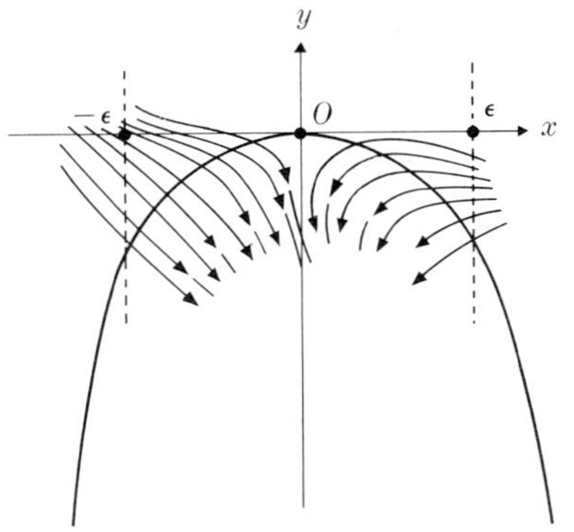

FIGURE 65

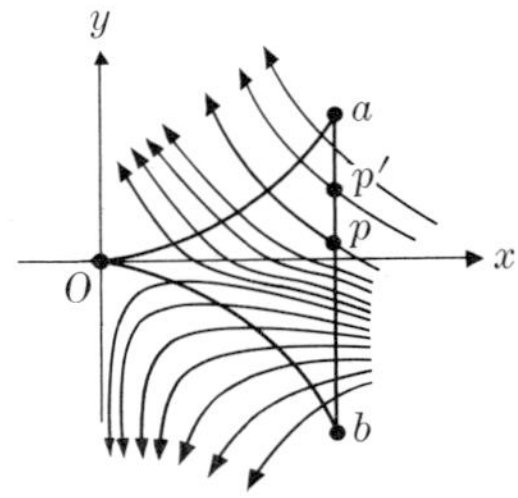

FIGURE 66

order terms in x and therefore a positive ϵ exists such that for $|x| \leq \epsilon$ we have

$$\dot{\omega}_{(7)}(x, \alpha x^2) \geq 0,$$

where the equality sign holds only for $x = 0$, that is, at the point O. From what has been proved it follows that all trajectories of system (1) except the state of equilibrium O intersect that section of the parabola (8) in the direction of increase of the function $\omega(x, y)$, that is, from the outside inwards.

In exactly the same way, it can be proved that the section $|x| \leq \epsilon$ of the parabola

$$y + \alpha x^2 = 0 \tag{9}$$

is intersected by all trajectories of the system (7), except the state of equilibrium O, from the outside inwards [the interior section of the parabola (9) contains the negative semiaxis of ordinates (Fig. 65)].

Let a and b be the points at which the straight line $x = \epsilon$ intersects the parabolas (8) and (9), respectively. We consider the triangle $[O, a, b]$ formed from two sections of the parabolas (8) and (9) and the line segment $[a, b]$. If ϵ is sufficiently small, then all trajectories of (1) passing through the triangle $[O, a, b]$ go from right to left (Fig. 66); in particular, they intersect the segment $[a, b]$ from right to left in entering the triangle $[O, a, b]$. This follows from the fact that the expression

$$\dot{x} = \lambda x + r(x, y)$$

[see (7)] is negative for $0 < x \leq \epsilon$, $|y| < \alpha x^2$, since $\lambda < 0$, and $r(x, y)$ is a "quadratic form" in x and y with bounded coefficients.

Let $\varphi(t, p)$ be a trajectory of system (7) starting at a certain point p of the interval (a, b) at $t = 0$. This trajectory enters the triangle $[O, a, b]$

through the side $[a, b]$. With an increase in t, it can either go out of the triangle through the parabolic arcs Oa, Ob, or not go out of the triangle at all. In the latter case the trajectory approaches the point O asymptotically as $t \to \infty$. It is geometrically obvious that if the trajectory $\varphi(t, p)$ leaves the triangle through the arc Oa, then the trajectory $\varphi(t, p')$, where p' is a point of the interval (a, p), also leaves the triangle through the arc Oa (Fig. 66). Furthermore, if the trajectory $\varphi(t, p)$ leaves the triangle through the arc Oa, then by the integral continuity theorem (Theorem 16) the trajectory $\varphi(t, p'')$, where p'' is a point sufficiently close to p', also leaves through the arc Oa. Thus the set of all such points p of the interval (a, b), for which the trajectory $\varphi(t, p)$ leaves the triangle through the arc Oa, forms a certain interval (a, a'). (This interval is nonempty, that is, $a' \neq a$, since trajectories starting in points p which are sufficiently close to a obviously intersect the arc Oa.) In exactly the same way, the set of all such points p for which the trajectory $\varphi(t, p)$ leaves the triangle through the side Ob forms an interval (b, b'). The intervals (a, a') and (b, b') cannot intersect, so that the point a' is located above the point b', or in the extreme case, coincides with it. (Actually, they coincide, but this requires a comparatively complicated proof.) Thus the segment $[a', b']$ contains at least one point, so that there exists a trajectory $\varphi(t, p_0)$, starting on the segment $[a', b']$ and approaching asymptotically the point O.

The tangent to the trajectory $\varphi(t, p_0)$ at the point (x, y) has the slope

$$k(x, y) = \frac{\mu y + s(x, y)}{\lambda x + r(x, y)}.$$

Since the point (x, y) of the trajectory $\varphi(t, p_0)$ belongs to the triangle $[O, a, b]$,

$$|y| < \alpha x^2, \qquad 0 < x < \epsilon, \tag{10}$$

whence it follows that the number $k(x, y)$ remains finite and tends to zero as $x \to 0$. On the other hand, the slope $l(x, y)$ of the secant drawn from the point O to the point (x, y) of the trajectory $\varphi(t, p_0)$ is equal to y/x, and since inequalities (10) are valid, we have $l(x, y) \to 0$ as $x \to 0$. Thus the curve $\varphi(t, p_0)$, based on the point O, has at O a continuous derivative and is tangent to the axis of abscissas. The trajectory $\varphi(t, p_0)$ represents the branch U_1. The branch U_2, which approaches the point O along the negative part of the axis of abscissas, is also tangent to the axis of abscissas at the point O. Both these branches together form a curve U with the equation

$$y = u(x), \tag{11}$$

where $u(x)$ is a continuous and continuously differentiable function of x, with $u'(0) = 0$.

Thus we have proved the existence of stable branches U_1 and U_2, which together with the point O form the curve U defined by equation (11). We shall now prove the uniqueness of these branches. For this we shall transform the coordinate system in the neighborhood of the origin of the xy-plane so that the curve (11) becomes the axis of abscissas. To do this, we substitute a new unknown function z in place of y by the formula

$$y = u(x) + z. \tag{12}$$

If we substitute (12) into (7), we obtain the new system of equations

$$\begin{aligned} \dot{x} &= f(x, u(x) + z) = F(x, z), \\ \dot{z} &= g(x, u(x) + z) - u'(x)f(x, u(x) + z) = G(x, z), \end{aligned} \tag{13}$$

where the unknown functions are x and z. Since $u(x)$ has a continuous derivative, the function $F(x, z)$ has continuous derivatives with respect to the two variables x and z, and $G(x, z)$ is continuous in x and has a continuous derivative with respect to z. However, the existence of a continuous derivative of the function $G(x, z)$ with respect to x has not been established. Thus it has not been established that our usual hypothesis concerning the continuous differentiability of the right-hand sides with respect to all the unknown variables holds for system (13). It is obvious, however, that to every solution of system (13) corresponds by (12) a solution of (7), and conversely. Thus the behavior of the trajectories of (7) can be determined by the behavior of the trajectories of (13).

The stable branches U_1 and U_2 of system (7) were transformed into segments of the axis of abscissas of the xz-plane, so that (13) has solutions in which the function x varies monotonically in a certain manner and approaches zero asymptotically, and the function z is identically zero. From this it follows that

$$G(x, 0) \equiv 0.$$

Below we shall show [see (C)] that the function $G(x, z)$ can be written in the form

$$G(x, z) = zH(x, z), \tag{14}$$

where $H(x, z)$ is a continuous function of x and z. It follows from (14) and the continuity of the function $H(x, z)$ that

$$\begin{aligned} \left.\frac{\partial G(x, z)}{\partial z}\right|_{\substack{x=0 \\ z=0}} &= \lim_{z\to 0} \frac{G(0, z) - G(0, 0)}{z - 0} = \lim_{z\to 0} \frac{G(0, z)}{z} \\ &= \lim_{z\to 0} H(0, z) = H(0, 0). \end{aligned}$$

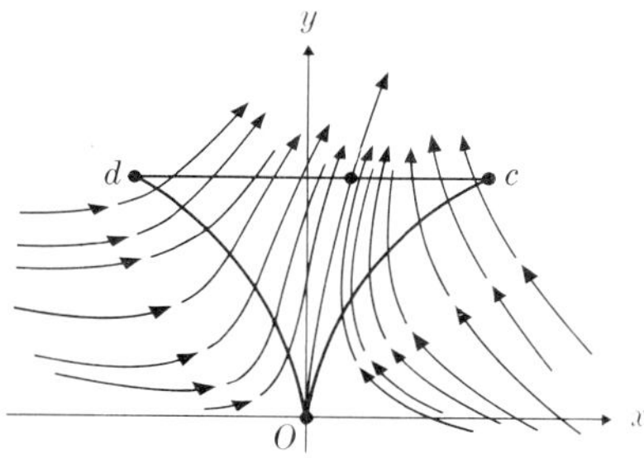

FIGURE 67

But by (7) and (13) we have $\partial G(x, z)/\partial z|_{\substack{x=0\\z=0}} = \mu$, so that

$$H(0, 0) = \mu.$$

Thus the second equation of (13) has the form

$$\dot{z} = zH(x, z),$$

where $H(x, z)$ is close to μ in the neighborhood of the origin and is therefore positive. Hence, in the neighborhood of the origin it follows that, along every trajectory which is distinct from the branches U_1 and U_2, the coordinate z does not change sign and its absolute value increases as t increases. Thus, no trajectory running outside the axis of abscissas of the xz-plane can approach the point O asymptotically, and the uniqueness of the stable branches U_1 and U_2 is proved.

It has been proved that on the interval (a, b) there exists only one point p_0 such that a trajectory of (7) starting from p_0 approaches the point O asymptotically as $t \to \infty$, forming the branch U_1. If the point p lies in the interval (a, p_0), then the trajectory starting from p intersects the arc Oa. If p lies in the interval (b, p_0), then the trajectory starting from p intersects the arc Ob.

From the parabolas

$$x - \alpha y^2 = 0, \tag{15}$$

$$x + \alpha y^2 = 0 \tag{16}$$

and the straight line

$$y = \epsilon$$

(Fig. 67), we can form a triangle $[O, c, d]$ with properties similar to those of the triangle $[O, a, b]$. There exists only one point q_0 in the interval (c, d) such that the trajectory from q_0 approaches asymptotically the point O for decreasing t and forms the unstable branch V_1. If the point q is located on the interval (c, q_0), then the trajectory from q intersects the arc Oc

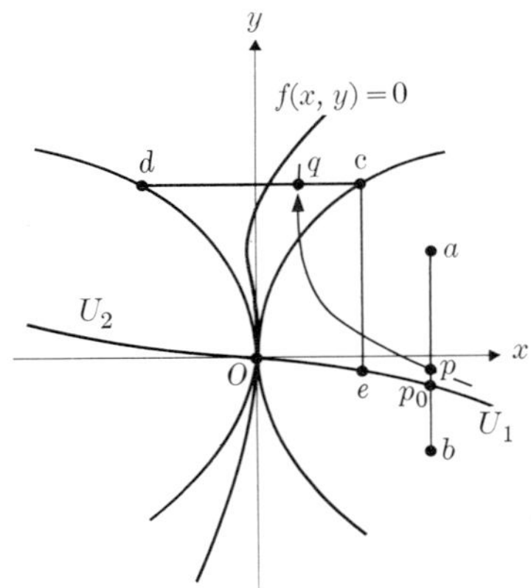

FIGURE 68

for decreasing t. If q is located on the interval (q_0, d), the trajectory from q intersects the arc Od for decreasing t.

Let us now consider the curve

$$f(x, y) = 0 \tag{17}$$

[see (7)]. It is easy to see that it is tangent to the axis of ordinates at the point O. Since the function $f(x, y)$ has continuous second derivatives, so that the curve (17) has a definite radius of curvature at the point O, we can find a number α so large and a number ϵ so small that on the interval $|y| \leq \epsilon$ the curve (17) passes between the parabolas (15) and (16) (Fig. 68). To the right of the curve (17) the function $f(x, y)$ is negative, so that the phase velocity vectors at points located to the right of the curve (17) are directed to the left. Let us draw a vertical segment $[c, e]$ from the point c, the lower end e of which is located on the branch U_1. Let p be a point of the interval (a, p_0). If the point p is sufficiently close to p_0, then by the integral continuity theorem (see §25, Theorem 16) the point, after leaving p, will pass sufficiently close to the origin of coordinates to intersect the segment $[c, e]$. As it moves farther, it must necessarily intersect the arc Oc. Indeed, if the moving point intersects the line (17), then it must intersect the arc Oc before this. If the moving point does not intersect the line (17), then it will always move to the left, and the vertical distance z from this point to the line (11) increases; in this case, therefore, the trajectory intersects the arc Oc. Thus, the trajectory enters the triangle $[O, c, d]$. After this, the trajectory will have to intersect the interval (c, q_0) at some point q. Conversely, if from the point q' of the interval (c, q_0) we follow the trajectory in the direction of decreasing t, then for sufficiently close points q and q_0 this trajectory intersects the interval

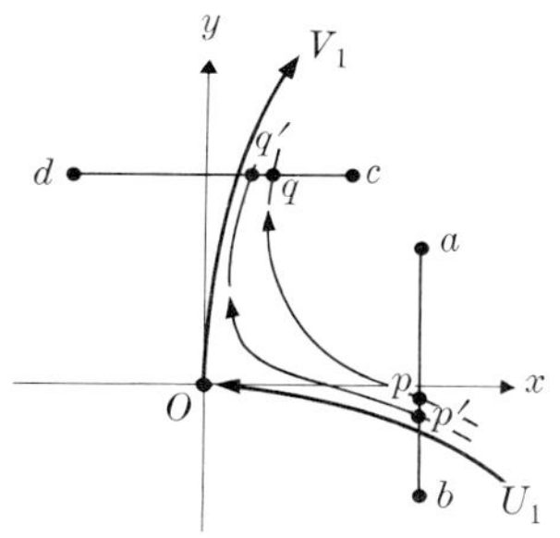

FIGURE 69

(a, p_0) at some point p' since it passes close to the origin (Fig. 69). Combining these two facts, it is easy to conclude that as $p \to p_0$ we have $q \to q_0$. This gives a complete qualitative idea concerning the behavior of trajectories in the neighborhood of a saddle. Thus Theorem 22 is proved.

We shall now prove property (14) of the function $G(x, z)$.

(C) Let $G(x, z)$ be a continuous function defined in the neighborhood of $x = z = 0$ and having a continuous derivative $(\partial/\partial z)G(x, z)$. If

$$G(x, 0) \equiv 0,$$

then

$$G(x, z) = zH(x, z),$$

where $H(x, z)$ is continuous.

To prove (C) we define the function $H(x, z)$ by setting

$$\begin{aligned} H(x, z) &= \frac{G(x, z)}{z} \qquad \text{for} \qquad z \neq 0, \\ H(x, z) &= \frac{\partial}{\partial z} G(x, z) \qquad \text{for} \qquad z = 0, \end{aligned} \tag{18}$$

and we show that this function is continuous. For $z \neq 0$ the function defined by (18) is obviously continuous. We shall prove that it is continuous at the point $(x_0, 0)$. We have

$$G(x, z) = G(x, z) - G(x, 0) = z \frac{\partial}{\partial z} G(x, \theta z),$$

where $0 \leq \theta \leq 1$. Since the function $(\partial/\partial z)G(x, z)$ is continuous, we have $G(x, z)/z = (\partial/\partial z)G(x, \theta z) \to (\partial/\partial z)G(x_0, 0)$ for $x \to x_0$, $z \to 0 (z \neq 0)$. Thus proposition (C) is proved.

Behavior of trajectories in the neighborhood of a node and of a focus. The study of the node and of the focus is considerably simpler than that of the saddle. Here it is sufficient to consider only the case of stability, since unstable nodes and foci are obtained from stable ones by changing the direction of the time flow. The principal method in the study of the node and the focus is the introduction of polar coordinates.

Theorem 23. Let $O = (0, 0)$ be a stable node of system (1) with eigenvalues λ and μ, where $\mu < \lambda < 0$. We draw through O a straight line P in the direction of an eigenvector with eigenvalue λ and a straight line Q in the direction of an eigenvector with eigenvalue μ. Then every trajectory starting sufficiently close to the point O approaches O asymptotically and has a tangent at the point O. Here only two trajectories are tangent to Q, and, in approaching the point O from the opposite side, all the remaining trajectories are tangent to P. When the node $(0 < \lambda < \mu)$ is unstable, the behavior of the trajectory as $t \to -\infty$ is similar.

We shall assume the existence of the third derivatives of the right-hand sides of (7). By proposition (B), system (1) can be written in the form (5); again denoting the variables ξ and η by x and y, we obtain the system

$$\begin{aligned} \dot{x} &= f(x, y) = \lambda x + r(x, y), \\ \dot{y} &= g(x, y) = \mu y + s(x, y). \end{aligned} \tag{19}$$

Here the functions $r(x, y)$ and $s(x, y)$ are three times continuously differentiable and vanish at the point O together with their first derivatives with respect to x and y.

Let us now introduce polar coordinates, i.e., we assume that

$$x = \rho \cos \varphi, \qquad y = \rho \sin \varphi. \tag{20}$$

Differentiating (20) and substituting the result into (19), we obtain

$$\begin{aligned} \dot{\rho} \cos \varphi - \rho \dot{\varphi} \sin \varphi &= \lambda \rho \cos \varphi + r(\rho \cos \varphi, \rho \sin \varphi), \\ \dot{\rho} \sin \varphi + \rho \dot{\varphi} \cos \varphi &= \mu \rho \sin \varphi + s(\rho \cos \varphi, \rho \sin \varphi). \end{aligned}$$

Solving the relations obtained for $\dot{\rho}$ and $\dot{\varphi}$, we obtain

$$\begin{aligned} \dot{\rho} &= \rho(\lambda \cos^2 \varphi + \mu \sin^2 \varphi) + F(\rho, \varphi), \\ \rho \dot{\varphi} &= (\mu - \lambda)\rho \sin \varphi \cos \varphi + G(\rho, \varphi), \end{aligned} \tag{21}$$

where the functions

$$\begin{aligned} F(\rho, \varphi) &= \cos \varphi \cdot r(\rho \cos \varphi, \rho \sin \varphi) + \sin \varphi \cdot s(\rho \cos \varphi, \rho \sin \varphi), \\ G(\rho, \varphi) &= -\sin \varphi \cdot r(\rho \cos \varphi, \rho \sin \varphi) + \cos \varphi \cdot s(\rho \cos \varphi, \rho \sin \varphi) \end{aligned}$$

are periodic in φ with period 2π, three times continuously differentiable with respect to ρ and φ, and vanish at $\rho = 0$ together with their first partial derivatives with respect to ρ, giving

$$F(0, \varphi) = G(0, \varphi) = \frac{\partial F(0, \varphi)}{\partial \rho} = \frac{\partial G(0, \varphi)}{\partial \rho} = 0. \tag{22}$$

By proposition (D) cited below, the function $G(\rho, \varphi)$ can be written in the form

$$G(\rho, \varphi) = \rho H(\rho, \varphi),$$

where $H(\rho, \varphi)$ is a function which is twice continuously differentiable with respect to ρ and φ and which vanishes at $\rho = 0$ for arbitrary φ [see (32) and (22)], giving

$$H(0, \varphi) = 0, \tag{23}$$

so that

$$\frac{\partial H(0, \varphi)}{\partial \varphi} = 0. \tag{24}$$

Dividing the second of relations (21) by ρ we obtain the system

$$\begin{aligned} \dot{\rho} &= \rho(\lambda \cos^2 \varphi + \mu \sin^2 \varphi) + F(\rho, \varphi), \\ \dot{\varphi} &= (\mu - \lambda) \sin \varphi \cos \varphi + H(\rho, \varphi). \end{aligned} \tag{25}$$

We shall consider the system (25) on the phase plane of the variables ρ and φ, plotting φ on the axis of abscissas and ρ on the axis of ordinates. The systems (19) and (25) are by no means equivalent, since the transformation (20) of the xy-plane into the $\rho\varphi$-plane is not one-to-one; nevertheless, from the behavior of the trajectories of (25) we can draw certain conclusions about the behavior of the trajectories of (19). The behavior of the trajectories of (25) will be considered only in the strip $|\rho| < \epsilon$. Let us first find the state of equilibrium of the system (25). From the first equation of (25) it is evident that for sufficiently small $\rho \neq 0$ the value of $\dot{\rho}$ is not zero [see (22)], and therefore in the band $|\rho| < \epsilon$ all states of equilibrium lie on the axis $\rho = 0$. After this we find all states of equilibrium from the second equation of (25) [see (23)]:

$$\rho = 0, \qquad \varphi = \frac{k\pi}{2}, \qquad k = 0, \pm 1, \pm 2, \ldots.$$

If we linearize the system (25) at the point $\rho = 0$, $\varphi = k\pi/2$, we obtain

$$\begin{aligned} \Delta\dot{\rho} &= \mu_k \, \Delta\rho, \\ \Delta\dot{\varphi} &= (\mu - \lambda) \cdot (-1)^k \, \Delta\varphi + \alpha_k \, \Delta\rho, \end{aligned}$$

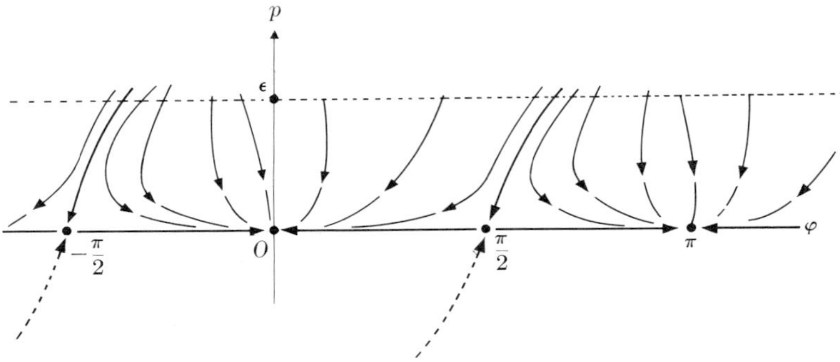

FIGURE 70

where μ_k (which is equal to λ for even values of k and to μ for odd) is a negative number. Thus the point $\rho = 0$, $\varphi = k\pi/2$ is a stable node of (25) whenever k is even and a saddle when k is odd (Fig. 70). Here the unstable branches of a saddle are directed along the φ-axis and the stable branches along the curves which approach the saddle from above and from below (see Theorem 22).

We shall now show that for sufficiently small positive ϵ, every solution of system (25) which starts in the strip $|\rho| < \epsilon$ either is a stable branch of one of the saddles of the system (25) or, by not leaving the strip $|\rho| < \epsilon$, approaches asymptotically one of the nodes of (25).

To every state of equilibrium $\rho = 0$, $\varphi = k\pi/2$ we shall make correspond its neighborhood U_k defined by the inequalities $|\rho| < \delta$, $|\varphi - k\pi/2| < \delta$, where δ is a positive number. If k is even, then the state of equilibrium under consideration is a stable node, and by virtue of its asymptotic stability there exists a positive number δ so small that every solution which starts in the neighborhood U_k approaches the node asymptotically. If k is odd, then the corresponding state of equilibrium is a saddle, and there exists a positive number δ so small that a solution different from the state of equilibrium and starting at U_k either describes the stable branch of a saddle or leaves the neighborhood U_k (see Theorem 22). Since the right-hand sides of (25) are periodic in φ, it is possible to select a positive δ, which is common for all neighborhoods U_k. Now it is possible to select a small positive number $\epsilon \leq \delta$ such that in the strip $|\rho| < \epsilon$ the right-hand side of the first of the equations (25) has a sign opposite to that of ρ, so that for every solution which starts in this strip, the quantity $|\rho|$ decreases. Further, for fixed δ we can choose a small positive number ϵ such that in the rectangle

$$|\rho| < \epsilon, \qquad \frac{k\pi}{2} + \delta \leq \varphi \leq \frac{(k+1)\pi}{2} - \delta$$

the right-hand side of the second of the equations (25) does not change sign, and its absolute value does not exceed some positive number α, so that the solution starting in this rectangle leaves it within a period of time not exceeding the number $\pi/2\alpha$ and enters that neighborhood U_k or U_{k+1} corresponding to the stable node. By the periodicity of the system (25) in φ, the number ϵ can be taken to be the same for all rectangles under consideration.

We now see that for this value of ϵ every solution which starts in the strip $|\rho| < \epsilon$ either passes through the stable branch of the saddle or approaches the stable node asymptotically.

To every solution of system (25) which starts in $|\rho| < \epsilon$ there corresponds a solution of system (19) which starts at a distance less than ϵ from the stable node O of this system. In order to obtain all such solutions of (19), it is sufficient to consider only solutions of (25) starting at $0 \leq \rho < \epsilon$. By the periodicity of the system (25) in φ and by the periodicity of the transformation (20), there exist only two solutions of (19) corresponding to stable branches of saddles of the system (25) which pass through for $\rho > 0$, and these solutions of (19) approach the state of equilibrium O asymptotically, being tangent to the straight line Q and approaching O from opposite sides. To those solutions of (25) which tend to stable nodes correspond solutions of (19) which tend to the state of equilibrium O and which are tangent to the straight line P as they approach O. Thus Theorem 23 is proved.

THEOREM 24. Let us assume that the origin O of system (1) is a focus, i.e., that the eigenvalues of the matrix (a^i_j) are the complex conjugate numbers

$$\lambda = \mu + i\nu, \qquad \bar{\lambda} = \mu - i\nu,$$

where $\mu \neq 0$, $\nu \neq 0$. Thus if $\mu < 0$, then as $t \to \infty$ all trajectories passing through some neighborhood of the point O spiral around the origin; if $\mu > 0$, then as $t \to -\infty$ all trajectories passing through a neighborhood of the point O spiral around the origin O.

Proof. For the proof we shall use the canonical form (6) by rewriting the variables ξ and η as x and y. Thus it is necessary to study the system of equations

$$\begin{aligned} \dot{x} &= f(x, y) = \mu x - \nu y + r(x, y), \\ \dot{y} &= g(x, y) = \nu x + \mu y + s(x, y). \end{aligned} \tag{26}$$

We introduce polar coordinates, i.e., we set

$$x = \rho \cos \varphi, \qquad y = \rho \sin \varphi. \tag{27}$$

Differentiating (27) and substituting the expressions obtained into (26), we have

$$\dot{\rho}\cos\varphi - \rho\dot{\varphi}\sin\varphi = \mu\rho\cos\varphi - \nu\rho\sin\varphi + r(\rho\cos\varphi, \rho\sin\varphi),$$
$$\dot{\rho}\sin\varphi + \rho\dot{\varphi}\cos\varphi = \nu\rho\cos\varphi + \mu\rho\sin\varphi + s(\rho\cos\varphi, \rho\sin\varphi).$$

Solving the system of equations obtained for $\dot{\rho}$ and $\dot{\varphi}$, we obtain

$$\begin{aligned} \dot{\rho} &= \mu\rho + \rho^2 p(\rho, \varphi), \\ \dot{\varphi} &= \nu + \rho q(\rho, \varphi), \end{aligned} \tag{28}$$

where $p(\rho, \varphi)$ and $q(\rho, \varphi)$ are functions which are bounded for small ρ and periodic in φ with period 2π. To be specific, we shall assume that $\mu < 0$. We shall consider the trajectory of system (28) which starts at the point (ρ_0, φ_0), where $0 < \rho_0 < \epsilon$ and ϵ is a sufficiently small number. From equations (28) it follows that this trajectory approaches asymptotically the axis $\rho = 0$, where φ tends either to $+\infty$ or to $-\infty$, depending on whether ν is positive or negative. Hence it follows that the corresponding trajectory in the xy-plane spirals around the origin. Thus Theorem 24 is proved.

The following proposition (D), which is essentially a generalization of proposition (C) proved above, is used only in proving Theorem 23.

(D) Let $G(\rho, \varphi)$ be a function defined in the domain W which is given by the inequalities $|\rho| < \epsilon$, $\beta_1 < \varphi < \beta_2$ and satisfies the condition

$$G(0, \varphi) = 0, \tag{29}$$

and which has the property that the function

$$\frac{\partial G(\rho, \varphi)}{\partial \rho}$$

exists and has continuous partial derivatives up to order r inclusive. Then in the domain W the function $G(\rho, \varphi)$ can be written in the form

$$G(\rho, \varphi) = \rho H(\rho, \varphi), \tag{30}$$

where the function $H(\rho, \varphi)$ is defined by the equations

$$\begin{aligned} H(\rho, \varphi) &= \frac{G(\rho, \varphi)}{\rho}, \qquad \text{for} \qquad \rho \neq 0, \\ H(0, \varphi) &= \frac{\partial G(0, \varphi)}{\partial \rho} \end{aligned} \tag{31}$$

and has continuous partial derivatives up to order r inclusive in the domain W. [For $r = 0$, proposition (D) reduces to proposition (C)].

To prove (D) we consider the function

$$K(\rho, \varphi) = \frac{\partial G(\rho, \varphi)}{\partial \varphi^{r-s}}, \qquad 0 \leq s \leq r, \tag{32}$$

which has continuous partial derivatives in the domain W with respect to ρ up to order $s + 1$ inclusive and satisfies the condition

$$K(0, \varphi) = 0 \tag{33}$$

[see (29)]. We shall prove that for $\rho \neq 0$ the inequality

$$\frac{\partial^k}{\partial \rho^k}\left(\frac{K(\rho, \varphi)}{\rho}\right) = \sum_{i=0}^{k} \gamma_i \frac{\partial^{k+1} K(\theta_i \rho, \varphi)}{\partial \rho^{k+1}}, \qquad 0 \leq k \leq s, \tag{34}$$

holds, where, for every fixed k, the numbers $\gamma_0, \gamma_1, \ldots, \gamma_k$ do not depend on the function $G(\rho, \varphi)$ and satisfy the condition

$$\gamma_0 + \gamma_1 + \cdots + \gamma_k = \frac{1}{k+1} \tag{35}$$

and the numbers $\theta_0, \theta_1, \ldots, \theta_k$ satisfy the inequalities

$$0 \leq \theta_i \leq 1, \qquad i = 0, 1, \ldots, k. \tag{36}$$

Calculating the derivative $(\partial^k/\partial \rho^k)(K(\rho, \varphi)/\rho)$ by the Leibniz formula, we obtain

$$\frac{\partial^k}{\partial \rho^k}\left(\frac{K(\rho, \varphi)}{\rho}\right) = \frac{1}{\rho^{k+1}} \sum_{i=0}^{k} a_i \rho^i \frac{\partial^i K(\rho, \varphi)}{\partial \rho^i}, \tag{37}$$

where, for every fixed k, the numbers $a_0, a_1, \ldots, a_k$ do not depend on the function $G(\rho, \varphi)$. Expanding each of the functions $\partial^i K(\rho, \varphi)/\partial \rho^i$, $i = 0, 1, \ldots, k$, in Taylor series in powers of ρ, we obtain

$$\frac{\partial^i K(\rho, \varphi)}{\partial \rho^i} = \frac{\partial^i K(0, \varphi)}{\partial \rho^i} + \frac{\rho}{1!} \frac{\partial^{i+1} K(0, \varphi)}{\partial \rho^{i+1}} + \cdots + \frac{\rho^{k-i}}{(k-i)!} \cdot \frac{\partial^k K(0, \varphi)}{\partial \rho^k} + \frac{\rho^{k-i+1}}{(k-i+1)!} \cdot \frac{\partial^{k+1} K(\theta_i \rho, \varphi)}{\partial \rho^{k+1}}. \tag{38}$$

Furthermore, on substituting (38) into (37), we obtain by virtue of (33)

$$\frac{\partial^k}{\partial \rho^k}\left(\frac{K(\rho, \varphi)}{\rho}\right) = \frac{1}{\rho^{k+1}}\left[\sum_{i=1}^{k} b_i \rho^i \frac{\partial^i K(0, \varphi)}{\partial \rho^i} + \sum_{j=0}^{k} \gamma_j \rho^{k+1} \frac{\partial^{k+1} K(\theta_j \rho, \varphi)}{\partial \rho^{k+1}}\right], \tag{39}$$

where b_i and γ_j are constants which, for every fixed k, do not depend on the choice of the function $G(\rho, \varphi)$.

To prove the relations (34) it is sufficient to establish that the constants $b_1, \ldots, b_k$ equal zero and that the constants $\gamma_0, \gamma_1, \ldots, \gamma_k$ satisfy condition (35). Since these constants do not depend on the choice of the function $G(\rho, \varphi)$, it is sufficient to establish these properties for any functions $G(\rho, \varphi)$ whatsoever. We consider the case when $G(\rho, \varphi)$ is the polynomial

$$G(\rho, \varphi) = \frac{\varphi^{r-s}}{(r-s)!} \cdot \sum_{i=1}^{k+1} \frac{\alpha_i}{i!} \rho^i. \tag{40}$$

By (32) we find that

$$\frac{\partial^k}{\partial \rho^k} \left(\frac{K(\rho, \varphi)}{\rho} \right) = \frac{\alpha_{k+1}}{k+1}. \tag{41}$$

On the other hand, the identity (39) for the polynomial (40) has the form

$$\frac{\partial^k}{\partial \rho^k} \left(\frac{K(\rho, \varphi)}{\rho} \right) = \frac{1}{\rho^{k+1}} \left[\sum_{i=1}^{k} b_i \rho^i \alpha_i + \alpha_{k+1} \rho^{k+1} \sum_{j=0}^{k} \gamma_j \right]. \tag{42}$$

The right-hand sides of (41) and (42) must coincide for $|\rho| < \epsilon$, $\rho \neq 0$, but since the numbers $\alpha_1, \ldots, \alpha_{k+1}$ are arbitrary, it follows that the numbers $b_1, \ldots, b_k$ are equal to zero, and (35) then follows. Thus formula (34) is proved.

We now consider the function $L_k(\rho, \varphi)$, $k = 0, 1, \ldots, s$, and set

$$\begin{aligned} L_k(\rho, \varphi) &= \frac{\partial^k}{\partial \rho^k} \left(\frac{K(\rho, \varphi)}{\rho} \right), \qquad \text{for} \qquad \rho \neq 0, \\ L_k(0, \varphi) &= \frac{1}{k+1} \frac{\partial^{k+1} K(0, \varphi)}{\partial \rho^{k+1}}. \end{aligned} \tag{43}$$

From (34) and (35) it follows that $L_k(\rho, \varphi)$ is a continuous function of both ρ and φ in the entire domain W. It is clear that for $\rho \neq 0$ the inequalities

$$L_{k+1}(\rho, \varphi) = \frac{\partial L_k(\rho, \varphi)}{\partial \rho}, \qquad k = 0, 1, \ldots, s-1, \tag{44}$$

hold. Let us prove that the equalities are also valid for $\rho = 0$. Let $0 < \rho_0 < \epsilon$, $0 < \rho < \epsilon$; then we have

$$L_k(\rho, \varphi) = L_k(\rho_0, \varphi) + \int_{\rho_0}^{\rho} L_{k+1}(\xi, \varphi) \, d\xi. \tag{45}$$

Since the functions on both sides are continuous, this equality is also valid for $\rho = 0$, so that we have

$$L_k(0, \varphi) = L_k(\rho_0, \varphi) + \int_{\rho_0}^{0} L_{k+1}(\xi, \varphi)\, d\xi. \tag{46}$$

Subtracting (46) from (45) and dividing the result by ρ, we obtain

$$\frac{L_k(\rho, \varphi) - L_k(0, \varphi)}{\rho} = \frac{\int_0^\rho L_{k+1}(\xi, \varphi)\, d\xi}{\rho}, \qquad \rho > 0.$$

Passing to the limit as $\rho \to 0$, we see that the right-hand derivative of the function $L_k(\rho, \varphi)$ with respect to ρ at the point $\rho = 0$ exists and is equal to $L_{k+1}(0, \varphi)$. In exactly the same way we can prove that the left-hand derivative is also equal to $L_{k+1}(0, \varphi)$. Thus (44) is valid in the entire domain W.

From (43) and (32) with $k = 0$ and $s = r$ we obtain (30) and (31), and from (44) and (32) it follows that the function $H(\rho, \varphi)$ has the continuous derivative

$$\frac{\partial^{r-s+k} H(\rho, \varphi)}{\partial \rho^k \partial \varphi^{r-s}}, \qquad 0 \le k \le s, \qquad 0 \le s \le r,$$

which means that the function $H(\rho, \varphi)$ has continuous partial derivatives up to order r inclusive. Thus proposition (D) is proved.

31. Stability of periodic solutions. In this paragraph we shall study the problem of stability of periodic solutions of autonomous systems and of systems with periodic right-hand sides.

The concept of stability. In §26 we gave the definition of Lyapunov stability of a state of equilibrium of an autonomous system. First of all, we shall give here a definition of Lyapunov stability of the solution of an arbitrary system of equations.

Let

$$\dot{\mathbf{x}} = \mathbf{f}(t, \mathbf{x}) \tag{1}$$

be the vector form of an arbitrary normal system of nth-order equations whose right-hand sides, together with their derivatives $\partial f^i(t, \mathbf{x})/\partial x^j$, are defined and continuous in a certain domain Γ of the space of the variables $t, \mathbf{x}$. The solution of equation (1) with initial values θ, $\boldsymbol{\xi}$ will be denoted by $\boldsymbol{\varphi}(t, \theta, \boldsymbol{\xi})$.

Definition. A solution $\boldsymbol{\varphi}(t)$ of equation (1) with initial values t_0, $\mathbf{x}_0$ is called *Lyapunov stable* if the following conditions are fulfilled: (1) There exists a positive number ρ such that, for $|\mathbf{x}_1 - \mathbf{x}_0| < \rho$, the solution $\boldsymbol{\varphi}(t, t_0, \mathbf{x}_1)$ is defined for all values $t \ge t_0$; in particular, the solution $\boldsymbol{\varphi}(t)$ itself is also defined for all $t \ge t_0$. (2) For every positive number ϵ, a positive number $\delta \leq \rho$ can be found such that for $|\mathbf{x}_1 - \mathbf{x}_0| < \delta$ we have $|\boldsymbol{\varphi}(t, t_0, \mathbf{x}_1) - \boldsymbol{\varphi}(t)| < \epsilon$ for $t \geq t_0$. The solution $\boldsymbol{\varphi}(t)$ of equation (1) which is Lyapunov stable with initial values t_0, $\mathbf{x}_0$ is called *asymptotically stable* if a positive number $\sigma \leq \rho$ can be found such that for $|\mathbf{x}_1 - \mathbf{x}_0| < \sigma$ we have

$$|\boldsymbol{\varphi}(t, t_0, \mathbf{x}_1) - \boldsymbol{\varphi}(t)| \to 0, \qquad \text{as} \qquad t \to \infty.$$

The definitions presented here of Lyapunov stability and of asymptotic stability are invariant with respect to the choice of initial values t_0, $\mathbf{x}_0$ of the solution $\boldsymbol{\varphi}(t)$. This can easily be derived from Theorem 18.

In the particular case when the system (1) is autonomous and the solution $\boldsymbol{\varphi}(t)$ is the state of equilibrium, the definitions of stability given here coincide with those given in §26.

We shall now consider systems (1) whose right-hand sides are periodic functions of t with period τ,

$$\mathbf{f}(t + \tau, \mathbf{x}) = \mathbf{f}(t, \mathbf{x}), \tag{2}$$

and also systems (1) which are autonomous

$$\mathbf{f}(t, \mathbf{x}) = \mathbf{f}(\mathbf{x}). \tag{3}$$

In both cases the problem of stability of a periodic solution with period τ,

$$\boldsymbol{\varphi}(t + \tau) = \boldsymbol{\varphi}(t), \tag{4}$$

will be investigated, and, in the case of an autonomous system, the periodic solution will be assumed to be different from the state of equilibrium. In the case of the periodic system [see (2)], sufficient conditions for the asymptotic stability of the solution (4) with period τ will be given. The autonomous system is a particular case of the periodic system, so that we might expect that these conditions apply also to a periodic solution of the autonomous system. It turns out, however, that they are not fulfilled (a periodic solution of an autonomous system cannot be asymptotically stable), so that for the Lyapunov stability of a periodic solution of the autonomous system, we shall give other weaker conditions.

(A) In order to study the behavior of solutions of (1) in the neighborhood of the solution $\boldsymbol{\varphi}(t)$, we shall introduce a new unknown vector function $\mathbf{y}$ by setting

$$\mathbf{x} = \boldsymbol{\varphi}(t) + \mathbf{y}. \tag{5}$$

In what follows we shall assume that the right-hand sides of (1) have continuous second-order derivatives in the domain with respect to the coordinates of the vector $\mathbf{x}$. Replacing the variables in system (1) by (5), using the fact that $\boldsymbol{\varphi}(t)$ is a solution of (1), and expanding the right-hand sides in powers of $\mathbf{y}$, we obtain

$$\dot{y}^i = \sum_j \frac{\partial f^i(t, \boldsymbol{\varphi}(t))}{\partial x^j} y^j + r^i(t, \mathbf{y}). \tag{6}$$

By linearizing this system, i.e., by discarding the terms r^i which are at least of the second order in $\mathbf{y}$, we obtain the linear system

$$\dot{\mathbf{y}} = A(t)\mathbf{y}, \tag{7}$$

where $A(t)$ is a matrix with elements

$$a^i_j(t) = \frac{\partial f^i(t, \boldsymbol{\varphi}(t))}{\partial x^j}.$$

We shall now assume that the right-hand side of (1) is a periodic function of t with period τ [see (2)] and that the solution $\boldsymbol{\varphi}(t)$ is also periodic with period τ. Under these assumptions, the linear system (7) is periodic with period τ:

$$a^i_j(t + \tau) = a^i_j(t), \qquad i, j = 1, \ldots, n,$$

so we can speak about its characteristic numbers [see §19, (E)]. Thus whenever (1) is autonomous [see (3)] and its periodic solution $\boldsymbol{\varphi}(t)$ is not a state of equilibrium, the linear system (7) necessarily has one characteristic number equal to unity.

Let us prove this last assertion. Let $\Psi(t)$ be a matrix which satisfies the matrix equation

$$\dot{\Psi} = A(t)\Psi$$

with the initial condition

$$\Psi(t_0) = E, \tag{8}$$

and let C be the fundamental matrix of the solution $\Psi(t)$ [see 19, (A)], so that

$$\Psi(t + \tau) = \Psi(t)C. \tag{9}$$

It is seen directly that any solution $\boldsymbol{\psi}(t)$ of (7) may be written in the form

$$\boldsymbol{\psi}(t) = \Psi(t)\boldsymbol{\psi}(t_0).$$

From this and from (8) and (9) it follows that

$$\boldsymbol{\psi}(t_0 + \tau) = C\boldsymbol{\psi}(t_0). \tag{10}$$

Remembering that (1) is autonomous, we have [see (3)]

$$\dot{\boldsymbol{\varphi}}(t) = \mathbf{f}(\boldsymbol{\varphi}(t)),$$

which, when differentiated with respect to t, gives

$$\ddot{\boldsymbol{\varphi}}(t) = A(t)\dot{\boldsymbol{\varphi}}(t).$$

Thus the vector function $\dot{\boldsymbol{\varphi}}(t)$ satisfies the vector equation (7). But $\dot{\boldsymbol{\varphi}}(t)$ is periodic with period τ, so that from (10) we obtain

$$\dot{\boldsymbol{\varphi}}(t_0) = \dot{\boldsymbol{\varphi}}(t_0 + \tau) = C\dot{\boldsymbol{\varphi}}(t_0), \tag{11}$$

and since $\dot{\boldsymbol{\varphi}}(t_0) \neq 0$ because $\boldsymbol{\varphi}(t)$ is not a state of equilibrium, it follows that the matrix C has an eigenvalue equal to unity and consequently one of the characteristic numbers of (7) is equal to unity.

The theorems of Lyapunov and Andronov-Witt. Now we can formulate sufficient conditions for the stability of the periodic solution $\boldsymbol{\varphi}(t)$ whenever system (1) is periodic and whenever it is autonomous.

THEOREM 25. *Let equation* (1) *be periodic in t with period τ [see* (2)*], and let $\boldsymbol{\varphi}(t)$ be a periodic solution of* (1) *also having period τ [see* (4)*]. If the absolute values of all characteristic numbers of* (7) *[see §19,* (E)*] are less than unity, then the solution $\boldsymbol{\varphi}(t)$ is asymptotically stable; moreover, there exists a number $\sigma > 0$ such that for $|\mathbf{x}_1 - \mathbf{x}_0| < \sigma$ the bound*

$$|\boldsymbol{\varphi}(t; t_0, \mathbf{x}_1) - \boldsymbol{\varphi}(t)| < re^{-\alpha t}|\mathbf{x}_1 - \mathbf{x}_0|, \qquad t \geq t_0, \tag{12}$$

holds, where r and α are two positive numbers which do not depend on $\mathbf{x}_1$.

THEOREM 26. *Let* (1) *be autonomous and let $\boldsymbol{\varphi}(t)$ be a periodic solution with period τ which is distinct from the state of equilibrium. If the characteristic number of* (7)*, which is equal to unity, has multiplicity one and if the absolute values of all remaining characteristic numbers of* (7) *are less than unity, then the solution $\boldsymbol{\varphi}(t)$ is Lyapunov stable.*

Theorem 25 was formulated by Lyapunov. Theorem 26 was derived by Andronov and Witt as a rather simple corollary of a rather obscure theorem of Lyapunov. We shall give here an alternative proof of Theorem 26, which will be based on Lyapunov's method.

We shall preface the proofs of Theorems 25 and 26 with certain constructions which will prove useful.

In §25 we gave the definition of the derivative of a certain function with respect to an autonomous system of equations. We shall present it here for the case of a nonautonomous system.

(B) Let

$$F(\mathbf{x}) = F(x^1, \ldots, x^n)$$

be a scalar function of the vector variable $\mathbf{x}$. We define the derivative

$\dot{F}_{(1)}(t_0, \mathbf{x}_0)$ of this function with respect to system (1) at the point t_0, $\mathbf{x}_0$ in the following manner. Let $\boldsymbol{\varphi}(t)$ be the solution of equation (1) with initial values t_0, $\mathbf{x}_0$. If we set

$$\dot{F}_{(1)}(t_0, \mathbf{x}_0) = \frac{d}{dt} F(\boldsymbol{\varphi}(t))\Big|_{t=t_0}$$

and perform the differentiation indicated on the right-hand side, we obtain

$$\dot{F}_{(1)}(t, \mathbf{x}) = \sum_{i=1}^{n} \frac{\partial F(\mathbf{x})}{\partial x^i} f^i(t, \mathbf{x}).$$

Whenever (1) is autonomous, the derivative $\dot{F}_{(1)}(t, \mathbf{x})$ of the function $F(\mathbf{x})$ with respect to (1) at the point t, $\mathbf{x}$ does not depend on t.

(C) Let

$$\dot{\mathbf{z}} = B\mathbf{z} + \mathbf{p}(t, \mathbf{z}) \tag{13}$$

be a normal system of differential equations in vector form, where $B = (b_j^i)$ is a constant matrix, all of whose eigenvalues have negative real parts, and $\mathbf{p}(t, \mathbf{z})$ is the remainder defined for $t \geq t_0$, $|z| < c$ $(c > 0)$ and possessing the bound

$$|\mathbf{p}(t, \mathbf{z})| \leq p|\mathbf{z}|^2, \tag{14}$$

where p is a positive number. Thus the solution $\mathbf{z} = 0$ of equation (13) is asymptotically stable; moreover, the estimate

$$|\boldsymbol{\chi}(t, \mathbf{z}_1)| \leq r|\mathbf{z}_1|e^{-\alpha t}, \qquad t \geq t_0, \tag{15}$$

is valid for the solution $\mathbf{z} = \boldsymbol{\chi}(t, \mathbf{z}_1)$ with initial values t_0, $\mathbf{z}_1$, $|\mathbf{z}_1| < c_1 < c$, where r and α are positive numbers which do not depend on $\mathbf{z}_1$.

Proposition (C) may be proved in exactly the same way as Lyapunov's theorem (see §26). We shall carry out this proof without going into excessive detail.

Let $W(\mathbf{z})$ be Lyapunov's function for the linear system

$$\dot{\mathbf{z}} = B\mathbf{z} \tag{16}$$

with constant coefficients [see §26, (E)], so that the inequality

$$\dot{W}_{(16)}(\mathbf{z}) = \sum_{i,j} \frac{\partial W(\mathbf{z})}{\partial z^i} b_j^i z^j \leq -2\beta W(\mathbf{z}), \qquad \beta > 0,$$

holds. From this inequality and from (14) we obtain for $W(\mathbf{z}) \leq c_2$ the inequality

$$\dot{W}_{(13)}(\mathbf{z}) = \sum_{i,j} \frac{\partial W(\mathbf{z})}{\partial z^i} b_j^i z^j + \sum_{i} \frac{\partial W(\mathbf{z})}{\partial z^i} p^i(t, \mathbf{z}) \leq -2\alpha W(\mathbf{z}),$$

where $\alpha < \beta$ and c_2 are certain positive numbers. Let us set

$$w(t) = W(\mathbf{x}(t, \mathbf{z}_1)) \quad \text{for} \quad W(\mathbf{z}_1) < c_2.$$

For the function $w(t)$, $t \geq t_0$, the inequality

$$\dot{w}(t) \leq -2\alpha w(t) \tag{17}$$

holds whenever the relation

$$w(t) \leq c_2$$

is valid. From (17) it follows that when the inequality $w(t) \leq c_2$ is valid, the function $w(t)$ decreases (or, more accurately, it does not increase), and since at the initial moment $t = t_0$ the inequality $w(t) < c_2$ is fulfilled, the point $\mathbf{x}(t, \mathbf{z}_1)$ cannot leave the closed set F defined by the inequality $W(z) \leq c_2$, so that the solution $\mathbf{x}(t, \mathbf{z}_1)$ is defined for all values $t \geq t_0$ [see §24, (B)] and (17) holds for all these values. Assuming now that $\mathbf{z}_1 \neq 0$, we may perform the following calculations, starting from (17):

$$\frac{\dot{w}(t)}{w(t)} \leq -2\alpha ,$$

or, by integrating, we obtain

$$\ln w(t) - \ln w(t_0) \leq -2\alpha(t - t_0),$$

and from this follows

$$w(t) \leq w(t_0)e^{-2\alpha(t-t_0)},$$

or, what is the same thing,

$$W(\mathbf{x}(t, \mathbf{z}_1)) \leq W(\mathbf{z}_1)e^{-2\alpha(t-t_0)}.$$

The estimate (15) follows directly from this estimate. Thus proposition (C) is proved.

Proof of Theorem 25. By Theorem 12 there exists a transformation

$$\mathbf{y} = T(t)\mathbf{z}, \tag{18}$$

where the matrix $T(t)$ is real and has period 2τ, which transforms (7) into the equation

$$\dot{\mathbf{z}} = B\mathbf{z}$$

with a constant real matrix B. The solution of the equation $\dot{\mathbf{z}} = B\mathbf{z}$ is the matrix e^{tB} [see §19(C)], so that the matrix $e^{2\tau B}$ is fundamental for this equation as well as for equation (7). Thus by the hypotheses of Theorem 25 the absolute value of each eigenvalue of the matrix $e^{2\tau B}$ is less than unity. But according to Theorem 27, the eigenvalues of the matrix $e^{2\tau B}$ have the form $e^{2\tau\lambda}$, where λ runs through all eigenvalues of B. Thus $|e^{2\tau\lambda}| < 1$, so that all eigenvalues of B have negative real parts. Under the transformation (18) equation (6) can be reduced to the form (13), and for its solution $\mathbf{z} = \boldsymbol{\chi}(t_1, \mathbf{z}_1)$ we obtain the estimate (15). From this estimate and from the fact that the matrix $T(t)$ is nondegenerate, we obtain the estimate (12). Thus Theorem 25 is proved.

Proof of Theorem 26. Proceeding from the assumption that equation (7) has a characteristic number equal to unity and of multiplicity one and that the absolute values of all remaining characteristic numbers are less than unity, we shall show that there exists a transformation

$$\mathbf{y} = T(t)\mathbf{z} \tag{19}$$

with a real matrix $T(t)$ of period 2τ, which transforms (7) into the equation

$$\dot{\mathbf{z}} = B\mathbf{z} \tag{20}$$

with a constant matrix B of the form

$$B = \left(\begin{array}{c|c} B^* & 0 \\ \hline 0 & 0 \end{array}\right), \tag{21}$$

where B^* is a square matrix of order $n - 1$, all of whose eigenvalues have negative real parts.

Let C be the fundamental matrix of a certain solution of the matrix equation [see (7)]

$$\dot{Y} = A(t)Y. \tag{22}$$

Since C has an eigenvalue equal to unity and of multiplicity one, then in some basis it has the form

$$\left(\begin{array}{c|c} C^* & 0 \\ \hline 0 & 1 \end{array}\right), \tag{23}$$

where C^* is a real square matrix of order $n - 1$, all of whose eigenvalues have an absolute value less than unity [see §32, (G)]. Since the matrix C and the matrix (23) may be obtained from each other by a linear transformation, the matrix (23) is fundamental for some solution of (22); we shall assume that C coincides with the matrix (23). By proposition (D) of §33 there exists a real matrix B^* which satisfies the condition

$$e^{2\tau B^*} = C^{*2},$$

where, by Theorem 27, all eigenvalues of the matrix B^* have negative real parts. It is clear that the matrix B [see (21)] satisfies the condition

$$e^{2\tau B} = C^2$$

[see (23)]. Thus (compare the proof of Theorem 12) there exists a transformation (19) which transforms (7) into (20).

We shall now determine those conditions which the matrix $T(t)$ must satisfy in order that the transformation (19) transforms (7) into (20). Differentiating (19), we obtain

$$\dot{\mathbf{y}} = \dot{T}(t)\mathbf{z} + T(t)\dot{\mathbf{z}} = \dot{T}(t)\mathbf{z} + T(t)B\mathbf{z}.$$

Replacing $\mathbf{z}$ in this equation by the formula $\mathbf{z} = T^{-1}(t)\mathbf{y}$, we obtain

$$\dot{\mathbf{y}} = (\dot{T}(t) + T(t)B)T^{-1}(t)\mathbf{y}.$$

Since this equation coincides with (7), we have

$$(\dot{T}(t) + T(t)B)T^{-1}(t) = A(t),$$

and, by multiplying this relation on the right by the matrix $T(t)$, we obtain

$$\dot{T}(t) + T(t)B = A(t)T(t). \tag{24}$$

This condition imposed upon the matrix $T(t)$ is necessary and sufficient for transformation (19) to transform (7) into (20). We shall decompose (24) into two relations by representing the matrix $T(t)$ in the form

$$T(t) = (T^*(t), \mathbf{t}(t)),$$

where $T^*(t)$ has n rows and $n - 1$ columns and where the matrix $\mathbf{t}(t)$ is the last row of the matrix $T(t)$ and therefore is a nonzero vector. We obtain [see (21)]

$$\dot{T}^*(t) + T^*(t)B^* = A(t)T^*(t), \tag{25}$$

$$\dot{\mathbf{t}}(t) = A(t)\mathbf{t}(t). \tag{26}$$

From the last relation it is clear that $\mathbf{t}(t)$ is a periodic solution with period 2τ of equation (7) so that the condition

$$\mathbf{t}(t_0) = \mathbf{t}(t_0 + 2\tau) = C^2\mathbf{t}(t_0)$$

is satisfied [compare (10)]. Thus the vector $\mathbf{t}(t_0)$ is an eigenvector of the matrix C^2 with an eigenvalue equal to unity. Since the matrix

$$C^2 = \left(\begin{array}{c|c} C^{*2} & 0 \\ \hline 0 & 1 \end{array}\right)$$

has unity as a simple eigenvalue and since we already know one vector $\dot{\boldsymbol{\varphi}}(t_0) \neq 0$ with this eigenvalue in the matrix C^2 [see (11)], we have

$$\mathbf{t}(t_0) = \gamma\dot{\boldsymbol{\varphi}}(t_0),$$

and therefore

$$\mathbf{t}(t) = \gamma\dot{\boldsymbol{\varphi}}(t),$$

since both vector functions $\mathbf{t}(t)$, $\dot{\boldsymbol{\varphi}}(t)$ are solutions of (7). From this it is clear that if we replace the last column $\mathbf{t}(t)$ of $T(t)$ by the vector $\dot{\boldsymbol{\varphi}}(t)$, then the matrix $\big(T^*(t), \dot{\boldsymbol{\varphi}}(t)\big)$ obtained again will satisfy conditions (25) and (26). Therefore we shall assume that

$$T(t) = \big(T^*(t), \dot{\boldsymbol{\varphi}}(t)\big). \tag{27}$$

Starting from (25) and (27), we shall transform the unknown function $\mathbf{x}$ of equation (1) in the autonomous case [see (3)] into the new unknown functions $\mathbf{z}^*$ and s, where $\mathbf{z}^* = (z^1, \ldots, z^{n-1})$ is an $(n - 1)$-dimensional vector (which will be regarded in the sequel as a matrix with one column) and where s is a new scalar variable. For this we set

$$\mathbf{x} = T^*(s)\mathbf{z}^* + \boldsymbol{\varphi}(s) = \mathbf{g}(\mathbf{z}^*, s). \tag{28}$$

This transformation is periodic in s with period 2τ. To each pair $\mathbf{z}^*$, s relation (28) places in correspondence for sufficiently small $|\mathbf{z}^*|$ the point $\mathbf{x}$, which is close to the point $\boldsymbol{\varphi}(s)$ of the periodic trajectory K defined by the solution $\mathbf{x} = \boldsymbol{\varphi}(t)$. In the neighborhood of each pair $\mathbf{z}^* = 0$, $s = s_0$ relation (28) is a one-to-one transformation, since the functional determinant of this transformation at the point $\mathbf{z}^* = 0$, $s = s_0$ is equal to the determinant of the matrix $T(s_0)$ [see (27)] and therefore is distinct from zero. The coordinate s of the pair $(\mathbf{z}^*, s)$ will be assumed to be a cyclic coordinate with period 2τ, i.e., we shall identify the pair $(\mathbf{z}^*, s)$ and $(\mathbf{z}^*, s + 2\tau)$. Since the pairs $(0, s_0)$ and $(0, s_0 + \tau)$ are mapped by (28) into the same point $\boldsymbol{\varphi}(s_0)$ of the trajectory K, some neighborhoods of the pairs $(0, s_0)$ and $(0, s_0 + \tau)$ are mapped one-to-one into the same neighborhood of the point $\boldsymbol{\varphi}(s_0)$ of K. Thus, transformation (28) maps the set of all pairs $(\mathbf{z}^*, s)$ (for sufficiently small $|\mathbf{z}^*|$) in a two-to-one manner onto some neighborhood of K. Here the closed curve consisting of all pairs $(0, s)$, $0 \leq s \leq 2\tau$ is mapped onto the line K twice.

We shall now substitute into equation (1) [see (3)] the unknown vector $\mathbf{x}$ according to formula (28). Substitution into the left-hand sides gives

$$\dot{\mathbf{x}} = T^{*\prime}(s)\mathbf{z}^*\dot{s} + T^*(s)\dot{\mathbf{z}}^* + \boldsymbol{\varphi}'(s)\dot{s}. \tag{29}$$

Substitution into the right-hand side gives

$$\mathbf{f}(\mathbf{x}) = \mathbf{f}\big(\boldsymbol{\varphi}(s)\big) + A(s)T^*(s)\mathbf{z}^* + R(s, \mathbf{z}^*), \tag{30}$$

where the remainder $R(s, \mathbf{z}^*)$ is periodic in s with period 2τ and is an infinitesimal of the second order with respect to the vector $\mathbf{z}^*$. Equating the right-hand sides of (29) and (30), we obtain

$$T^*(s)\dot{\mathbf{z}}^* + \boldsymbol{\varphi}'(s)\dot{s} + T^{*\prime}(s)\mathbf{z}^*\dot{s} = \mathbf{f}(\boldsymbol{\varphi}(s)) + A(s)T^*(s)\mathbf{z}^* + R(s, \mathbf{z}^*).$$

Replacing the matrix $A(s)T^*(s)$ according to formula (25) and substituting $\boldsymbol{\varphi}'(s)$ for $\mathbf{f}[\boldsymbol{\varphi}(s)]$, we obtain

$$T^*(s)\dot{\mathbf{z}}^* + \boldsymbol{\varphi}'(s)\dot{s} + T^{*\prime}(s)\mathbf{z}^*\dot{s} = \boldsymbol{\varphi}'(s) + (T^{*\prime}(s) + T^*(s)B^*)\mathbf{z}^* + R(s, \mathbf{z}^*),$$

whence

$$T^*(s)(\dot{\mathbf{z}}^* - B^*\mathbf{z}^*) + (\boldsymbol{\varphi}'(s) + T^{*\prime}(s)\mathbf{z}^*)(\dot{s} - 1) = R(s, \mathbf{z}^*). \tag{31}$$

We now consider two new auxiliary variables, the vector

$$\mathbf{u}^*(u^1, \ldots, u^{n-1})$$

and the scalar u^n. In the n-dimensional space of the variables $(\mathbf{u}^*, u^n) = (u^1, u^2, \ldots, u^n)$ we consider the linear transformation M which depends on the parameters s and $\mathbf{z}^*$, and set

$$M(\mathbf{u}^*, u^n) = T^*(s)\mathbf{u}^* + (\boldsymbol{\varphi}'(s) + T^{*\prime}(s)\mathbf{z}^*)u^n.$$

For $\mathbf{z}^* = 0$ the transformation M reduces to $T(s)$, so that close to zero M is nondegenerate. Thus the equation

$$M(\mathbf{u}^*, u^n) = R(s, \mathbf{z}^*)$$

can be solved uniquely (for $\mathbf{z}^*$ close to zero) for the unknowns $\mathbf{u}^*$, u^n, and the solution

$$\mathbf{u}^* = \mathbf{q}^*(s, \mathbf{z}^*),$$
$$u^n = q(s, \mathbf{z}^*)$$

is periodic in s with period 2τ and is an infinitesimal of the second order with respect to $\mathbf{z}^*$. Since (31) can be rewritten in the form

$$M(\dot{\mathbf{z}}^* - B^*\mathbf{z}^*, \dot{s} - 1) = R(s, \mathbf{z}^*),$$

we obtain

$$\dot{\mathbf{z}}^* - B^*\mathbf{z}^* = \mathbf{q}^*(s, \mathbf{z}^*), \qquad \dot{s} - 1 = q(s, \mathbf{z}^*).$$

Thus in terms of the variables $\mathbf{z}^*$, s, equation (1) may be written in the form

$$\dot{\mathbf{z}}^* = B^*\mathbf{z}^* + \mathbf{q}^*(s, \mathbf{z}^*), \tag{32}$$

$$\dot{s} = 1 + q(s, \mathbf{z}^*). \tag{33}$$

There now exists a positive number ϵ such that for $|\mathbf{z}^*| < \epsilon$ the remainder $q(s, \mathbf{z}^*)$ satisfies the inequality $|q(s, \mathbf{z}^*)| < 1$. When this inequality is satisfied for every solution $\mathbf{z}^* = \mathbf{z}^*(t)$, $s = s(t)$, then s may be taken in place of t as the independent variable, and equations (32) and (33) may be rewritten in the form

$$\frac{d\mathbf{z}^*}{ds} = \frac{B^*\mathbf{z}^* + \mathbf{q}^*(s, \mathbf{z}^*)}{1 + q(s, \mathbf{z}^*)},$$

$$\frac{dt}{ds} = \frac{1}{1 + q(s, \mathbf{z}^*)},$$

or

$$\frac{d\mathbf{z}^*}{ds} = B^*\mathbf{z}^* + \mathbf{k}^*(s, \mathbf{z}^*), \tag{34}$$

$$\frac{dt}{ds} = 1 + k(s, \mathbf{z}^*), \tag{35}$$

where the remainders $\mathbf{k}^*(s, \mathbf{z}^*)$ and $k(s, \mathbf{z}^*)$ are periodic in s with period 2τ and are infinitesimals of the second order with respect to the vector $\mathbf{z}^*$.

In the system (34), (35), s is the independent variable, but $\mathbf{z}^*$ and t are considered to be unknown functions of s. Equation (34) does not contain the unknown function t and can be solved separately. Thus, in order to find the solution of the system (32) and (33) with initial values t_0, $\mathbf{z}_1^*$, s_1, it is first necessary to find the solution $\mathbf{z}^*(s, \mathbf{z}_1^*, s_1)$ of (34) with the initial values $\mathbf{z}_1^*$, s_1 which by proposition (C) is defined for all values $s \geq s_1$ for sufficiently small $|\mathbf{z}_1^*|$ and has the estimate

$$|\mathbf{z}^*(s, \mathbf{z}_1^*, s_1)| \leq r|\mathbf{z}_1^*|e^{-\alpha s}. \tag{36}$$

After this we must find the solution of (35) with the initial values t_0, $\mathbf{z}_1^*$, s_1; this solution is given by the obvious formula

$$t = t_0 + \int_{s_1}^{s} [1 + k(s, \mathbf{z}^*(s, \mathbf{z}_1^*, s_1))]\, ds$$

$$= t_0 - s_1 + s + \int_{s_1}^{s} k(s, \mathbf{z}^*(s, \mathbf{z}_1^*, s_1))\, ds. \tag{37}$$

The last equation can be solved for s whenever $|\mathbf{z}^*|$ is sufficiently small, so that we obtain

$$s = s(t, \mathbf{z}_1^*, s_1). \tag{38}$$

Substituting this expression for s in the solution $\mathbf{z}^*(s, \mathbf{z}_1^*, s_1)$ of equation (34), we have

$$\mathbf{z}^*(t) = \mathbf{z}^*(s(t, \mathbf{z}_1^*, s_1), \mathbf{z}_1^*, s_1). \tag{39}$$

Formulas (38) and (39) together give the solution of the system (32), (33) with the initial values t_0, $\mathbf{z}_1^*$, s_1. From (37) it follows that for $t \geq t_0$ we have

$$|s(t, \mathbf{z}_1^*, s_1) - t| \leq |s_1 - t_0| + l|\mathbf{z}_1^*|^2, \tag{40}$$

where l is some positive constant. In the particular case when $\mathbf{z}_1^* = 0$, $s_1 = t_0$, the solution (38), (39) has the form

$$\mathbf{z}^*(t) = 0, \qquad s(t) = t.$$

From the estimates (36) and (40) it follows that this solution of the system (32), (33) is Lyapunov stable.

If we substitute the solution (38), (39) into the transformation formula (28), we obtain the solution $\mathbf{x} = \boldsymbol{\varphi}(t, \mathbf{x}_1)$ of equation (1) with initial values $t = t_0$, $\mathbf{x} = \mathbf{x}_1 = g(\mathbf{z}_1^*, s_1)$. Since (28) is a one-to-one mapping onto some neighborhood of the pair $\mathbf{z}^* = 0$, $s = t_0$, any solution $\boldsymbol{\varphi}(t, \mathbf{x}_1)$ of equation (1) with initial values $t_0, \mathbf{x}_1$ can be obtained in this way from a certain solution (38), (39) of the system (32), (33) for sufficiently small $|\mathbf{x}_1 - \mathbf{x}_0|$. Here the solution $\mathbf{x} = \boldsymbol{\varphi}(t)$ is obtained from the solution $\mathbf{z}^* = 0$, $s \equiv t$. Now the Lyapunov stability of the original periodic solution $\mathbf{x} = \boldsymbol{\varphi}(t)$ follows from the Lyapunov stability of the solution $\mathbf{z}^* = 0$, $s \equiv t$ and from the uniform continuity of (28). Thus Theorem 26 is proved.

Let us apply the results obtained here to the case of a limit cycle.

(D) We shall assume that the autonomous system (1) [see (3)] is of the second order:

$$\dot{x}^i = f^i(x^1, x^2) = f^i(\mathbf{x}), \qquad i = 1, 2,$$

and let

$$\mathbf{x} = \boldsymbol{\varphi}(t)$$

be its periodic solution with period τ. Here system (7) has the form

$$\dot{y}^i = \frac{\partial f^i(\boldsymbol{\varphi}(t))}{\partial x^1} y^1 + \frac{\partial f^i(\boldsymbol{\varphi}(t))}{\partial x^2} y^2, \qquad i = 1, 2.$$

By proposition (A) one characteristic number of this system is equal to unity; the second will be denoted by λ. Then

$$\lambda = \exp\left[\int_0^\tau \left(\frac{\partial f^1(\boldsymbol{\varphi}(t))}{\partial x^1} + \frac{\partial f^2(\boldsymbol{\varphi}(t))}{\partial x^2}\right) dt\right]. \tag{41}$$

Thus if

$$\int_0^\tau \left(\frac{\partial f^1(\boldsymbol{\varphi}(t))}{\partial x^1} + \frac{\partial f^2(\boldsymbol{\varphi}(t))}{\partial x^2}\right) dt < 0,$$

then the periodic solution $\mathbf{x} = \boldsymbol{\varphi}(t)$ is Lyapunov stable. In fact (see example) there exists a succession function $\chi(u)$ of the periodic solution $\mathbf{x} = \boldsymbol{\varphi}(t)$ (see §28) for which

$$\chi'(u_0) = \lambda, \tag{42}$$

so that for $\lambda \neq 1$ the periodic solution $\mathbf{x} = \boldsymbol{\varphi}(t)$ is a rough limit cycle. It is stable for $\lambda < 1$ and unstable for $\lambda > 1$.

Let us prove the inequality (41). The fundamental matrix C of the solution $Y = \Psi(t)$ of the equation $\dot{Y} = A(t)Y$ with initial value $\Psi(0) = E$ [see (A)] is given by the equality

$$C = \Psi(\tau).$$

By Liouville's formula we have

$$\text{Det}\,\Psi(\tau) = \text{Det}\,\Psi(0)\cdot\exp\left(\int_0^\tau S(t)\,dt\right),$$

where

$$S(t) = a_1^1(t) + a_2^2(t) = \frac{\partial f^1(\boldsymbol{\varphi}(t))}{\partial x^1} + \frac{\partial f^2(\boldsymbol{\varphi}(t))}{\partial x^2}$$

[see §17, (G)]. In our case, when the matrix C is of second order, and one of its eigenvalues is equal to unity and the other is equal to λ, we have

$$\lambda = \text{Det}\,C = \exp\left[\int_0^\tau \left(\frac{\partial f^1(\boldsymbol{\varphi}(t))}{\partial x^1} + \frac{\partial f^2(\boldsymbol{\varphi}(t))}{\partial x^2}\right)dt\right].$$

Example

Let $\boldsymbol{\varphi}(t)$ be the periodic solution of the autonomous system (1) [see (3)] with period τ with the initial values t_0, $\mathbf{x}_0$. The solution of this system with initial values t_0, $\boldsymbol{\xi}$ will be denoted by $\boldsymbol{\varphi}(t, \boldsymbol{\xi})$. We shall construct for the solution $\boldsymbol{\varphi}(t)$ an analogue of the succession function (see §28), which here will be a mapping of the $(n-1)$-dimensional space of variables into itself.

Let

$$\mathbf{x} = \mathbf{g}(\mathbf{u}); \qquad \mathbf{u} = (u^1, \ldots, u^{n-1}) \tag{43}$$

be the equation of a surface which intersects the trajectory $\boldsymbol{\varphi}(t)$ at the unique point

$$\mathbf{x}_0 = \boldsymbol{\varphi}(t_0, \mathbf{x}_0) = \mathbf{g}(\mathbf{u}_0) \tag{44}$$

and which is not tangent to the trajectory $\boldsymbol{\varphi}(t)$ at this point, so that the vectors

$$\dot{\boldsymbol{\varphi}}(t_0), \quad \frac{\partial \mathbf{g}(\mathbf{u}_0)}{\partial u^1}, \ldots, \frac{\partial \mathbf{g}(\mathbf{u}_0)}{\partial u^{n-1}} \tag{45}$$

are linearly independent. Let us find the intersection of the trajectory $\boldsymbol{\varphi}[t, \mathbf{g}(\mathbf{u})]$ with the surface (43) at t close to $t_0 + \tau$, assuming that $|\mathbf{u} - \mathbf{u}_0|$ is small. Let $\mathbf{g}(\mathbf{v})$ be the point of intersection; then the relation

$$\boldsymbol{\varphi}(t, \mathbf{g}(\mathbf{u})) - \mathbf{g}(\mathbf{v}) = 0 \tag{46}$$

holds. For $\mathbf{u} = \mathbf{u}_0$, we have the trivial solution of equation (16):

$$t = t_0 + \tau, \qquad \mathbf{v} = \mathbf{u}_0$$

[see (4) and (44)]. Here we consider $\mathbf{u}$ as an independent variable and t, $\mathbf{v}$ as unknowns. Since the functional determinant of (46) at $t = t_0 + \tau$, $\mathbf{u} = \mathbf{u}_0$, $\mathbf{v} = \mathbf{u}_0$ with respect to the unknown functions t and $\mathbf{v}$ does not vanish because of the linear independence of the vectors (45), then for small $|\mathbf{u} - \mathbf{u}_0|$, there exists a solution

$$t = t(\mathbf{u}), \qquad \mathbf{v} = \boldsymbol{\chi}(\mathbf{u})$$

of (46) with small $|t(\mathbf{u}) - (t_0 + \tau)|$ and $|\boldsymbol{\chi}(\mathbf{u}) - \mathbf{u}_0|$. The mapping $\boldsymbol{\chi}(\mathbf{u})$ of the space of variables $u^1, \ldots, u^{n-1}$ into itself (which is defined for small $|\mathbf{u} - \mathbf{u}_0|$) will be called the *succession mapping*. To each solution $\mathbf{u} = \mathbf{u}_1$ of the equation

$$\boldsymbol{\chi}(\mathbf{u}) - \mathbf{u} = 0 \tag{47}$$

corresponds the periodic solution $\boldsymbol{\varphi}[t, \mathbf{g}(\mathbf{u}_1)]$ of the autonomous equation (1) [see (3)] with a period close to τ; in particular, to the solution $\mathbf{u} = \mathbf{u}_0$ corresponds the original periodic solution $\boldsymbol{\varphi}(t) = \boldsymbol{\varphi}(t, \mathbf{g}(\mathbf{u}_0))$. If the functional matrix

$$M = \left(\frac{\partial \chi^i(\mathbf{u}_0)}{\partial u^j}\right), \qquad i, j = 1, \ldots, n-1,$$

does not have eigenvalues equal to unity, then the solution $\mathbf{u} = \mathbf{u}_0$ of (47) is isolated. Indeed, the functional matrix of (47) at $\mathbf{u} = \mathbf{u}_0$ is equal to

$$M - E^*.$$

For the determinant of this matrix not to vanish, it is necessary and sufficient that the matrix M not have an eigenvalue equal to unity.

We shall now answer the question of whether every periodic trajectory K_1 passing close to the trajectory K described by the solution $\boldsymbol{\varphi}(t)$ can be described by the solution $\boldsymbol{\varphi}[t, \mathbf{g}(\mathbf{u}_1)]$, where $\mathbf{u}_1$ is some solution of equation (47). This was just the situation in the plane case ($n = 2$). It turns out that for $n \geq 3$ the situation is different. Let us examine this question. We shall assume that τ is a minimal period of the solution $\boldsymbol{\varphi}(t)$, i.e., that the equality

$$\boldsymbol{\varphi}(t_0 + t) = \boldsymbol{\varphi}(t_0)$$

can be valid only when $t = k\tau$, where k is an integer [see §15 (C)]. If the trajectory K_1 is close to the trajectory K, then it intersects the surface (43) at some point $\mathbf{g}(\mathbf{u}_1)$, where $|\mathbf{u}_1 - \mathbf{u}_0|$ is close to zero. We shall set

$$\mathbf{u}_2 = \boldsymbol{\chi}(\mathbf{u}_1), \quad \mathbf{u}_3 = \boldsymbol{\chi}(\mathbf{u}_2), \quad \ldots, \quad \mathbf{u}_{i+1} = \boldsymbol{\chi}(\mathbf{u}_i), \quad \ldots .$$

Since the trajectory K_1 is closed, we can find a point in this sequence coinciding with the point $\mathbf{u}_1$; let $\mathbf{u}_{k+1}$ be the first such point. Then the trajectory K_1 is described by the solution $\boldsymbol{\varphi}[t, \mathbf{g}(\mathbf{u}_1)]$, where its minimal period is close to the number $k\tau$; the solution $\boldsymbol{\varphi}[t, \mathbf{g}(\mathbf{u}_1)]$ closes only after it goes around the trajectory $\boldsymbol{\varphi}(t)$ k times. In the plane case only the case $k = 1$ is possible. We shall call the number k the *multiplicity* of the trajectory K_1. For determining *double* trajectories it is necessary to solve, not equation (47), but the equation

$$\boldsymbol{\chi}(\boldsymbol{\chi}(\mathbf{u})) - \mathbf{u} = 0;$$

for determining triple trajectories it is necessary to solve the equation

$$\boldsymbol{\chi}[\boldsymbol{\chi}(\boldsymbol{\chi}(\mathbf{u}))] - \mathbf{u} = 0,$$

and so on. The functions $\boldsymbol{\chi}(\boldsymbol{\chi}(\mathbf{u}))$, $\boldsymbol{\chi}[\boldsymbol{\chi}(\boldsymbol{\chi}(\mathbf{u}))]$, ... are called *iterations* of the function $\boldsymbol{\chi}(\mathbf{u})$; the kth iteration will be denoted by $\boldsymbol{\chi}^k(\mathbf{u})$. Thus to find all k-tuple periodic solutions close to the solution $\boldsymbol{\varphi}(t)$, the equation

$$\boldsymbol{\chi}^k(\mathbf{u}) - \mathbf{u} = 0 \tag{48}$$

must be solved, but from all solutions of equation (48), only those are to be taken which are not solutions of equations of multiplicities encountered previously; the solution $\mathbf{u} = \mathbf{u}_0$ of equation (47) is also the solution of any equation (48). The functional matrix of equation (48) at $\mathbf{u} = \mathbf{u}_0$ is obviously equal to $M^k - E^*$; thus in order that equation (48) have only one solution $\mathbf{u} = \mathbf{u}_0$ close to $\mathbf{u}_0$, it is sufficient that the determinant of the matrix $M^k - E^*$ not vanish or, what is the same thing, that the matrix M^k not have eigenvalues equal to unity, or finally, that the matrix M not have eigenvalues equal to $\sqrt[k]{1}$. Thus, in order that there not be periodic trajectories of a given multiplicity k in the neighborhood of the trajectory K, it is sufficient that the matrix M not have eigenvalues equal to $\sqrt[k]{1}$. In particular, the matrix M does not have such eigenvalues if the absolute values of all its eigenvalues are less than unity.

From what has been said, it is evident how important a role the matrix M plays in the study of trajectories of the autonomous equation (1) [see (3)] which are close to the periodic solution $\boldsymbol{\varphi}(t)$. We shall now show that if equation (7) has a characteristic number equal to unity of multiplicity one, then for some choice of the surface (43) the matrix M coincides with the matrix C^* [see (23)]. Let us set

$$\psi^i_j(t) = \left.\frac{\partial\varphi^i(t, \xi)}{\partial\xi^j}\right|_{\xi=\mathbf{x}_0}, \qquad \Psi(t) = (\psi^i_j(t)).$$

By Theorem 15 we have

$$\dot{\Psi}(t) = A(t)\Psi(t), \tag{49}$$

where the initial condition

$$\Psi(t_0) = E$$

is fulfilled. Thus the matrix $\Psi(t)$ represents the solution of the matrix equation (49), which is the matrix form of equation (7), and therefore

$$\Psi(t_0 + \tau) = C.$$

Since the matrix C has an eigenvalue equal to unity of multiplicity one, we can choose a basis in the space of vectors $\mathbf{y}$ [see (A)] such that the matrix C can be written in the form (23). Let us now take for the coordinates in the phase space of equation (1) [see (3)] the components of the vector $\mathbf{y}$, by setting

$$\mathbf{x} = \boldsymbol{\varphi}(t_0) + \mathbf{y}$$

[compare (5)]. The coordinates thus obtained in the phase space will again be denoted by $x^1, \ldots, x^n$, and the surface (43) will be defined by the equations

$$x^1 = u^1, \quad \ldots, \quad x^{n-1} = u^{n-1}, \quad x^n = 0.$$

If we differentiate (46) with respect to $u^1, \ldots, u^{n-1}$ at $\mathbf{u} = 0$, $t = t_0 + \tau$, $\mathbf{v} = 0$ under the assumption that $t = t(\mathbf{u})$ and $\mathbf{v} = \boldsymbol{\chi}(\mathbf{u})$ are functions of the variables $u^1, \ldots, u^{n-1}$, we obtain the equality

$$C^* = M. \tag{50}$$

In the case $n = 2$, the matrix C^* is a scalar λ, and the relation (50) gives (42).

If the absolute values of all eigenvalues of the matrix C^* are less than unity, then there are no periodic solutions of any multiplicity in the neighborhood of the trajectory K. This follows from the estimate (36).

CHAPTER 6

LINEAR ALGEBRA

This chapter is intended to supplement the basic material of the book. We present here those results in the area of linear algebra which have been used in some of the more advanced sections. It should be noted that §34 is based only on the results of §32 and does not use the results of §33 at all.

32. The minimal annihilating polynomial. *Eigenvalues and eigenvectors.* (A) To each nth-order square matrix

$$A = (a^i_j), \qquad i, j = 1, 2, \ldots, n,$$

whose elements are real or complex numbers there corresponds a linear transformation $\mathbf{A}$ of the n-dimensional vector space R; namely, to the vector

$$\mathbf{x} = (x^1, \ldots, x^n)$$

of the space R corresponds the vector

$$\mathbf{A}\mathbf{x} = \mathbf{y} = (y^1, \ldots, y^n),$$

which is defined by the relation

$$y^i = \sum_j a^i_j x^j.$$

To the zero matrix 0 (all of whose elements are zero) corresponds the zero transformation $\mathbf{0}$ which transforms every vector into zero. To the unit matrix

$$E = (\delta^i_j), \qquad \delta^i_j = \begin{cases} 0, & i \neq j, \\ 1, & i = j, \end{cases}$$

corresponds the unit or identity transformation $\mathbf{E}$ of R:

$$\mathbf{E}\mathbf{x} = \mathbf{x}.$$

If in the space R we introduce the new coordinates $x'^1, \ldots, x'^n$ which are related to the old coordinates $x^1, \ldots, x^n$ by the transformation formulas

$$x'^j = \sum s^j_i x^i,$$

or, in matrix form,

$$\mathbf{x}' = S\mathbf{x},$$

then the matrix

$$A' = SAS^{-1} \tag{1}$$

corresponds to the transformation **A** in the new coordinate system. Let us prove relation (1). We have

$$\mathbf{y}' = S\mathbf{y} = SA\mathbf{x} = SAS^{-1}\mathbf{x}'.$$

(B) Let **A** be the linear transformation and A the matrix corresponding to the transformation **A** in some coordinate system. A nonzero vector **h** is called an *eigenvector* of the transformation **A**, and the number λ is called an *eigenvalue* of this transformation corresponding to the vector **h**, if the relation

$$\mathbf{Ah} = \lambda\mathbf{h} \tag{2}$$

holds. The determinant of the matrix $(a_j^i - z\delta_j^i)$,

$$D(z) = |a_j^i - z\,\delta_j^i| = |A - zE|,$$

is called the *characteristic polynomial* of the matrix A. The coefficients of the polynomial $D(z)$ do not depend on the choice of the system of coordinates but are completely determined by the transformation **A**. Therefore the polynomial $D(z)$ is called the characteristic polynomial of the transformation **A**. Furthermore, the number λ is an eigenvalue of the transformation **A** if and only if it is a root of the polynomial $D(z)$.

We shall prove that the polynomial $D(z)$ is independent of the choice of the coordinate system. In the new system of coordinates, the matrix SAS^{-1} corresponds to the transformation **A**, where S is a certain nonsingular matrix [see (1)]. We have

$$\begin{aligned}|SAS^{-1} - zE| &= |SAS^{-1} - zSES^{-1}| = |S(A - zE)S^{-1}| \\ &= |S| \cdot |A - zE| \cdot |S^{-1}| = |S| \cdot |A - zE| \cdot |S|^{-1} = |A - zE|.\end{aligned}$$

Let us now write the relation $(\mathbf{A} - \lambda\mathbf{E})\mathbf{h} = 0$, which is equivalent to relation (2), in terms of the coordinates

$$\sum_{j=1}^{n} (a_j^i - \lambda\,\delta_j^i)h^j = 0, \qquad i = 1, \ldots, n.$$

This system of homogeneous equations has a nontrivial solution $h^1, \ldots, h^n$ if and only if the determinant $D(\lambda)$ of this system is equal to zero. Thus every root λ of $D(z)$ is an eigenvalue of the transformation **A**, and conversely.

(C) If the eigenvalues $\lambda_1, \ldots, \lambda_k$ of the transformation **A** are pairwise distinct, then the corresponding eigenvectors $\mathbf{h}_1, \ldots, \mathbf{h}_k$ are linearly independent.

The proof is by induction on the number k. For $k = 1$ this assertion is obvious. Let us assume that it is true for $k - 1$ vectors, and prove it for k vectors. Let us assume that $a_1\mathbf{h}_1 + \cdots + a_k\mathbf{h}_k = 0$. Applying the transformation $\mathbf{A}$ to this relation, we obtain

$$a_1\lambda_1\mathbf{h}_1 + \cdots + a_k\lambda_k\mathbf{h}_k = 0;$$

on the other hand,

$$\lambda_k(a_1\mathbf{h}_1 + \cdots + a_k\mathbf{h}_k) = 0.$$

If we take the difference of these relations, we obtain

$$a_1(\lambda_1 - \lambda_k)\mathbf{h}_1 + \cdots + a_{k-1}(\lambda_{k-1} - \lambda_k)\mathbf{h}_{k-1} = 0.$$

Hence, by the induction hypothesis, it follows that $a_1 = \cdots = a_{k-1} = 0$.

Thus if all roots of the characteristic polynomial $D(z)$ are distinct, we can take the eigenvectors $\mathbf{h}_1, \ldots, \mathbf{h}_n$ of $\mathbf{A}$ as a basis of the space R. In this basis a diagonal matrix corresponds to $\mathbf{A}$. In the general case, the reduction of the transformation matrix to diagonal form is impossible, and we are forced to construct a comparatively complicated theory, which we now proceed to do.

The minimal annihilating polynomial. (D) By well-known rules we can add and multiply square matrices of order n, as well as multiply them by given numbers; to these operations on matrices correspond the same operations on transformations. Thus if

$$f(z) = a_0z^m + a_1z^{m-1} + \cdots + a_m$$

is a polynomial with real or complex coefficients in z, then by substituting the matrix A for the variable z in this polynomial, we obtain the matrix

$$f(A) = a_0A^m + a_1A^{m-1} + \cdots + c_mE,$$

which is a polynomial in A. The polynomial $f(\mathbf{A})$ of the transformation $\mathbf{A}$ is defined similarly. If $f(z) \not\equiv 0$ and if $f(A)$ is the zero matrix (in which case it is also, obviously, a zero transformation), then the polynomial $f(z)$ is said to *annihilate* the matrix A and the transformation $\mathbf{A}$. Thus the characteristic polynomial $D(z)$ of the matrix A annihilates A:

$$D(A) = 0.$$

For the proof let us consider the n-dimensional vector space R with the basis

$$\mathbf{e}_1, \mathbf{e}_2, \ldots, \mathbf{e}_n,$$

and let us study the coordinate system which corresponds to this basis, so that

$$\mathbf{e}_j = (0, \ldots, 1, \ldots, 0),$$

where the coordinate 1 is in the j-th place. We shall denote by $\mathbf{A}$ the transformation corresponding to the matrix A in the basis chosen. Then we have

$$\mathbf{A}\mathbf{e}_j = \sum_s a_j^s \mathbf{e}_s$$

or, what is the same thing,

$$\sum_s (a_j^s \mathbf{E} - \delta_j^s \mathbf{A})\mathbf{e}_s = 0. \tag{3}$$

Let us set

$$L_j^s(z) = a_j^s - \delta_j^s z.$$

Here $L_j^s(z)$ is a polynomial in z of degree zero or one, and

$$(L_j^s(z))$$

is the matrix formed from the polynomials. The cofactor of the element $L_j^s(z)$ in this matrix will be denoted by $M_s^j(z)$, so that the relation

$$\sum M_i^j(z) L_j^s(z) = \delta_i^s D(z) \tag{4}$$

holds. Multiplying (3) on the left by the polynomial $M_i^j(\mathbf{A})$ and summing over j, we obtain by (4):

$$\sum_{s,j} M_i^j(\mathbf{A})(a_j^s \mathbf{E} - \delta_j^s \mathbf{A})\mathbf{e}_s = \sum_{s,j} M_i^j(\mathbf{A}) L_j^s(\mathbf{A})\mathbf{e}_s$$

$$= \sum_s \delta_j^s D(\mathbf{A})\mathbf{e}_s = D(\mathbf{A})\mathbf{e}_i = 0.$$

Thus $D(\mathbf{A})$ transforms all basis vectors of the space R into zero, so that it is the zero transformation, which means that the corresponding matrix $D(A)$ is also zero:

$$D(A) = 0.$$

(E) In the set of all polynomials annihilating the matrix A (or the transformation $\mathbf{A}$), there exists a unique polynomial $\Delta(z)$ of minimal degree, which is unique up to a numerical factor; this polynomial $\Delta(z)$ is a divisor of all other polynomials which annihilate the matrix A and is called the *minimal* polynomial which annihilates the matrix A. We shall adopt the convention that the coefficient of the term of highest degree of the polynomial $\Delta(z)$ is equal to unity. Whenever the matrix A is real, the polynomial $\Delta(z)$ is real.

To prove (E), we recall that if $f(z)$ and $g(z)$ are two arbitrary polynomials and $d(z)$ is their greatest common divisor, then the identity

$$d(z) = p(z)f(z) + q(z)g(z) \tag{5}$$

holds, where $p(z)$ and $q(z)$ are suitably chosen polynomials. The existence of the identity (5) can be proved by the division algorithm for polynomials. From (5) it follows that if the polynomials $f(z)$ and $g(z)$ annihilate the matrix A, then their greatest common divisor $d(z)$ also annihilates A. From (D) it follows that there exist polynomials which annihilate the matrix A. Now let $\Delta(z)$ be a polynomial of minimal degree which annihilates A, and let $f(z)$ be an arbitrary polynomial also annihilating A. If the polynomial $f(z)$ is not divisible by $\Delta(z)$, then the greatest common divisor of these polynomials would have a degree smaller than that of $\Delta(z)$ and also would annihilate A, but this is impossible by hypothesis. Now if A is real, then

$$0 = \overline{\Delta(A)} = \overline{\Delta}(\overline{A}) = \overline{\Delta}(A).$$

Thus, the polynomial $\overline{\Delta}(z)$ annihilates A and therefore is divisible by $\Delta(z)$, but this is possible only if $\Delta(z) = \overline{\Delta}(z)$. Thus, proposition (E) is proved.

(F) Let $\Delta(z)$ be the minimal annihilating polynomial of the matrix A. The number λ is an eigenvalue of the matrix A if and only if it is a root of the polynomial $\Delta(z)$.

To prove this we denote by $\mathbf{A}$ the transformation of the n-dimensional vector space which corresponds to the matrix A. We remark that if $f(z)$ is an arbitrary polynomial, then

$$\mathbf{Ah} = \lambda\mathbf{h} \quad \text{implies} \quad f(\mathbf{A})\mathbf{h} = f(\lambda)\mathbf{h}. \tag{6}$$

Indeed, we have

$$\mathbf{Eh} = \mathbf{h}, \quad \mathbf{Ah} = \lambda\mathbf{h}, \quad \mathbf{A}^2\mathbf{h} = \mathbf{A}\lambda\mathbf{h} = \lambda^2\mathbf{h}, \quad \ldots, \quad \mathbf{A}^m\mathbf{h} = \lambda^m\mathbf{h}.$$

Multiplying these relations by the coefficients of the polynomial $f(z)$ and summing, we obtain (6).

Let us assume that λ is an eigenvalue of the matrix A; then there exists a vector $\mathbf{h} \neq 0$, such that $\mathbf{Ah} = \lambda\mathbf{h}$, and (6) implies $\Delta(\mathbf{A})\mathbf{h} = \Delta(\lambda)\mathbf{h}$; but $\Delta(\mathbf{A}) = 0$, so that $\Delta(\lambda) = 0$. Conversely, if λ is a root of the polynomial $\Delta(z)$, then $\Delta(z) = (z - \lambda)\Gamma(z)$. Since $\Delta(z)$ is the minimal annihilating polynomial of the matrix A, the polynomial $\Gamma(z)$ does not annihilate it, so that the matrix $\Gamma(A)$ and consequently the transformation $\Gamma(\mathbf{A})$ are nonzero. Thus there exists a vector $\mathbf{x}$, for which $\Gamma(\mathbf{A})\mathbf{x} = \mathbf{h}' \neq 0$, and we have $0 = \Delta(\mathbf{A})\mathbf{x} = (\mathbf{A} - \lambda\mathbf{E})\Gamma(\mathbf{A})\mathbf{x} = (\mathbf{A} - \lambda\mathbf{E})\mathbf{h}'$, so that

$\mathbf{Ah}' = \lambda\mathbf{h}'$, whence λ is an eigenvalue of the matrix A. Hence, proposition (F) is proved.

If necessary, the vector space R can be taken to be real whenever it has only vectors with real coordinates, or as complex whenever it has vectors with complex coordinates. If $\mathbf{x} = (x^1, \ldots, x^n)$ is a vector of a complex space R, then $\bar{\mathbf{x}} = (\bar{x}^1, \ldots, \bar{x}^n)$ is the vector which is the complex conjugate of $\mathbf{x}$. If S is a subspace of a complex space R, then the subspace $\bar{S}$, consisting of all vectors which are complex conjugate to the vectors of S, is defined as the complex conjugate of the subspace S.

The complex (or real) space R is said to be decomposed into a direct sum of its subspaces S_1 and S_2 if every vector $\mathbf{x}$ from R can be written uniquely in the form of a sum:

$$\mathbf{x} = \mathbf{x}_1 + \mathbf{x}_2,$$

where the vector $\mathbf{x}_i$ belongs to the subspace S_i, $i = 1, 2$.

(G) Let $\Delta(z) = \Delta_1(z)\,\Delta_2(z)$ be a factorization of the minimal polynomial annihilating the matrix A into two relatively prime factors. We denote by S_i, $i = 1, 2$, the linear subspace of the space R consisting of all vectors $\mathbf{x}$ of R which satisfy the condition $\Delta_i(\mathbf{A})\mathbf{x} = 0$, where $\mathbf{A}$ is the transformation with matrix A. It turns out that the space R can be decomposed into a direct sum of the subspaces S_1 and S_2. (If the matrix A is complex, then in the statement formulated here the space R must be considered complex.) Let us now assume that the matrix A is real; then it is necessary to distinguish two important cases. (1) The factors $\Delta_1(z)$ and $\Delta_2(z)$ are real; then the space R and its subspaces S_1 and S_2 can be considered real. (2) The factors $\Delta_1(z)$ and $\Delta_2(z)$ are complex conjugate; then the space R should be considered complex and its subspaces S_1 and S_2 turn out to be complex conjugate.

Let us prove proposition (G). Because the factors $\Delta_1(z)$ and $\Delta_2(z)$ are relatively prime, the identity

$$1 = p_1(z)\,\Delta_1(z) + p_2(z)\,\Delta_2(z) \tag{7}$$

is valid, where $p_1(z)$ and $p_2(z)$ are suitably chosen polynomials [see (5)]. We note that if the factors $\Delta_1(z)$ and $\Delta_2(z)$ are real, then the polynomials $p_1(z)$ and $p_2(z)$ can be chosen as real, since they are obtained by applying the division algorithm to the polynomials $\Delta_1(z)$ and $\Delta_2(z)$. Now let $\mathbf{x}$ be an arbitrary vector of R; by (7) we have

$$\mathbf{x} = p_1(\mathbf{A})\,\Delta_1(\mathbf{A})\mathbf{x} + p_2(\mathbf{A})\,\Delta_2(\mathbf{A})\mathbf{x}.$$

Setting

$$\mathbf{x}_1 = p_2(\mathbf{A})\,\Delta_2(\mathbf{A})\mathbf{x}, \qquad \mathbf{x}_2 = p_1(\mathbf{A})\,\Delta_1(\mathbf{A})\mathbf{x},$$

we obtain the decomposition $\mathbf{x} = \mathbf{x}_1 + \mathbf{x}_2$, where

$$\Delta_1(\mathbf{A})\mathbf{x}_1 = \Delta_1(\mathbf{A})p_2(\mathbf{A})\,\Delta_2(\mathbf{A})\mathbf{x} = p_2(\mathbf{A})\,\Delta(\mathbf{A})\mathbf{x} = 0,$$
$$\Delta_2(\mathbf{A})\mathbf{x}_2 = \Delta_2(\mathbf{A})p_1(\mathbf{A})\,\Delta_1(\mathbf{A})\mathbf{x} = p_1(\mathbf{A})\,\Delta(\mathbf{A})\mathbf{x} = 0,$$

so that the vector $\mathbf{x}_i$ belongs to the subspace S_i. Now if $\mathbf{x} = \mathbf{x}_1' + \mathbf{x}_2'$ is any decomposition of the vector $\mathbf{x}$ into a sum in which $\mathbf{x}_i'$ belongs to S_i, $i = 1, 2$, then by (7) we have

$$\mathbf{x}_1' = p_1(\mathbf{A})\,\Delta_1(\mathbf{A})\mathbf{x}_1' + p_2(\mathbf{A})\,\Delta_2(\mathbf{A})\mathbf{x}_1' = p_2(\mathbf{A})\,\Delta_2(\mathbf{A})(\mathbf{x}_1' + \mathbf{x}_2') = \mathbf{x}_1;$$

in exactly the same way $\mathbf{x}_2' = \mathbf{x}_2$, and the uniqueness of the decomposition is proved.

If the matrix A is real and the factors $\Delta_1(z)$ and $\Delta_2(z)$ are real, then, starting from the real vector $\mathbf{x}$, we obtain real vectors $\mathbf{x}_1$ and $\mathbf{x}_2$. However, if the matrix A is real and the factors $\Delta_1(z)$ and $\Delta_2(z)$ are complex conjugate, then the subspaces S_1 and S_2 are complex conjugate by definition. Thus proposition (G) is proved.

(H) Let $\mathbf{A}$ be a linear transformation of the n-dimensional space R, let

$$\lambda_1, \lambda_2, \ldots, \lambda_r$$

be the set of all eigenvalues of this transformation, let

$$\Delta(z) = (z - \lambda_1)^{k_1}(z - \lambda_2)^{k_2} \cdots (z - \lambda_r)^{k_r}$$

be the minimal annihilating polynomial of the transformation $\mathbf{A}$, and let

$$D(z) = (-1)^n(z - \lambda_1)^{q_1}(z - \lambda_2)^{q_2} \cdots (z - \lambda_r)^{q_r} \tag{8}$$

be the characteristic polynomial of the transformation $\mathbf{A}$. Since the polynomial $D(z)$ is divisible by the polynomial $\Delta(z)$, we have

$$q_1 \geq k_i, \qquad i = 1, \ldots, r.$$

By (G) the space R can be decomposed into a direct sum of the subspaces $S_1, S_2, \ldots, S_r$, where S_i consists of all vectors $\mathbf{x}$ which satisfy the condition

$$(\mathbf{A} - \lambda_i\mathbf{E})^{k_i}\mathbf{x} = 0.$$

Thus the space S_i is of dimension q_i, and the number q_i is called the *multiplicity* of the eigenvalue λ_i.

We shall prove that the dimension of the space S_i is equal to q_i. The space S_i is *invariant* with respect to the transformation $\mathbf{A}$, that is, $\mathbf{A}S_i$ is contained in S_i. Thus, if in the space S_i a certain basis is chosen, then to the transformation $\mathbf{A}$ on S_i there corresponds a certain matrix A_i

of order p_i, where p_i is the dimension of the space S_i. If the basis of R consists of the bases of all the spaces S_i, then in the basis obtained, there will correspond to the transformation **A** a matrix A consisting of matrices $A_1, \ldots, A_r$ located along the diagonal of the matrix A. From this it is clear that the characteristic polynomial of the transformation **A** of R will be equal to the product

$$D_1(z)D_2(z) \ldots D_r(z),$$

where $D_i(z)$ is the characteristic polynomial of the transformation **A** considered on the subspace S_i. Since $\Delta_i(z) = (z - \lambda_i)^{k_i}$ is the annihilating polynomial of **A** on the subspace S_i, the transformation **A** on S_i has only one eigenvalue λ_i, so that $D_i(z)$ has the form $(-1)^{p_i}(z - \lambda_i)^{p_i}$, because its degree is equal to the order of the matrix A_i, that is, to the dimension p_i of the space S_i. Consequently $D(z) = (-1)^n(z - \lambda_1)^{p_1}(z - \lambda_2)^{p_2} \ldots (z - \lambda_r)^{p_r}$, and therefore $p_i = q_i$ [see (8)]. Thus proposition (H) is proved.

33. Matrix functions. In this section we shall make no distinction between the transformation **A** and the corresponding matrix A, since we shall not change the coordinate system. In addition, in this section we shall also use some elementary facts from the theory of functions of a complex variable (see, for example, Ahlfors, *Complex analysis*, McGraw-Hill, New York, 1953).

Matrix power series. (A) Let

$$\begin{gathered} \Delta(z) = (z - \lambda_1)^{k_1}(z - \lambda_2)^{k_2} \cdots (z - \lambda_r)^{k_r}, \\ k_i > 0, \quad i = 1, \ldots, r, \quad k_1 + k_2 + \cdots + k_r = k, \end{gathered} \tag{1}$$

be a minimal annihilating polynomial of the matrix A, where

$$\lambda_1, \lambda_2, \ldots, \lambda_r \tag{2}$$

are its pairwise distinct roots. By proposition (F) of §32 the numbers (2) constitute the set of all eigenvalues of A. We say that *the function W is defined on the spectrum of the matrix A* if there is a correspondence between the eigenvalue λ_i of the matrix A and the sequence of numbers

$$W^{(0)}(\lambda_i), \quad W^{(1)}(\lambda_i), \ldots, W^{(k_i-1)}(\lambda_i), \qquad i = 1, \ldots, r. \tag{3}$$

If $W(z)$ is a function of the complex variable z which is holomorphic at the points λ_i, then, if the numbers (3) are regarded as the value of the function itself and of its derivatives up to order $k_i - 1$ at the point λ_i, we obtain a function defined on the spectrum of the matrix A. If for two functions of z, the values (3) coincide, respectively, then we shall say that the two functions *coincide on the spectrum of the matrix A*. Thus two polynomials $f(z)$ and $g(z)$ coincide on the spectrum of A if and only if $f(A) = g(A)$. Furthermore, it turns out that for any arbitrarily given numbers (3), there always exists a unique polynomial $\varphi(z)$ of degree not larger than $k - 1$, whose values on the spectrum of A coincide with the numbers (3), i.e.,

$$\varphi^{(j)}(\lambda_i) = W^{(j)}(\lambda_i), \qquad j = 0, \ldots, k_i - 1, \qquad i = 1, \ldots, r. \tag{4}$$

Here the coefficients of $\varphi(z)$ are linear functions of the quantities (3) and therefore depend continuously on them.

We shall prove these statements. If we set $h(z) = f(z) - g(z)$, and if $f(A) = g(A)$, then $h(A) = 0$. Furthermore, if the values of the polynomials $f(z)$ and $g(z)$ coincide on the spectrum of A, then the function $h(z)$ vanishes on the spectrum of A. Thus, in order to prove that part of the assertion of (A) concerning $f(z)$ and $g(z)$, it is sufficient to prove that the polynomial $h(z)$ annihilates the matrix A if and only if it vanishes on the spectrum of this matrix. To prove this, let us assume that $h(z)$ annihilates A; then by (E) of §32 it is divisible by the polynomial $\Delta(z)$, and therefore λ_i is a root of multiplicity not less than k_i [see (1)], and from this it follows that it vanishes on the spectrum of A. If the polynomial $h(z)$ vanishes on the spectrum of A, then the number λ_i is a root of multiplicity not less than k_i, and therefore $h(z)$ is divisible by the polynomial $\Delta(z)$ [see (1)]. Hence it follows that $h(A) = 0$.

We shall now prove that part of proposition (A) concerning the function $\varphi(z)$. The set of relations (4) can be regarded as a system of linear equations in the coefficients of the polynomial $\varphi(z)$. This system has k equations with k unknowns. To prove statement (A) it is sufficient to show that the determinant of this system is distinct from zero, and for this in turn it is sufficient to prove that whenever the right-hand sides of these equations vanish, there exists only the zero polynomial $\varphi(z)$ satisfying conditions (4). Whenever the right-hand sides of (4) vanish, the polynomial $\varphi(z)$ vanishes on the spectrum of the matrix A so that it is divisible by $\Delta(z)$, and, since it is of degree not greater than $k - 1$, it is identically zero. Thus proposition (A) is proved.

(B) Let A be a real matrix; then its minimal annihilating polynomial $\Delta(z)$ is real [see §32, (E)], so that to each root λ_i [see (1)] corresponds a complex conjugate root $\bar{\lambda}_i$ of the same multiplicity. Thus, if the numbers (3) satisfy the conditions

$$W^{(j)}(\bar{\lambda}_i) = \overline{W^{(j)}(\lambda_i)}, \qquad j = 0, \ldots, k_i - 1, \qquad i = 1, \ldots, r, \tag{5}$$

the polynomial $\varphi(z)$ defined by (4) is real, so that the matrix $\varphi(A)$ is also real.

To prove proposition (B), we shall denote the coefficients of the polynomial $\varphi(z)$ by $\varphi^1, \ldots, \varphi^k$. Without going into details, the system (4) in the unknowns $\varphi^1, \ldots, \varphi^k$ may now be written in the form

$$\sum_{\beta=1}^{k} c^\alpha_\beta \varphi^\beta = d^\alpha, \qquad \alpha = 1, \ldots, k. \tag{6}$$

By (5) this system has the property that, for each of its equations, there is also a complex conjugate equation, i.e., the equation

$$\sum_{\beta=1}^{k} \overline{c^\alpha_\beta} \varphi^\beta = \overline{d^\alpha}.$$

Let us now pass from the equalities (6) to their conjugates

$$\sum_{\beta=1}^{k} \overline{c^\alpha_\beta \varphi^\beta} = \overline{d^\alpha}. \tag{7}$$

The set of relations (7) represents a system of linear equations in the unknowns $\overline{\varphi}^1, \ldots, \overline{\varphi}^k$. However, in view of the property which has been formulated for (6), it must coincide with (7), the only possible difference being the numbering of the equations. Since system (6) has a nonzero determinant, its two solutions $\varphi^1, \ldots, \varphi^k$ and $\overline{\varphi}^1, \ldots, \overline{\varphi}^k$ coincide, that is, $\varphi^\alpha = \overline{\varphi}^\alpha$, $\alpha = 1, \ldots, k$, so that the numbers $\varphi^1, \ldots, \varphi^k$ are real. Thus proposition (B) is proved.

Let

$$f(z) = a_0 + a_1 z + a_2 z^2 + \cdots + a_m z^m + \cdots \tag{8}$$

be an analytic function of the complex variable z defined by the series (8) with radius of convergence ρ, so that for $|z| < \rho$ the series (8) converges, and for $|z| > \rho$ it diverges.

For the sequel we recall that the series

$$f'(z) = a_1 + 2a_2 z + \cdots + m a_m z^{m-1} + \cdots,$$

obtained from (8) by means of formal differentiation, has the same radius of convergence as (8) and converges inside the circle of convergence to the derivative $f'(z)$ of $f(z)$.

It can happen that by substituting the matrix A for z in (8), we obtain a convergent matrix series

$$f(A) = a_0 E + a_1 A + \cdots + a_m A^m + \cdots. \tag{9}$$

(A matrix series is called "convergent" if the numerical series consisting of the elements in the i-th row and j-th column converge for arbitrary

$i, j = 1, \ldots, n$.) In this case we say that the function $f(z)$ is defined on the matrix A.

THEOREM 27. We shall retain the notation of proposition (A). If all eigenvalues of the matrix A lie inside the circle of convergence of (8), i.e.,

$$|\lambda_i| < \rho, \qquad i = 1, \ldots, r,$$

then the matrix series (9) converges, so that the matrix $f(A)$ is defined. The numbers

$$f(\lambda_i), \qquad i = 1, \ldots, r, \tag{10}$$

which need not be distinct, comprise the set of all eigenvalues of the matrix $f(A)$. In addition, if the eigenvalues λ_i of the matrix A lie inside the circle of convergence of the series defining a certain function $g(z)$, so that the matrix $g(A)$ is defined, then, in order that the matrices $f(A)$ and $g(A)$ coincide, it is necessary and sufficient that the functions $f(z)$ and $g(z)$ coincide on the spectrum of the matrix A.

Proof. Let us form the partial sum

$$f_m(z) = a_0 + a_1 z + \cdots + a_m z^m$$

of the series (8); then for $|z| < \rho$ we have

$$f^{(j)}(z) = \lim_{m\to\infty} f_m^{(j)}(z).$$

In addition, let $\varphi_m(z)$ be a polynomial of a degree not exceeding $k - 1$ which coincides with the polynomial $f_m(z)$ on the spectrum of the matrix A [see (A)]. Since the eigenvalues (2) of A satisfy the condition $|\lambda_i| < \rho$, $i = 1, \ldots, r$, we have

$$\lim_{m\to\infty} \varphi_m^{(j)}(\lambda_i) = f^{(j)}(\lambda_i), \qquad j = 0, \ldots, k_i - 1, \qquad i = 1, \ldots, r.$$

From this it follows by proposition (A) that the sequence of polynomials $\varphi_m(z)$ converges coefficient-wise to some polynomial $\varphi(z)$ of degree $\leq k - 1$, where the polynomial $\varphi(z)$ and the function $f(z)$ coincide on the spectrum of A. Since the polynomials $\varphi_m(z)$ and $f_m(z)$ coincide on the spectrum of A, we have

$$f_m(A) = \varphi_m(A).$$

As $m \to \infty$ the right-hand side tends to $\varphi(A)$, and this means that the left-hand side also converges as $m \to \infty$. Thus the series (9) converges to the matrix $f(A) = \varphi(A)$.

We shall now prove that the polynomial

$$\Gamma(z) = [z - f(\lambda_1)]^{k_1}[z - f(\lambda_2)]^{k_2} \cdots [z - f(\lambda_r)]^{k_r}$$

annihilates the matrix $f(A)$. For this we consider the polynomial

$$\Phi_m(z) = [\varphi_m(z) - \varphi_m(\lambda_1)]^{k_1}[\varphi_m(z) - \varphi_m(\lambda_2)]^{k} \cdots [\varphi_m(z) - \varphi_m(\lambda_r)]^{k_r} \tag{11}$$

and show that it annihilates A. The polynomial $\varphi_m(z) - \varphi_m(\lambda_i)$ vanishes at $z = \lambda_i$, so that it is divisible by the expression $z - \lambda_i$. Thus, the polynomial (11) can be written in the form

$$\Phi_m(z) = \Psi_m(z)\,\Delta(z),$$

so that the polynomial $\Phi_m(z)$ annihilates A:

$$[\varphi_m(A) - \varphi_m(\lambda_1)E]^{k_1}[\varphi_m(A) - \varphi_m(\lambda_2)E]^{k_2} \ldots [\varphi_m(A) - \varphi_m(\lambda_r)E]^{k_r} = 0.$$

Passing to the limit in this relation as $m \to \infty$, we obtain

$$[f(A) - f(\lambda_1)E]^{k_1}[f(A) - f(\lambda_2)E]^{k_2} \cdots [f(A) - f(\lambda_r)E]^{k_r} = 0,$$

which means that the polynomial $\Gamma(z)$ annihilates the matrix $f(A)$.

In particular, it follows from what has been proved that all eigenvalues of the matrix $f(A)$ are contained among the numbers (10) [see §32, (F)]. We shall prove that every number (10) is an eigenvalue of $f(A)$. Let $\mathbf{h}_i$ be the eigenvector of A corresponding to the eigenvalue λ_i, so that

$$A\mathbf{h}_i = \lambda_i\mathbf{h}_i.$$

From this it follows, by formula (6) of §32, that $f_m(A)\mathbf{h}_i = f_m(\lambda_i)\mathbf{h}_i$. Passing to the limit in this relation as $m \to \infty$, we obtain

$$f(A)\mathbf{h}_i = f(\lambda_i)\mathbf{h}_i.$$

Thus the number $f(\lambda_i)$ is an eigenvalue of the matrix $f(A)$.

Let us now assume that the circle of convergence of the function $g(z)$ also contains all the eigenvalues of A. It then follows from what we have proved that the matrix $g(A)$ is defined and that there exists a polynomial $\psi(z)$ of degree not exceeding $k - 1$ which coincides with the function $g(z)$ on the spectrum of A, where $\psi(A) = g(A)$. Now if $f(A) = g(A)$, then $\varphi(A) = \psi(A)$, and by proposition (A) the polynomials $\varphi(z)$ and $\psi(z)$ coincide on the spectrum of A, so that $f(z)$ and $g(z)$ must also coincide on the spectrum of A. Conversely, if the functions $f(z)$ and $g(z)$ coincide on the spectrum of A, then the polynomials $\varphi(z)$ and $\psi(z)$ also coincide on this spectrum, so that, by proposition (A), $\varphi(A) = \psi(A)$, which implies $f(A) = g(A)$. Thus Theorem 27 is proved.

Implicit functions of matrices. Let $F(z, w)$ be a function of two complex variables defined by the series

$$F(z, w) = a + bz + cw + dz^2 + ezw + fw^2 + \cdots. \tag{12}$$

When the order of the factors in the terms of this series changes (for example, when the product $z^\alpha w^\beta$ is replaced by $w^\beta z^\alpha$), the function $F(z, w)$ does not change. Therefore, if we replace the arguments z and w by *matrices* A and B in (12), it is natural to confine ourselves to the case when the matrices A and B commute. If (12) converges for arbitrary values of the variables z and w, then it can be proved that by substituting for z and w in this series any matrices A and B which commute, we obtain a convergent matrix series which will define a certain matrix which we shall denote by $F(A, B)$. However, we shall not prove the convergence of this series in the general case, since we shall consider below only particular cases in which there is a *finite number* of terms depending on z, so that in fact the question is really that of the convergence of a series of the *one* complex variable w.

(C) Let $F(z, w)$ be an analytic function of two variables defined by (12), which is convergent for all values z, w, and let A be a given matrix. Further, suppose that we have a correspondence between every eigenvalue λ_i of A and a number μ_i which satisfies the conditions

$$F(\lambda_i, \mu_i) = 0, \qquad \frac{\partial}{\partial w} F(\lambda_i, \mu_i) \neq 0, \qquad i = 1, \ldots, r. \tag{13}$$

Then there exists a matrix B, which commutes with A, satisfying the condition

$$F(A, B) = 0. \tag{14}$$

Furthermore, if the coefficients of (12) and the matrix A are real and if for every two complex conjugate eigenvalues λ_i and $\lambda_j = \bar{\lambda}_i$ of the matrix A the corresponding numbers μ_i and μ_j are also complex conjugate ($\mu_i = \bar{\mu}_j$), then there exists a real matrix B, which commutes with A and which satisfies condition (14).

We shall prove proposition (C). It follows from (13) and the implicit function theorem in the complex case that for any $i = 1, \ldots, r$ there exists a function $W(z) = W_i(z)$ which is defined for z close to λ_i and which satisfies the conditions

$$F(z, W(z)) = 0, \tag{15}$$

$$W(\lambda_i) = \mu_i, \qquad i = 1, \ldots, r. \tag{16}$$

To determine the derivatives $W^{(j)}(\lambda_i)$ of $W(z)$ at the point $z = \lambda_i$, it is necessary to differentiate successively relation (15) with respect to z and then to substitute in it $z = \lambda_i$:

$$\frac{d^j}{dz^j} F(z, W(z))|_{z=\lambda_i} = 0. \tag{17}$$

For these relations, the numbers

$$W^{(j)}(\lambda_i), \qquad j = 1, \ldots, k_i - 1, \qquad i = 1, \ldots, r, \tag{18}$$

can be determined successively. Starting from (16) and (18), we shall construct the polynomial $\varphi(z)$ which satisfies conditions (4). We shall show that the matrix $B = \varphi(A)$, which obviously commutes with A, satisfies (14).

For the proof we shall substitute in (12) the value $w = \varphi(z)$. We then obtain the function $\Phi(z) = F[z, \varphi(z)]$ of z. To prove (14) it is sufficient to show that $\Phi(z)$ is equal to zero on the spectrum of the matrix A (see Theorem 27). By calculating the derivatives $\Phi^{(j)}(\lambda_i)$ of $\Phi(z)$ at the point λ_i, $j = 0, 1, \ldots, k_i - 1$, we can substitute the function $W(z)$ for the polynomial $\varphi(z)$, since the respective derivatives of these functions of orders $0, 1, \ldots, k_j - 1$, at the point λ_i are equal. But by replacing the polynomial $\varphi(z)$ in $F(z, \varphi(z))$ by the function $W(z)$ defined in the neighborhood of λ_i, we obtain an identity [see (15)]. Thus the function $\Phi(z)$ vanishes on the spectrum of A.

We shall now prove that if the coefficients of (12) and the matrix A are real and the numbers μ_i satisfy the requirement of conjugacy, i.e., if

$$W(\overline{\lambda_i}) = \overline{W(\lambda_i)}, \qquad i = 1, \ldots, r,$$

then the polynomial $\varphi(z)$, as well as the matrix $B = \varphi(A)$, is real. Indeed, under these hypotheses the numbers $W^{(j)}(\lambda_i)$, coming from (17), satisfy (5), so that the polynomial $\varphi(z)$ is real [see (B)]. Thus, proposition (C) is proved.

(D) The analytic function e^z of z is defined by the series

$$e^z = 1 + z + \frac{z^2}{2!} + \cdots + \frac{z^m}{m!} + \cdots, \tag{19}$$

which converges for all values of z. As is known, the identity $e^{z+w} = e^z \cdot e^w$, which follows from the properties of (19), is valid for two arbitrary complex numbers z and w. Hence it follows that the identity

$$e^{A+B} = e^A \cdot e^B \tag{20}$$

is valid for two commuting square matrices A and B. Thus, for any nonsingular matrix A there exists a matrix B, commuting with A, which satisfies the condition

$$e^B = A. \tag{21}$$

In addition, we find that for any real nonsingular matrix A there exists a real matrix B_1, which commutes with A and satisfies the condition

$$e^{B_1} = A^2. \tag{22}$$

To prove that (21) can be solved for B it is sufficient to apply proposition (C) to the function $F(z, w) = e^w - z$. Indeed, since the matrix A is nonsingular, all its eigenvalues λ_i are distinct from zero, so that there exist numbers μ_i which satisfy the condition $e^{\mu_i} - \lambda_i = 0$ [see the first of the relations (13)]; the second of the relations (13) are obviously satisfied here.

To prove the existence of a real matrix B_1 satisfying condition (22), it is sufficient to apply the second part of proposition (C) to the function $F(z, w) = e^w - z^2$. Indeed, if λ_i is a real positive or negative number, then we set $\mu_i = \ln \lambda_i^2$, where we take the real branch of the logarithm. If, however, λ_i is a complex number, then we can take complex conjugate numbers for $W(\lambda_i)$ and $W(\bar{\lambda}_i)$. Thus, proposition (D) is proved.

34. The Jordan form of a matrix. (A) The sequence of vectors

$$\mathbf{h}_1, \ldots, \mathbf{h}_m \tag{1}$$

of the space R is called a *basis set* or a *series* with eigenvalue λ for the transformation $\mathbf{A}$ if the relations

$$\mathbf{A}\mathbf{h}_1 = \lambda\mathbf{h}_1, \quad \mathbf{A}\mathbf{h}_2 = \lambda\mathbf{h}_2 + \mathbf{h}_1, \quad \ldots, \quad \mathbf{A}\mathbf{h}_m = \lambda\mathbf{h}_m + \mathbf{h}_{m-1}, \quad \mathbf{h}_1 \neq 0,$$

are fulfilled. If the matrix A of the transformation $\mathbf{A}$ is real, then the sequence

$$\bar{\mathbf{h}}_1, \ldots, \bar{\mathbf{h}}_m \tag{2}$$

obviously forms a series with eigenvalue $\bar{\lambda}$. The series (1) and (2) will be called *complex conjugate*. If the number λ and the vectors (1) are real, then the series is considered *real*.

THEOREM 28. There exists a basis of the space R consisting of all the vectors of one or more series for the transformation $\mathbf{A}$. If the matrix A is real, then the series constituting the basis can be chosen in such a way that a series with real eigenvalues is real and series with complex eigenvalues are pairwise conjugate.

Proof. Let

$$\Delta(z) = (z - \lambda_1)^{k_1} \ldots (z - \lambda_r)^{k_r} \tag{3}$$

be the minimal polynomial annihilating the matrix A, where

$$\lambda_1, \ldots, \lambda_r$$

are distinct eigenvalues of A. By proposition (G) of §32, the space R can be decomposed into a direct sum of the subspaces $S_1, \ldots, S_r$ corresponding to the factors (3), so that the space S_i consists of all vectors $\mathbf{x}$ which satisfy the condition $(\mathbf{A} - \lambda_i\mathbf{E})^{k_i}\mathbf{x} = 0$. This means that the annihilating polynomial of the transformation $\mathbf{A}$ taken on the space S_i is the polynomial $(z - \lambda_i)^{k_i}$. It is easy to see that this polynomial is minimal.

Let us assume that A is real. First we combine all the factors of (3) with real λ_i into the factor $\Delta_1(z)$ and all the other factors into the factor $\Delta_2(z)$. Then $\Delta(z) = \Delta_1(z)\,\Delta_2(z)$ is a factorization into real relatively prime factors, and the corresponding decomposition of the space R into a direct sum of R_1 and R_2 may be considered real. The space R_1 is now decomposed into a sum of real summands corresponding to real eigenvalues λ_i, and in these real direct summands we shall later construct bases consisting of real series. The space R_2 will be decomposed into pairwise complex conjugate direct summands corresponding to the complex conjugate eigenvalues, and in these complex conjugate spaces we shall later construct bases consisting of complex conjugate series; here it is sufficient to construct the basis from a series in one of the two complex conjugate spaces, and in the other space to take the complex conjugate basis.

Thus it is sufficient for us to prove that if the linear transformation $\mathbf{A}$ operating in the vector space S has the minimal annihilating polynomial $(z - \lambda)^k$, then in this space we can choose a basis consisting of series for the transformation $\mathbf{A}$, and, moreover, of real series if the space S, the matrix A, and the eigenvalue λ are all real.

We shall go on to the proof of this statement. For brevity we assume that $\mathbf{C} = \mathbf{A} - \lambda\mathbf{E}$, and we denote by T_i the set of all vectors $\mathbf{x}$ of S which satisfy the condition

$$\mathbf{C}^i\mathbf{x} = 0.$$

Then we have

$$S = T^k \supset T^{k-1} \supset \cdots \supset T^1 \supset T^0 = 0.$$

Let

$$\mathbf{h}_i^1, \ldots, \mathbf{h}_i^r, \qquad i = 1, \ldots, k,$$

be a system of vectors from T_i which are linearly independent with respect to the space T_{i-1}; this means that the vector

$$a_1\mathbf{h}_i^1 + \cdots + a_r\mathbf{h}_i^r$$

can belong to space T_{i-1} only if

$$a_1 = \ldots = a_r = 0.$$

We shall show that for fixed i and j the vectors

$$\mathbf{h}_{i-j}^{\alpha} = \mathbf{C}^j \mathbf{h}_i^{\alpha}, \qquad j < i, \tag{4}$$

belong to the space T_{i-j} and are linearly independent with respect to the space T_{i-j-1}. We have

$$\mathbf{C}^{i-j}\mathbf{h}_{i-j}^{\alpha} = \mathbf{C}^i \mathbf{h}_i^{\alpha} = 0, \qquad \alpha = 1, \ldots, r,$$

and consequently the vectors (4) belong to the space T_{i-j}. Let us assume now that the vector

$$a_1 \mathbf{h}_{i-j}^1 + \cdots + a_r \mathbf{h}_{i-j}^r = \mathbf{x}$$

belongs to the space T_{i-j-1}. Then we have

$$0 = \mathbf{C}^{i-j-1}\mathbf{x} = \mathbf{C}^{i-1}(a_1 \mathbf{h}_i^1 + \cdots + a_r \mathbf{h}_i^r),$$

and this means that the vector $a_1 \mathbf{h}_i^1 + \cdots + a_r \mathbf{h}_i^r$ belongs to the space T_{i-1} so that the numbers $a_1, \ldots, a_r$ are zero.

Let us select a maximal system of vectors

$$\mathbf{h}_k^1, \ldots, \mathbf{h}_k^{r_k} \tag{5}$$

of T_k which are linearly independent with respect to T_{k-1}. According to what has been proved, the vectors

$$\mathbf{h}_{k-1}^{\alpha} = \mathbf{C}\mathbf{h}_k^{\alpha}, \qquad \alpha = 1, \ldots, r_k, \tag{6}$$

belong to T_{k-1} and are linearly independent with respect to T_{k-2}; thus system (6) can be augmented to a maximal system

$$\mathbf{h}_{k-1}^1, \ldots, \mathbf{h}_{k-1}^{r_{k-1}}, \qquad r_{k-1} \geq r_k, \tag{7}$$

of vectors of T_{k-1} which are linearly independent with respect to T_{k-2}. Continuing this process, we shall construct in the space $T_i (i > 0)$ a maximal system of vectors

$$\mathbf{h}_i^1, \ldots, \mathbf{h}_i^{r_i}, \tag{8}$$

which are linearly independent with respect to T_{i-1}; here the relations

$$\mathbf{h}_i^{\alpha} = \mathbf{C}\mathbf{h}_{i+1}^{\alpha}, \qquad \alpha = 1, \ldots, r_{i+1}; \quad r_i \geq r_{i+1},$$

will be fulfilled.

We shall now prove that the set $\sum_j$ of all vectors which belong to all systems (8), $i = j, j-1, \ldots, 1$, constitutes a basis of the space T_j. The proof will be carried out by induction on the number j. For $j = 1$ the system $\sum_1$ coincides with the system (8) and therefore is a basis of the space T_1, $T_0 = 0$. Let us assume that our assertion has been proved for the system $\sum_j$, and we shall prove it for the system $\sum_{j+1}$. Let us assume that the relation

$$a_1\mathbf{h}_{j+1}^1 + \cdots + a_{r_{j+1}}\mathbf{h}_{j+1}^{r_{j+1}} + b_1\mathbf{h}_j^1 + \cdots + b_{r_j}\mathbf{h}_j^{r_j} + \cdots = 0 \tag{9}$$

holds. Applying the transformation $\mathbf{B}^j$ to (9) we obtain

$$a_1\mathbf{h}_1^1 + \cdots + a_{r_{j+1}}\mathbf{h}_1^{r_{j+1}} = 0,$$

and this is possible only if $a_1 = \cdots = a_{r_{j+1}} = 0$; thus relation (9) can contain only vectors of the system $\sum_j$, so that by the induction hypothesis, relation (9) is trivial. Now let $\mathbf{x}$ be an arbitrary vector of the space T_{j+1}. Since the system (8) for $i = j + 1$ forms a maximal linearly independent system for the space T_j, there exists a vector

$$\mathbf{y} = a_1\mathbf{h}_{j+1}^1 + \cdots + a_{r_{j+1}}\mathbf{h}_{j+1}^{r_{j+1}}$$

such that $\mathbf{x} - \mathbf{y}$ belongs to T_j and, by the induction hypothesis, may be expressed linearly in terms of the vectors of the system $\sum_j$, so that the vector $\mathbf{x}$ may be expressed linearly in terms of the vectors of $\sum_{j+1}$.

Thus we have shown that $\sum_k$ is a basis of the space $S = T_k$.

If the space S, the matrix A, and the number λ are real, then by taking the vectors of (5) as real, we obtain a real system (6), which can be augmented to the real system (7). Continuing in this way, we obtain the *real* system $\sum_k$.

We shall now show that the system $\sum_k$ consists of series, i.e., we shall show that the vectors $\mathbf{h}_1^\alpha, \mathbf{h}_2^\alpha, \ldots$ form a series with eigenvalue λ. We have

$$0 = \mathbf{C}\mathbf{h}_1^\alpha = (\mathbf{A} - \lambda\mathbf{E})\mathbf{h}_1^\alpha,$$

so that $\mathbf{A}\mathbf{h}_1^\alpha = \lambda\mathbf{h}_1^\alpha$; further,

$$\mathbf{h}_1^\alpha = \mathbf{C}\mathbf{h}_2^\alpha = (\mathbf{A} - \lambda\mathbf{E})\mathbf{h}_2^\alpha,$$

so that $\mathbf{A}\mathbf{h}_2^\alpha = \lambda\mathbf{h}_2^\alpha + \mathbf{h}_1^\alpha$, and so on. Thus Theorem 28 is proved.

In the basis constructed by means of Theorem 28, there corresponds to the transformation $\mathbf{A}$ not the original matrix $A = (a_j^i)$, but some new matrix $B = (b_j^i)$ which has a particularly simple form, called the *Jordan* form. Thus Theorem 28 is a *theorem on the reduction of a matrix to Jordan form.* Let us examine this question in detail.

(B) By a Jordan block of the mth-order with eigenvalue λ we shall mean a square matrix (g_j^i) of mth-order which is defined by the relations

$$g_i^i = \lambda, \qquad i = 1, \ldots, m; \qquad g_{i+1}^i = 1, \qquad i = 1, \ldots, m-1;$$
$$g_j^i = 0, \qquad j - i < 0 \qquad \text{and} \qquad j - i > 1;$$

i.e., the matrix

$$\begin{pmatrix} \lambda & 1 & 0 & \cdots & 0 & 0 \\ 0 & \lambda & 1 & \cdots & 0 & 0 \\ 0 & 0 & \lambda & \cdots & 0 & 0 \\ \vdots & & & & & \\ 0 & 0 & 0 & \cdots & \lambda & 1 \\ 0 & 0 & 0 & \cdots & 0 & \lambda \end{pmatrix}.$$

It turns out that for every nth-order square matrix A, a nonsingular square matrix S may be chosen such that the matrix $B = SAS^{-1}$ which is obtained from the matrix A by a transformation of the matrix S is of *Jordan form,* i.e., consists of one or more Jordan blocks located along its principal diagonal, while all the elements not contained in the Jordan blocks are equal to zero.

Let us prove this. Let R be an n-dimensional vector space and $\mathbf{A}$ a linear transformation corresponding (in a certain coordinate system) to the matrix A. Now let $\mathbf{f}_1, \ldots, \mathbf{f}_n$ be a basis of the space R consisting of series (see Theorem 28). We shall assume that the vectors $\mathbf{f}_1, \ldots, \mathbf{f}_n$ are arranged in such an order that the vectors of each series follow one another in the sequence $\mathbf{f}_1, \ldots, \mathbf{f}_n$. We shall denote by $B = (b_j^i)$ the matrix of the transformation $\mathbf{A}$ in the basis $\mathbf{f}_1, \ldots, \mathbf{f}_n$. Let

$$\mathbf{h}_1 = \mathbf{f}_1, \quad \ldots, \quad \mathbf{h}_m = \mathbf{f}_m$$

be the first series appearing in the sequence $\mathbf{f}_1, \ldots, \mathbf{f}_n$, and let λ be the corresponding eigenvalue. Then, as follows directly from the definition of a series, we have

$$b_i^i = \lambda, \qquad i = 1, \ldots, m; \qquad b_{i+1}^i = 1, \qquad i = 1, \ldots, m-1;$$
$$b_j^i = 0, \qquad i + 1 < j \leq m \qquad \text{and} \qquad i > j \leq m.$$

Thus the first Jordan block in the matrix B corresponds to the first series of the sequence $\mathbf{f}_1, \ldots, \mathbf{f}_n$. In exactly the same way the second Jordan block in the matrix B corresponds to the second series of the basis $\mathbf{f}_1, \ldots, \mathbf{f}_n$, and so on. Since the matrix A can be transformed into B [see §32, (A)], we have $B = SAS^{-1}$, which proves proposition (B).

INDEX

ABCDE698765432

Date Due			
MY 24 '65			
FE 19 '69			

Pontriagin, L.

Ordinary differential equations.